Fundamentals of Ground Combat System Ballistic Vulnerability/Lethality

Fundamentals of Ground Combat System Ballistic Vulnerability/Lethality

Paul H. Deitz
U.S. Army Research Laboratory, Aberdeen Proving Ground, Maryland

Harry L. Reed, Jr.
Aberdeen, Maryland

J. Terrence Klopcic
Kenyon College, Gambier, Ohio

James N. Walbert
SURVICE Engineering Company, Dumfries, Virginia

Edited by
Eric W. Edwards
SURVICE Engineering Company, Belcamp, Maryland

William L. Hacker
Applied Research Associates, Albuquerque, New Mexico

William L. Kincheloe
D. R. Kennedy & Associates, Mountain View, California

Dennis C. Bely
SURVICE Engineering Company, Belcamp, Maryland

Volume 230
PROGRESS IN
ASTRONAUTICS AND AERONAUTICS

Frank K. Lu, Editor-in-Chief
University of Texas at Arlington
Arlington, Texas

Published by the
American Institute of Aeronautics and Astronautics, Inc.
1801 Alexander Bell Drive, Reston, Virginia 20191-4344

American Institute of Aeronautics and Astronautics, Inc., Reston, Virginia

1 2 3 4 5

The intent of this book is to provide general information on how ground combat and other systems are analyzed. All quantitative results shown are purely notional and are included for illustrative purposes only. They are *not* reflective of the vulnerability/lethality of any specific platform, configuration, threat, hit location, or other variables.

The findings in this book are not to be construed as an official position of the Department of Army or the Department of Defense unless so designated by other authorized documents.

Citation of manufacturers' or trade names does not constitute an official endorsement or approval of the use thereof.

Copyright © 2009 by the American Institute of Aeronautics and Astronautics, Inc. All rights reserved. Printed in the United States of America. No part of this publication may be reproduced, distributed, or transmitted, in any form or by any means, or stored in a database or retrieval system, without the prior written permission of the publisher.

The U.S. Government has rights under the guidance of Defense Federal Acquisition Regulation Supplement (DFAR) Clause 252.227-7013. AIAA shall thereby grant to the U.S. Government, a nonexclusive, irrevocable, unrestricted copyright license solely for government purposes only. The U.S. Government has these rights to the "technical data," in whole or in part, and in any manner for U.S. Government purposes only. These rights do not include the right to have or permit others to use the copyright of the Work for commercial purposes.

ISBN 978-1-60086-015-7

Progress in Astronautics and Aeronautics

Editor-in-Chief
Frank K. Lu
University of Texas at Arlington

Editorial Board

David A. Bearden
The Aerospace Corporation

Eswar Josyula
U.S. Air Force Research Laboratory

John D. Binder
viaSolutions

Gail Klein
Jet Propulsion Laboratory

Steven A. Brandt
U.S. Air Force Academy

Konstantinos Kontis
University of Manchester

Jose Camberos
U.S. Air Force Research Laboratory

Richard C. Lind
University of Florida

Richard Curran
Delft University of Technology

Ning Qin
University of Sheffield

Sanjay Garg
NASA Glenn Research Center

Oleg Yakimenko
U.S. Naval Postgraduate School

Christopher H. Jenkins
Montana State University

Foreword

THE process of assessing conventional weapons and their effects is still far from a science. It is more a patchwork of empirical fits based on test data and quasi-physical relationships, anecdotal combat data, and hands-on experience.

Prompted by the initiation of the Joint Live Fire (JLF) test program in 1984 and passage in 1986 of the legislation mandating Live Fire testing, the extensive growth of tools and methods for assessing the vulnerability/lethality (V/L) of combat systems to the effects of conventional weapons makes the need for a book such as this one apparent. Classical methods and predictive models have played a vital role in this process and continue to undergird the lessons coming from these tests. These two activities have prompted great interest in, and activity related to, improving the nation's analytical approaches for assessing our combat vulnerability, lethality, and ultimate survivability.

This publication presents the historical basis and describes some of the most widely used methods of assessing combat V/L and, ultimately, conventional weapons effectiveness. The complexity of the material has been kept to a modest level to be understandable to those who are entering the discipline as well as to serve as a handy reference for those who regularly practice these disciplines. An attempt has been made not only to share those areas where significant progress has been made but also to identify those areas where methodology is weak or currently nonexistent.

I would like to express my sincere appreciation to all those who have given unselfishly of themselves to bring this text into being and to those many more who have worked tirelessly over the years to gather the data necessary to derive these methods. Furthermore, this book is dedicated to those in our fighting forces who have put, and continue to put, their lives on the line defending freedom around the world.

James F. O'Bryon
Former Director, Live Fire Testing
Office of the Secretary of Defense
March 2009

Table of Contents

Preface xi
Contributors xv

Chapter 1. Introduction 1
Overview and Definitions 2
A Brief History of V/L Analysis 4
Roles of V/L Analysis 8

Chapter 2. Vulnerability/Lethality Analysis Process 13
The Missions and Means Framework 13
Initial Representation and Damage Mechanisms 19
Induced Dysfunction of Components 44
Target Response 58
Tactical Utility 62

**Chapter 3. Vulnerability Assessment and Measures
   of Effectiveness** 65
Traditional Measures of Mission Effectiveness 65
Fault Trees and Degraded States 89
An Example of Fault-Tree Development for Traditional Metrics 93
An Example of Fault-Tree Development for Degraded States Methodology 110
Networks and Networked Combat Systems 117

Chapter 4. Modeling and Simulation Tools and Methods 127
Background 127
Empirical vs Semi-Empirical Models 128
Phenomenological Models 129
Engineering Models 143
Geometric Representation of Targets 144
System Models 167
Computer Environments 195
Force-Level Modeling 205
Verification, Validation, and Accreditation for V/L Assessment 206

Chapter 5. Applications 217
System Acquisition 217
System Life Cycle 226
Vulnerability Reduction 232
Tactics and Doctrine 239

References 241

Appendix A. Penetration from Fragmentation Munitions 253
Simple Methods 255
Compact Fragment Penetration Characteristics 257
FATEPEN Penetration Model Overview 260
References 264

Appendix B. Behind-Armor Debris Characterization 267
Witness Pack 267
Analysis of Witness Plates 269
Description of the Mathematical Model 270
Summary 274
Reference 274

**Appendix C. Estimating Component Probability of
Damage Given a Hit** 275
Graphical Method 278
Multiple Fragment Effects 279
References 281

**Appendix D. Case Study: MUVES–SQuASH VV&A for the
Bradley Fighting Vehicle (BFV)** 283
MUVES Functionality and Code 283
V&V of MUVES–SQuASH Submodels and Program Inputs 285
Overall MUVES–SQuASH Model Validation 286
Statistical Comparison to LFT 286
Model Input 287
Grading Criteria 287
Box Plot 288
Comparisons of Component Capability (MMF Level 2) 288
Comparisons of Vehicle Capability (MMF Level 3) 290
Comparisons of Vehicle Utility (MMF Level 4) 294
IDA and VAST Comparisons 295
Conclusions of the MUVES–SQuASH VV&A for the BFV 296
Reference 296

**Appendix E. Details and Developments in the Missions and
Means Framework** 297
Paper 1: The Military Missions and Means Framework 299
Paper 2: The MMF Formal Process 319
Paper 3: The Connection Between Functions and Capabilities 324
References 345

Appendix F. Acronyms and Abbreviations 347

Index 353

Supporting Materials 367

Preface

THAT the past is prologue as history is written is certainly evident in the pages that follow. In large measure, the progress in ballistics and vulnerability/lethality (V/L) reported in these pages shares a significant nexus over the past decade with the establishment and development of Aberdeen Proving Ground (APG) in northeastern Maryland.

By the close of World War I, the gun facilities support at Sandy Hook, New Jersey, had been greatly exceeded, and it was clear to the Army that expansive new real estate would be critical to the prosecution of future ballistic testing. Accordingly, the lands now known as APG were identified as ideal and purchased by the Army from various private owners. Soon after experimental gun testing was established at APG, a theoretical gun section was formed, which supported studies in areas such as interior and exterior effects. In addition, the scientific discipline of "ballistics" was developing as fire-control solutions were codified and computed repetitively under parametric exercise. The limits of human dedication were tested as many thousands of hours of manual computation were performed in the development of ballistic firing tables.

During the 1930s, new means of automated computation, including the use of the Bush differential analyzer, were examined. In 1938, the Ballistic Research Laboratory (BRL) was formed at APG and became a focus of research in theoretical and experimental ballistics. As World War II began, the gap between ballistic computation needs and capabilities was so great that a contract was made with the University of Pennsylvania to serve as a computing adjunct to BRL. From that activity came the proposal for construction and delivery of ENIAC (Electronic Numerical Integrator and Computer), the world's first general-purpose, programmable computing machine. ENIAC and the machines that were to follow Electronic Discrete Variable Computer (EDVAC), Ordnance Variable Automatic Computer (ORDVAC), BRL Electronic Scientific Computer (BRLESC-I, BRLESC-II, etc.) set the vital automation backdrop that, with the confluence of ballistic theory and experimentation, established the foundations for today's V/L studies.

Another important development in the study of V/L involves the detailed interaction of a threat and a target, in particular, the causes and extent of damage and the reduction of target capability and utility. V/L testing began in earnest during World War II with firings against aircraft. Tests by the hundreds continued after the war as many old aircraft were brought to APG. By the 1950s, tests on tanks and other ground vehicles began as well. At each step, warhead and armor algorithms were blended with vehicle representations to understand design tradeoffs. About 1960, the results of full-scale tank testing in Canada were codified in a lumped-parameter computer program called the compartment model. Ten years later, the Computation of Vulnerable Areas and Repair Times (COVART) model, a somewhat higher resolution model, was developed to support aircraft assessment.

Then by 1972, the Vulnerability Analysis for Surface Targets (VAST) model, the first spall-handling model, was developed for armored ground targets.

By this time, the prosecution of V/L modeling was largely limited by two critical issues: 1) the daunting task of generating voluminous amounts of geometric, material, ballistic, and engineering data as input; and 2) the availability of copious amounts of computing cycles and storage, and requisite graphics support. (Progress in computing has essentially eliminated the latter constraint, but the former continues as a critical resource limitation in the application of state-of-the-art V/L modeling in today's studies.)

Accordingly, by the 1980s, the stage had been set for a number of key developments that resulted in an extremely dynamic period: 1) the Director of BRL established a V/L methodology effort that ultimately led to a rich set of tools [e.g., Ballistic Research Laboratory – Computer-Aided Design (BRL-CAD, a registered trademark of the U.S. Army Research Laboratory), behind-armor spall modeling, stochastic V/L modeling, degraded states analysis, and shareable computer environments]; 2) the vulnerability experimental facilities and techniques were upgraded and expanded; and 3) the Live Fire law provided a critical stimulus to test and model to ever more exacting standards. Added to this mix are the contributions of the Navy and Air Force modeling and testing programs, the support of the Joint Technical Coordinating Group for Munitions Effectiveness (JTCG/ME) and the Joint Technical Coordinating Group on Aircraft Survivability (JTCG/AS) [now an element of the Joint Aircraft Survivability Program (JASP)], and the leadership of the Department of the Army and Office of the Secretary of Defense.

Thus, the work that follows builds on the efforts of many scores of military and civilian workers. Their efforts span numerous technical fields, including mechanics, physics, engineering, computer science, statistics, and operations research. The developments include some of the most recent in the fields of ballistics and computer science but, just as often, rely on concepts and practices established and improved over the previous decade. This work is, therefore, a testament to the skill and dedication of numerous workers, past and present.

The focus of this book, as its title implies, is on ground combat system vulnerability. However, the reader will note that many of the principles, methodologies, and tools discussed here are also applicable to the air and sea system communities. Although much has been made about the distinctions that exist among these communities, when it comes to vulnerability analysis, there are more similarities than differences.

Chapter 1 begins by introducing the basic language, history, and uses of V/L analysis. Chapter 2 then discusses elements of the V/L analysis process, including the Missions and Means Framework (MMF), initial representations and damage mechanisms, component and personnel dysfunction, target response, and tactical utility. The focus of Chapter 3 is on vulnerability assessment and measures of effectiveness, including mission effectiveness, fault trees, degraded states, and networked systems. Chapter 4 discusses V/L modeling and simulation, including empirical/semi-empirical, phenomenological, engineering, system, and force-level modeling; geometric representation; computer environments; and verification, validation, and accreditation (VV&A). Finally, Chapter 5 focuses on V/L application topics such as system acquisition, system life cycle, vulnerability reduction, and tactics and doctrine.

In addition, more detailed discussions on selected topics are provided in the appendices. Appendices A and B discuss the characterization of penetrating fragments and behind-armor debris (BAD), respectively. Appendix C provides details on estimating component probability of damage. Appendix D presents a case study of an actual VV&A implementation. Appendix E provides more detailed discussions and more recent developments/applications of the MMF.

Finally, because no volume of this type could begin to cover the depth and breadth of this constantly changing subject, an extensive set of references has been included at the end of the text and after each appendix.

In conclusion, the time is right for a book on the fundamentals of ground combat system vulnerability. We are especially grateful to James O'Bryon, former Director, Live Fire Testing, Office of the Secretary of Defense, who served as the catalyst for the initiation of this project and who has, through the years, continued to stimulate its progress.

Paul H. Deitz
March 2009

In addition, more detailed discussions on selected topics are provided in the appendices. Appendices A and B discuss the characterization of penetrating fragments and behind-armor debris (BAD), respectively. Appendix C provides details on estimating component probability of damage. Appendix D presents a case study of an actual VV&A implementation. Appendix E provides more detailed discussions and more recent developments/applications of the MMF.

Finally, because no volume of this type could begin to cover the depth and breadth of this constantly changing subject, an extensive set of references has been included at the end of the text and after each appendix.

In conclusion, the time is right for a book on the fundamentals of ground combat system vulnerability. We are especially grateful to James O'Bryon, former Director, Live Fire Testing, Office of the Secretary of Defense, who served as the catalyst for the initiation of this project and who has, through the years, continued to stimulate its progress.

Paul H. Deitz
March 2009

Contributors

THIS work represents the collective efforts of many individuals in government and industry who have contributed to the content and/or presentation of this text over its numerous years of development. The names of these individuals, along with their organizational affiliations at the time of their contribution, are given in alphabetical order in the following list. It is recognized that some names may have been inadvertently overlooked over the years. The authors regret any such omissions and sincerely thank all contributors, named and unnamed, who have worked to bring this book to fruition.

Applin, K. A., U.S. Army Research Laboratory
Bingham, B. L., Applied Research Associates
Bodt, K. A., U.S. Army Research Laboratory
Bray, B. E., Dynamics Research Corporation
Butler, L. A., U.S. Army Research Laboratory
Davis, E. G., U.S. Army Research Laboratory
Dietrich, A. M., U.S. Army Research Laboratory
Eberius, N. L., U.S. Army Research Laboratory
Farenwald, D. B., U.S. Army Research Laboratory
Fedele, P. D., U.S. Army Research Laboratory
Finnerty, A. E., U.S. Army Research Laboratory
Frey, R. B., U.S. Army Research Laboratory
Grote, R. L., U.S. Army Research Laboratory
Hoffman, A. J., U.S. Army Ballistic Research Laboratory
Hunt, J. E., Applied Research Associates
Jacobson, J. R., U.S. Army Research Laboratory
Johnson, S. A., Olgoonik Logistics, LLC
Kennedy, D. R., D. R. Kennedy & Associates, Inc.
Kinsler, R. E., The Analytical Sciences Corporation
Kirby, R. L., U.S. Army Research Laboratory
Kiwan, A. R., U.S. Army Research Laboratory
Lottero, R. E., U.S. Army Research Laboratory
Maestas, F. A., Applied Research Associates
Minchew, M. L., Dynamics Research Corporation
Muuss, M. J., U.S. Army Research Laboratory
Neades, D. N., U.S. Army Research Laboratory
Nelson, M. K., Franklin and Marshall College
Ozolins, A., U.S. Army Research Laboratory
Petty, D. W., U.S. Army Research Laboratory
Ploskonka, J. J., U.S. Army Research Laboratory
Polesne, J. T., U.S. Army Research Laboratory
Ritondo, M. E., U.S. Army Research Laboratory

Roach, L. K., U.S. Army Research Laboratory
Ruyle, P. A., The SURVICE Engineering Company
Sandmeyer, R. S., U.S. Army Research Laboratory
Saucier, R., U.S. Army Research Laboratory
Schmidt, E. M., U.S. Army Research Laboratory
Sheehan, J. H., Defense Modeling and Simulation Office
Shnidman, R., U.S. Army Research Laboratory
Sperrazza, J., U.S. Army Materiel Systems Analysis Activity
Tanenbaum, P. J., U.S. Army Research Laboratory
Vitali, R., U.S. Army Research Laboratory
Ward, B. S., U.S. Army Research Laboratory
Watson, J. L., U.S. Army Research Laboratory
Winner, W. A., U.S. Army Research Laboratory
Wong, A. B. H., U.S. Army Materiel Systems Analysis Activity
Yatteau, J. D., Applied Research Associates
Young, L. A., Applied Research Associates

Chapter 1

Introduction

THIS text is intended to serve as a reference on the subject of ground combat system vulnerability/lethality (V/L) analysis, not only for the novice but also for those familiar with the subject. The Missions and Means Framework (MMF), one of the most significant developments in military operations research analysis in the recent past, is the unifying theory that has brought more rigor to this discipline. MMF is an organizing structure that codifies the V/L analysis process and provides a powerful and unifying method to characterize the numerous activities and measures that result. Accordingly, this text is organized largely along the lines of the MMF, which is later described in detail in Chapter 2 (and further discussed in Appendix E). In this chapter, we present some background information and a historical perspective of V/L analysis.

Many years ago, at the start of weapon system analysis, the work most often had to be carried out with a minimum of unifying theory, a limited amount of empirical data, and only rudimentary computational capability. The early technology grew as an arcane art form practiced by experienced analysts who were qualified by on-the-job training. But much has happened in the last two decades to mature this technology into a more rigorous engineering discipline. Developments in digital computers, with parallel processing and sophisticated software systems, now provide computational power never dreamed of when serious analysis was introduced. Many of those advances are summarized in this book.

Although this book has been written primarily for the analysis of ballistic threats [e.g., kinetic energy (KE) penetrators and high-explosive (HE) warheads], the underlying methodology can be applied to other threat mechanisms, such as directed energy (DE); nuclear, biological, and chemical (NBC); and electronic. Discussions of these other mechanisms have been included where appropriate. In addition, because the desire that this book capture some of the rules of thumb and insights that have been developed as analysts have gone about the daily business of getting numbers to meet deadlines, we have included discussions of the practical lessons learned over the years.

More recently, so-called network-centric warfare has required analysts to consider the vulnerability of groups of combat systems taken as ensembles, rather than simply analyzing individual platforms. Whereas, in the past, platforms have been largely responsible for their own offensive and defensive capabilities, the network allows capabilities to be distributed among a number of platforms and over a wide geographical area. In this context, any number of networked platforms can contribute to both the survivability and the lethality of one another and the unit as a whole. Analysts are just now beginning to understand how to adapt existing platform-centric methods and tools to this emerging class of problem.

Finally, the face of warfare is also changing. Asymmetric warfare and the emergence of rapid deployment as a cornerstone of U.S. defense have led to lighter-weight systems and a reliance on less conventional survivability measures. Consequently, these factors necessitate a shift in the techniques used to analyze them. The last section of this text is devoted to looking at what is in store for the future of V/L analysis.

I. Overview and Definitions

As with any technical discipline, there is unique terminology associated with different aspects of ballistic V/L testing and analysis. Thus, when first encountering the field, acquiring a firm grasp of this terminology is essential for developing a clear understanding of the subject. Throughout the book, terminology is presented in a list, with a term (in italics) followed by its definition. The following definitions are provided for some of the principal terms used throughout this book:

Survivability The total capability of a system (resulting from the synergism among personnel, materiel, design, tactics, and doctrine) to avoid, withstand, or recover from damage to a system or crew in hostile (man-made or natural) environments without suffering an abortive impairment of its ability to accomplish its designated mission.

Susceptibility The characteristics of a system that make it unable to avoid being engaged by threats on the battlefield. This covers being detected, tracked, targeted, and engaged, up to the point of being hit.

Vulnerability The characteristics of a system that cause it to suffer degradation (loss or reduction of capability to perform the designated mission) as a result of having been subjected to a hostile environment on the battlefield. It is generally an assumption in vulnerability studies that the threat has engaged the target.

Lethality The ability of a weapon system to cause the loss of, or degradation in, the ability of a target system to complete its designated mission. Often for direct-fire weapons, the delivery of the threat from launch to target impact is integral to the lethality analysis. For indirect-fire weapons, studies often begin with warhead initiation in the neighborhood of the target.

Battle damage repair The ability of a system to be reconstituted to a full, or partial, capability after suffering damage from a threat in a hostile environment. This term represents a system's return to some level of its functionality prior to a previous threat engagement.

Target A platform or materiel that performs a military mission(s). This could be on the battlefield itself or behind the lines in a logistics, or a command, communications, or control role. Typically, targets have been tanks, trucks, missile launchers, and other ground-mobile vehicles, as well as fixed- and rotary-wing aircraft, missiles, and spacecraft. Ground-fixed targets such as buildings and buried bunkers are also often considered targets.

Threat Either the weapon system or its effects that impinge on, and interact with, a target, potentially causing damage to the target. This can be a specific threat, such as a particular shaped charge (SC) warhead, or a class of threats such as SCs in general. This volume focuses primarily on penetrating threats, blast threats, threats that cause in-structure shock, and threats that generate fumes or fire inside the target.

Survivability is a function of susceptibility, vulnerability, and repairability. The greater a system's susceptibility and/or vulnerability, the less is its survivability. Likewise, the greater a system's ability to repair or reconstitute itself, the greater is its survivability. Thus, the term *survivability* covers a much larger area than the term *vulnerability*. For instance, the designer can enhance the survivability of a system, without reducing its vulnerability, by making it more difficult to find (e.g., by adding a stealth capability), more difficult to hit (e.g., increasing its agility), or more easily repairable.

Vulnerability is usually considered to be "target centric" in that the response of the target to various weapons effects is the primary concern. Lethality, on the other hand, is considered "weapon centric" because the performance of the weapon system against a target (or a set of targets) is studied. The physical interaction of the weapon effects with the target and the resulting response of the target are common to both areas. Because of this commonality, efforts accomplished in these two areas are often referred to together as V/L studies, tests, analysis, or assessments. Differences between vulnerability analysis and lethality analysis are mostly a matter of application and emphasis. Vulnerability analysis is most commonly used to characterize the ability of a friendly system to withstand enemy threats, although lethality analysis generally characterizes the ability of a friendly weapon to disable enemy systems. There are two important implications: 1) to be conservative, the analyst tends to prefer approximations that *overestimate* the effect of the weapon or threat for vulnerability analysis, although preferring approximations that *underestimate* the effect of the threat for lethality analysis; and 2) the description of the threat is more likely to be speculative, and this limitation is typically a less serious drawback in the case of vulnerability analysis. Dually, it is both more likely and more acceptable for the target to be speculative in the case of lethality analysis.

Because of these approximations and speculations, V/L analysis can be a complex engineering discipline. It can be viewed as involving four stages:

1) *Determining the relevant threat-engagement parameters and values.* These parameters typically include the type of threat, the engagement geometry, and, if appropriate, the functioning of the threat in the vicinity of the target (e.g., fuze functioning).

2) *Analyzing the elementary ballistic processes (e.g., penetration, spall, blast deformation).* Each of these processes is complicated in itself and is not completely describable in adequate first-principles models.

3) *Evaluating the response of major subsystems (e.g., mobility, firepower, communications) to damage.* This stage involves considerable systems engineering.

4) *Estimating the implications of these subsystem responses for tactical utility and mission success.* These implications are often viewed as the province of the military operations research analyst and beyond the actual V/L analysis, but they must at least be understood by the V/L analyst as an ultimate goal.

These stages correspond to some of the MMF levels described in Chapter 2.

As one might expect, there are numerous fuzzy boundaries between V/L analysis and other related engineering disciplines. For example, consider the following: 1) fragmentation data might be developed by a (warhead) system analyst, and the V/L analyst might be asked to develop the implications of the impacts of representative fragments; the system analyst might then combine the fragment effects to define a warhead's effectiveness; 2) conversely, the V/L analyst might be asked to determine the effects of an artillery barrage on a group of tanks; 3) ballistic data can be the responsibility of terminal ballisticians (at least for the more general characterizations), or the V/L analyst may need to develop data for a particular munition (based on data for similar situations or from the more general characterizations); 4) the implications of the damage might be defined by the V/L analyst (as in the case of the generic kill categories), or more functional data (such as degraded-states data) might be used by the military analyst in combat simulations.

The reader should be aware of these nuances but should also understand that V/L analysis involves all the aforementioned considerations regardless of who actually performs the parts thereof. The good V/L analyst may not actually do all of these things, but he or she must understand the contributions of the other players at the edges of a V/L analysis and must assume (at least some) responsibility for the quality of the input data supplied by others and the appropriateness of the results of the V/L analysis to other applications.

II. A Brief History of V/L Analysis

Major resources for the history presented in this section include texts such as Volumes I, II, and III of *Ballisticians in War and Peace* (Schmidt a; Schmidt b; Reed 1992a); the *Vulnerability Day* proceedings (Reed 1992b); and a number of Ballistic Research Laboratory (BRL) reports, including "Historical Perspectives on Vulnerability/Lethality Analysis" (Klopcic and Reed 1999) and "Current Simulation Methods in Military Systems Vulnerability Assessment" (Deitz et al. 1990).

A. Early History

Although the technology associated with V/L analysis goes back a long way, this type of study was not generally viewed as a unified discipline until the 1950s, and the analyses were, for the most part, ad hoc. By today's standards, the older methods might be viewed as crude, but in actuality they were often quite ingenious. For example, consider the following description of a surprisingly complete wound ballistics analysis from a century ago (Ezell 1981):

> In 1904 Colonels John T. Thompson (Ordnance Corps) and Louis A. LaGarde (Medical Corps) formed a board to select the caliber for a new Army pistol that was to replace the .38 caliber revolver. There was a particular concern for enough power to stop such threats as a fanatic Philippine Moro or a charging horse in their tracks. To this end, they tested a variety of calibers and sizes of bullets. Firings against 10 human cadavers, 16 steers, and 2 horses were conducted in the Philadelphia Polyclinic followed by firings at the Nelson Morris slaughterhouse at the Chicago Stockyards against live steers. Their tests and analyses led to the selection of the .45-caliber round and eventually to the .45 Colt Automatic Pistol.

In general, wound ballistics concerned itself with the empirical development of energy parameters to characterize the killing power of projectiles or fragments given hits in certain body areas. This led to such rules of thumb as the widely accepted standard that 58 ft-lbs of energy is needed to incapacitate a human. In addition, the vulnerability of materiel targets such as aircraft was, in the early days, largely based on presented areas of critical subsystems (engine, crew, etc.). These areas were often determined from drawings (or photographs or shadowgraphs) of the targets or by measurement on the target itself with devices such as planimeters. Sometimes techniques such as the weighing of paper cutouts were also used.

Of the many significant advances in vulnerability analysis during these early years, two stand out as milestones in formalizing the collection of data for analyzes. First, at Aberdeen Proving Ground (APG), around 1925, extensive live-fire testing of aircraft targets was performed. Anti-aircraft shells were evaluated by firing against aircraft until sufficient damage was done to require major repairs, machine-gun ammunition was fired against aircraft with running engines, bombs were detonated beneath the wings of aircraft, and bombs were fired upon. Second, in 1934, R. H. Kent (one of the founders of the BRL) suggested to the Office of the Chief of Ordnance (OCO) that, in order to evaluate the ability of a bursting munition to damage a target, the ballistician must know the number of fragments, their velocity as a function of size, the location of the shell from which they were expelled, and their drag characteristics.

The previously mentioned Volume I of "Ballisticians in War and Peace" (Schmidt a) provides a good deal of information on the activities in ballistics from 1914 through 1956.

B. World War II Through the Cold War

World War II was a turning point in military research. There was total dedication throughout the United States to defeating the Axis threat. Leading scientists and engineers (including Nobel laureates) rushed to support the war effort, and areas such as ballistics reached new levels of sophistication—feeding applications such as V/L analysis. Many of the luminaries of the time and their contributions are described by Goldstine (1993).

During World War II, numerous tests were conducted to observe the effects of various munitions against armored vehicles, including German Tiger and Panther tanks and the U.S. M26 General Pershing heavy tank. This effort was summarized in a collection of 1946 reports of the Office of Scientific Research and Development/National Defense Research Council, entitled *Effects of Impact and Explosion*. Unfortunately, this fascinating collection of data and methodology is long out of print.

By the end of World War II, the need for systematic collection of data on a wide variety of targets and munitions was well recognized. In an unpublished paper, Arthur Stein noted the following (Stein 1946; Klopcic and Reed 1999):

> In July 1945, The Office, Chief of Ordnance (OCO) directed that investigations be carried out to determine the optimum caliber for aircraft weapons, and this was supplemented in March 1946 to include anti-aircraft weapons. In

April 1947 a project was initiated to determine the vulnerability of combat aircraft and guided missiles to ordnance weapons.

How was this program, the most extensive of its kind in the world, conducted? The 1200 aircraft received, some brand new and still in crates, consisted of essentially all types of WWII Air Force and Navy combat aircraft. In addition, a large number of off-line tests were conducted. Most of the work was for use of the Air Force; a much lesser amount for the Navy. First, procedures had to be developed for measuring the vulnerability of various aircraft components to different types of anti-air weapons. Second, these procedures were used in experimental and test work to obtain the detailed data. Hypotheses developed during the course of the program were tested using additional firings. Then, on the basis of the findings design changes were recommended to improve aircraft passive defense; these were made available to the Air Force, Navy and to aircraft designers. In addition, the results were used as inputs to weapons effectiveness studies, particularly by BRL itself but also to new emerging government and industry operations research groups. Although obsolete WWII aircraft were used to obtain data, the major objective was the formulation of design principles and predicted capabilities of sufficient universality that they could be applied to combat aircraft generally, especially to aircraft still in the design stage. Recommendations for passive defense of specific aircraft were regarded as secondary.

In the latter part of 1948, a program was initiated to study the vulnerability of armored vehicles. Around 1954, trials were conducted in which tank-fired armor-piercing (AP), armor-piercing capped (APC), high-velocity armor-piercing (HVAP), high-explosive anti-tank (HEAT), and high-explosive plastic (HEP) projectiles were used to attack M4, M5, M24, M26, T29, M46, and T34 tanks, and hundreds of static firings of HEAT rounds were fired against stacks of steel and other armors.

Analysts documented the materials used in the tests, the conditions of the tests, the observed physical damage to vehicle components, and the values of three metrics called mobility kill (M-Kill), firepower kill (F-Kill), and catastrophic kill (K-Kill), as scored by persons experienced in tank operations and officially designated as "damage assessors." (See Chapter 3 for detailed descriptions of subsequent refinements of these kill metrics.)

Tank vulnerability analysts worked over large blueprints of tanks spread on drafting tables. Much of their time was spent trying to figure out what the tank was really like at some particular point of interest—what was the armor thickness, what was the armor obliquity, and into what compartment might residual jet, projectile fragments, or spall impinge.

The emergence of the digital computer greatly helped bring together the processing and storage of all of this interaction information, allowing the detailed representation of the configuration of targets and the tracing of myriad projectile paths into those targets. Even so, the computers of the 1950s (and considerably later) were unable to handle the large amounts of detail that would be associated with anything that might be called first-principle modeling of V/L phenomena (Goldstine 1993; Fritz 1994). For a detailed history of Army computing, the reader is encouraged to consult texts such as the previously cited *The Computer from Pascal to von Neumann* (Goldstine 1993) as well as "50 Years of Army Computing: From ENIAC to MSRC" (Bergin 2000).

In the decades following World War II, the V/L analysis process was enhanced through greater unification, better data, and more efficiency. The unification and efficiency came from ever-increasing computing capability and the creation of increasingly complex and powerful computer models for the analyses. These models allowed detailed geometric modeling of the target, ray-tracing of penetrators and fragments through the target, and assessment of component damage. Several references provide insight into the issues associated with the development of such models [Mathematical Applications Group, Inc. (MAGI) 1967; Muuss et al. 1983; Muuss 1991; Hoyt 1969; and Rapp 1983].

V/L models for armored ground systems went through a series of models from the early compartment models (in which, for example, the crew compartment and engine compartment are the components) through point-burst models (in which detailed component damage is modeled for penetrators and individual spall fragments) (Nail et al. 1979; Nail 1982). These were to be the progenitors of the stochastic models created later to evaluate Live Fire tests (Deitz and Ozolins 1989).

Better data came from a number of major data-gathering efforts. For example, in 1958, at the Third Tripartite Conference on Tank Armament held in the United Kingdom, a working group on anti-tank trials and evaluation was established. A year later, the group planned and conducted an extensive series of anti-tank tests on the M47 and M48 at the Canadian Armament Research and Development Establishment (CARDE). The original correlations between the penetration parameters and the damage were based largely on this test series.

C. More Recent Times

There was a dearth of full-scale testing of the vulnerability of armored vehicles from the CARDE tests in the late 1950s until the middle 1980s, when concerns about the Bradley Fighting Vehicle (BFV) led to Live Fire testing legislation (Title 10, U.S. Code Section 2366, Chapter 139). [Within the V/L community, initial capitalization is commonly used to distinguish between the term for the official, statutorily mandated test program ("Live Fire") and the more generic term used to describe a type of test or experiment ("live fire").] On September 19, 1984, defense officials briefed a joint session of Congress on the need for realistic, full-scale testing of the vulnerability or lethality of all major U.S. combat systems. Such Live Fire testing was mandated in the FY86 Department of Defense (DOD) Authorization Act, which has been elaborated on since but is still in effect. This law has made possible the full-scale (or at least large-scale) testing of expensive and scarce hardware that otherwise would probably not be made available.

A related program [the Joint Live Fire (JLF) test series] was chartered in March of 1984 by the Director of Defense Test and Evaluation, Office of the Secretary of Defense (OSD), and provided the impetus for reviving the databases from earlier studies. The test series was established to develop data on the lethality of U.S. weapons against Soviet systems and the vulnerability of U.S. systems to Soviet weapons. The responsibility for tests against armored systems was assigned to the

Joint Technical Coordinating Group for Munitions Effectiveness (JTCG/ME), and the responsibility for tests against aircraft was assigned to the Joint Technical Coordinating Group on Aircraft Survivability (JTCG/AS).

In the mid- to late-1980s, analysts began to concern themselves with developing a rigorous framework for describing the V/L process. Spurred by a reconsideration of the probabilistic metrics of V/L analysis of tanks, the concept of the vulnerability levels, or spaces, was born. Out of this first taxonomy grew the MMF, the conceptual linkage between survivability and mission effectiveness (Sheehan et al. 2003).

An adjunct to analysis of Live Fire tests of the BFV and the Abrams tank was the development of computer environments that coordinate a variety of V/L analysis processes. The environments brought together items such as target data files, threat data files, analysis programs, and output files, and proved to be particularly useful in providing a user-friendly way to perform V/L analysis. Two such environments are the Modular Unix-based Vulnerability Estimation Suite (MUVES) and Modular Effectiveness Vulnerability Analysis (MEVA), which are both discussed in detail in Chapter 4 (Hanes et al. 1991; Dunn et al. 1998).

Not surprisingly, modeling a large-scale military system, such as an air-defense unit with radar vans, command center, and missile launchers, poses special problems. In the early 1980s, the Army Unit Resiliency Analysis (AURA) model was created for such analyses (Klopcic 1984). AURA was a large, interconnected collection of analysis models that provided a detailed evaluation of the ability of an Army unit to accomplish a series of missions in a combat scenario. Although AURA is no longer used, the need for this type of multi-mission evaluation is even greater today than it was 25 years ago. Recent advances in network technology and the demand for lighter, more deployable combat systems have required analysts to develop a number of new strategies for conducting vulnerability analyses on systems of systems (SoS) (Walbert 2004a).

One significant change brought about by networking and the sharing of tasks by multiple systems is an increased emphasis on susceptibility for ground combat systems. Long-range sensors and smart munitions help reduce the burden of the direct-fire engagement and add signature management and threat avoidance as key elements of survivability. Although platform vulnerability and crew survivability remain as vital concerns, the analyst must now look at a broader spectrum than ever before. In particular, signature management and threat avoidance are mission-dependent parameters, forcing vulnerability analysts to look beyond the traditional boundaries of their field and venture into force-level simulations.

III. Roles of V/L Analysis

A. Fielding New Materiel

The Army acquisition process begins with the perceived need for a military capability and results in the fielding of an item of materiel to provide that capability. The need for a particular capability may come from any of such sources as experience in actual combat, field exercises, and paper studies. During peacetime, it most often comes from a study of force performance based on a series of scenarios considered representative of the threats to be faced by U.S. forces in the near future.

1. User Needs and Technology Opportunities

The objective here is to determine the characteristics and capabilities needed to fight successfully against a set of hypothesized future threats in a representative set of scenarios that may be faced by future U.S. forces. A series of war games or other combat simulations is run using various mixes of current and notional future systems [e.g., vehicles; weapons; munitions; sensors; and command, control, communications, computers, and intelligence (C4I) equipment]. Proper evaluation of the contribution of the various systems to the battle outcome, as well as the correct play of the threat forces in the scenario, requires accurate V/L data and methodologies for the systems in the scenario. Often, a first cut at the characterization of the notional system requirements is made by noting the deficiencies in U.S. force performance in a base case composed only of current systems.

In the case of the current U.S. systems and known foreign and threat systems, actual V/L data resulting from measurements, experimentation, and testing are often available. For notional U.S. systems, the analyst is required to use data based on the best estimates of what will be available in the time frame of the scenario. These estimates often cover a range from conservative, minor improvements in current materiel to more daring extrapolations. In the case of notional future threat systems, the data are nearly all estimates based on the best available intelligence. Surrogates are used frequently in actual model runs.

By varying the force mix used in the war games and combat simulations, the performers of the study will find that certain weapon mixes are eliminated because they are unsuccessful in the scenario relative to other mixes, because they are too costly or too speculative. Examining those force mixes that are successful in the scenarios, that seem to be within reasonable cost constraints, and that appear to be technologically achievable (with acceptably high risk), decision makers identify areas for further consideration.

New systems included in the successful force mixes are typically subjected to sensitivity analyses: how the battle outcome would be affected if certain changes in capability were made. By iterating on this process, participants then identify those systems that fill a need not met by other systems in the time frame of the scenario. The capabilities of these systems then become needs.

After considering several alternative ways of satisfying those needs, individuals conducting further studies may down-select to one or more system concepts for which requirement and other acquisition documents are written and staffed. These documents often require further refinement to get a precise, usable definition of the V/L requirements.

In addition, developing meaningful V/L requirements is iterative in nature. For Army materiel, a Training and Doctrine Command (TRADOC) element leads this process. TRADOC is supported with studies by the relevant Research, Development, and Engineering Centers (RDECs), the U.S. Army Research Laboratory (ARL), the U.S. Army Materiel Systems Analysis Activity (AMSAA), the U.S. Army Test and Evaluation Command (ATEC), and other organizations, with the intent that requirements will be brought into line with feasible technological capabilities.

2. Research, Development, and Acquisition

As specific design concepts emerge, V/L models can be used to support advanced design efforts. In the best case, a detailed description of the conceptual vehicle or munition is made available for use in the V/L models early on. In other cases, only structural or subsystem concepts are available at first. In any case, the hypothesized designs can be analyzed with V/L models not only to determine whether the V/L requirements are likely to be met but also to help the designer achieve these requirements in the most efficient manner.

If the model runs determine that the V/L requirements are not likely to be met with the current design configuration of the system or munition, then either the design must change or the proponent for the system or munition must obtain permission to relax the requirements (or both). The preferable solution is to find a modification to the design that allows retention of the desired performance, that keeps cost and weight within bounds, and that increases survivability so that V/L requirements are met. However, sometimes the desirability of fielding the system or munition is judged to outweigh the need to meet the original V/L requirements, and so they are relaxed.

As the development nears completion, the system may be required by the previously mentioned Live Fire Law (Title 10, U.S. Code Section 2366, Chapter 139) to undergo Live Fire tests. These tests are intended to demonstrate the V/L characteristics of the system or munition via actual tests of live munitions against full-up vehicles. Perhaps more important, they also identity situations where vulnerability reduction (or lethality improvement) is required. V/L methodologies can be used in planning the sequence of Live Fire shots, not only by helping determine which shots should be included in the test, and in what order they should be performed to gain the most information and minimize the number of test vehicles destroyed, but by ensuring the data from the shots will address critical evaluation issues.

3. Estimating Logistics Burden

As a new system is developed for introduction into the force, it must meet a complex series of integrated logistics system (ILS) requirements. Among these are studies of the logistical implications of the new system. The V/L connection is the requirement for studies of manning and training and of expected spare parts stocking requirements.

War games, combat simulations, and logistics models may run typical scenarios to estimate numbers and types of systems damaged. The results of the killer–victim scoreboards (i.e., number of systems of each type killed/damaged by each weapon type) are used in analyses to determine the distribution of components killed for each system viewed as a target. This distribution, along with estimated rates of component failure from normal usage, feeds the requirements for spare parts stocks, for field and depot maintenance requirements, and for numbers of personnel with various repair and maintenance military occupational specialties (MOS). These models also estimate crew casualties and, consequently, the expected medical workload. It is thus clear that accurate V/L data and methodologies are required for good estimation of the impact of a new materiel system on the logistics burden.

4. Training and Mission Rehearsal

Ideally, soldiers are trained in realistic exercises or simulators wherein V/L effects are accurately accounted for. Of course, safety of the soldiers usually requires that field exercises not be conducted with live ammunition, and security requirements for handling and processing classified data also limit the realism of exercises and simulators.

For a number of years, combat results in field exercises for direct-fire weapons have been used to score the Multiple Integrated Laser Equipment System (MILES) low-power laser system. Each weapon (typically a vehicle) in the exercise is equipped with a low-power laser in lieu of live ammunition and a laser alarm that goes off when lased. When the laser registers a hit on the alarm, a kill occurs with a specified probability. Aside from the deficiency of playing all kills as K-Kills (target removed from further participation in exercise), this system has even more deficiencies because the effective range of the low-power laser is typically much less than that of a tank main gun or an anti-tank guided missile (ATGM).

One would like to evolve a training system in which realistic V/L data and methodologies could be used to simulate live shots. The use of distributed interactive simulation (DIS) simulators (which provide a consistent format for conducting real-time, platform-level war-gaming over numerous host computers) is a step in the right direction, but the V/L methodologies used in those models are typically a generation or so behind the current best. Also, the requirement to use unclassified data in many exercises compromises the accuracy and realism of the training.

The issues for mission rehearsal are similar to those for training except that, in the former, the scenario is a situation that the trainees are expected to face during a real mission they are about to undertake. Mission rehearsal is more likely to be permitted to use actual classified data though those data are not necessarily available on call.

B. Actual Combat

In real warfare, the results of V/L analyses are used [typically from Joint Munitions Effectiveness Manual (JMEM) products] to estimate optimal munition loads for aircraft, volumes and types of ammunition for artillery fire missions, survivability countermeasures to reduce the lethality of enemy fires, etc. Clearly, the use of accurate V/L data here can increase both the efficiency of munition usage and the survivability of equipment, and it can reduce the loss of life from incorrectly estimated threats.

In addition, V/L estimates often reflect changing battlefield conditions, based on actual experience, intelligence data, and/or captured equipment. The Vietnam War and the more recent wars in Southwest Asia have demonstrated the need for rapid vulnerability analyses to protect U.S. soldiers from an adaptive, asymmetric threat. As we address this ever-changing threat, measures of effectiveness (MOE) continue to evolve. We can no longer attempt to apply a single number, developed under conventional assumptions, to describe the vulnerability of a class of combat systems to a class of munitions in a highly specialized setting. In the end, each war brings with it a new set of assumptions, applications, and analyses.

Chapter 2

Vulnerability/Lethality Analysis Process

I. The Missions and Means Framework

THE previously mentioned Live Fire test legislation brought increased attention to the testing and prediction of ballistic live-fire phenomenology. An outgrowth of the efforts to rationalize the comparison of tests and predictions was the V/L Taxonomy (Deitz et al. 1990; Deitz and Ozolins 1989; Klopcic et al. 1992; Deitz 1996; Deitz and Starks 1995; Klopcic 1999). This taxonomy was developed from a platform (target)-centric perspective, with a numbering system related to characteristics of the platform being described. As shown in Fig. 2.1, the taxonomy characterizes the platform (tank, aircraft, or other target) in several levels (or sets) of measurable characteristics. These levels represent the "spaces" of potential configurations of outcomes associated with each level.

In terms of a ballistic threat–target interaction, level 1 describes the initial state of components before the event. Level 2 describes the state of those same components after the event (i.e., which components function and which do not). Level 3 describes the corresponding capability of the platform after the event (i.e., what is the vehicle's ability to move, shoot, communicate, etc.). Level 4 describes the corresponding state of combat utility or mission effectiveness after the event (i.e., what is the vehicle's ability to achieve its mission). Moreover, the "operators" between the levels represent the transformations from a fully operational, undamaged platform to a damaged target with reduced capabilities.

The most difficult aspect of applying this taxonomy is the transition from level 3 (platform capability) to level 4 (combat utility). The traditional role of the vulnerability analyst stopped with platform capability; combat utility was left to the military specialists, who tried to generalize over all missions and conditions. Unfortunately, because probabilities dominated the set of measures of vulnerability inferring higher-level, mission-specific measures from aggregated probabilities was a tenuous, if not downright incorrect, process.

Accordingly, the realization that vulnerability analyses had to permit introduction of specific sets of missions and combat tasks led to a significant extension of the V/L Taxonomy to what is now the MMF (Sheehan et al. 2003). Incorporating the four levels of the V/L Taxonomy, the MMF integrated additional elements, such as mission purpose, tasks, context/environment, and a time/location index, to more comprehensively specify military operations (see Fig. 2.2). Also included were the three previously mentioned operators, as well as a fourth operator, the

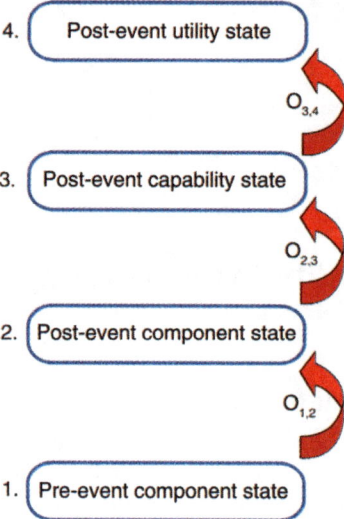

Fig. 2.1 The V/L Taxonomy.

$O_{4,1}$, that transforms level 4 task sequences into level 1 interaction conditions. This connection of the level 4 to level 1 effectively "resets the trigger" and enables the MMF to expand its focus beyond a single event and address issues associated with multiple interactions and sustained operations (e.g., repair, replenishment,

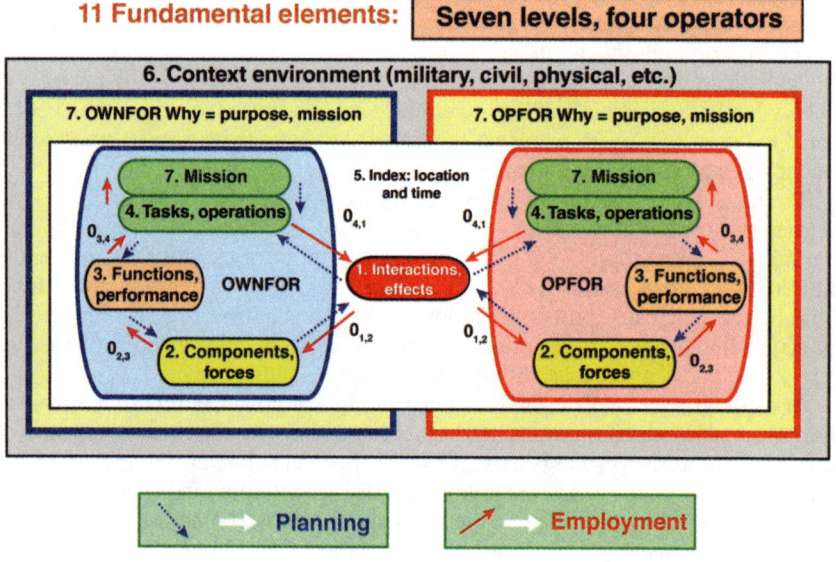

Fig. 2.2 The MMF.

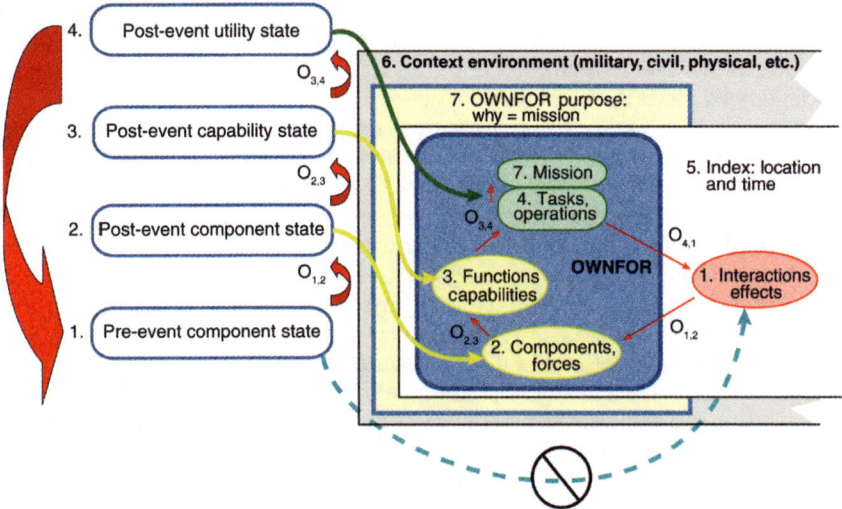

Fig. 2.3 Mapping of the V/L Taxonomy to the MMF.

reliability, etc.). This is the reason why the level 1s are not equivalent in the V/L Taxonomy-to-MMF mapping in Fig. 2.3.

In one sense, the upper levels of the MMF collectively represent the "missions" part of the framework (i.e., a specification of what needs to be accomplished), whereas the lower levels and operators collectively represent the "means" by which those missions are accomplished. In so doing, the MMF is intended to provide a compatible extension not only to the V/L Taxonomy but also to the DOD architecture framework and the Functional Descriptions of the Mission Space (FDMS) (Wolfowitz 2003; Haddix 2003).

Also of note is the MMF's two-sided nature. As illustrated in Fig. 2.2, the missions, tasks, and components of the opposing forces (OPFOR) coalition can be seen to influence the mission prosecution and outcome of the friendly forces (OWNFOR) coalition and vice versa. As such, the MMF provides a symmetric representation of an asymmetric (perhaps decidedly asymmetric) conflict.

Observability is also an important property of the framework. Because each weapon–target interaction event is a single, physical process, the quantities involved in most of the levels are inherently and directly observable. That is, one can (at least in principle) observe and/or measure the quantities of, say, the encounter conditions (level 1), the results of physical processes (level 2), and the residual capability of the system (level 3). Unfortunately, V/L practitioners have not always been able to incorporate these types of observable metrics in their analyses.

As indicated in Chapter 1, this text is organized along the lines of the MMF structure. In the rest of this chapter, we discuss level 1 encounter conditions (initial representation) and damage mechanisms (Sec. 2.II), level 2 component and/or personnel functional states as a consequence of damage mechanisms (Sec. 2.III), the functional states at level 2 to the performance (capability states) at level 3 (Sec. 2.IV), and the implications of the capability states at level 3 to the utility states at level 4 and beyond (Sec. 2.V). In addition, Appendix E provides a more in-depth look at MMF components, processes, and more recent developments.

Chapter 3 of this book details the various measures used to describe conditions at each of these levels; Chapter 4 describes the modeling and simulation tools and methods that implement the methods described in Chapter 3; and Chapter 5 presents applications of these methods to various parts of the life cycle of combat systems. In addition, several topics requiring more detailed technical discussion are presented in Appendices A–D. Note that these appendices are in no way comprehensive and are not required for a basic understanding of the topic of V/L analysis. They do, however, provide the interested reader with an in-depth look at how vulnerability analysts treat these particular topics.

Present-day V/L methodologies compute the effects of the physical damage mechanisms and then consider components as being either functional or nonfunctional. For these methodologies, the elements of the level 2 state vector are binary. Similarly, the capabilities of a target can be expressed as a capability state vector, which is an array of M elements whose values reflect the capability possessed by the associated target. For example, a truck might have the capability to move and support a load. This (exceptionally coarse) capability state is describable by a length-two vector in which the first element might be the truck's maximum speed capability and the second element the maximum load.

From the preceding discussion, the issue of granularity (fineness of detail) arises in at least two ways in the expression of state vectors. First, it is obviously present in the values allowed within each element. In the previous example of the truck, the representation of its ability to move could be restricted to a binary element, as is customarily done in damage states: the truck is assessed as either able to move (element = 1) or unable to move (element = 0). Alternately, the element might be expressed with infinite granularity with a real number expressing the maximum speed, or a range of capabilities, such as shown in Table 2.1.

There is, however, a deeper question of granularity, namely the number and associations of elements in a state vector. In the truck example, rather than representing the capability to move by the single element for maximum speed, a number of elements could have been used, quantifying the truck's ability to turn, stop, climb, or pull.

To date, no general method has been found to quantify, let alone specify, the level of granularity that is required for a particular analysis. Nonetheless, it is

Table 2.1 Sample capability

Max speed, s (mph)	Element value
s = 0	0
0 < s ≤ 10	1
10 < s ≤ 30	2
s > 30	3

highly desirable to have a consistent level of granularity throughout an analysis. It would be wasteful to specify radios down to the circuit board at level 2 if the capability to communicate were accounted for as a single binary element at level 3. Conversely, the capability to communicate (level 3) at various frequencies could not be separately evaluated if communication equipment were physically represented as a single black box (level 2).

Of course, a level of granularity is often imposed by available data. Because the details of the interior of a foreign target may not be known, it makes little sense to specify hypothesized equipment to the subcomponent level. Hence, the granularity throughout a foreign target analysis may be less than that employed in a U.S. target analysis.

It is also essential to recognize that more detail is not equivalent to finer granularity. For example, a truck contributes its movement capability to its unit. Quantifying movement in terms of the finer elements (speed, agility, stopping) is a matter of granularity. However, the truck can also be said to have, for example, the capability to pump fuel to its engine or lubricate its moving parts. In this case, the increased detail is not a description of movement itself but rather of the factors that support movement. Pumping fuel to itself is not a capability that the truck contributes to its unit; to include it in a capability vector would be to change the meaning of the vector, and thus is not a matter of granularity.

In addition, there are environmental factors (such as terrain in the case of ground vehicles or atmospheric factors in the case of aircraft) that, although not part of the platform being analyzed, certainly interact with the capabilities of the platform. At first, it would seem that a factor such as mud, which could affect the top speed of a tracked vehicle, would be considered a part of the scenario. Upon further consideration, one notes that there is an essential difference between a capability and a requirement. The requirement to travel in mud depends upon the scenario in which the system is being analyzed, while the capability of a platform to move in mud depends on the motive power of the system, the configuration of the tracks, the ground clearance, and other engineering factors. Thus, the analysis of the capability to move in mud is an engineering analysis (carried out in level 3). With regard to environmental factors in the V/L process, the following are noted: 1) capabilities to perform in various environmental conditions are, as mentioned previously, included as separate elements in the capability state vector; and 2) V/L analysis output values for these elements can be combined as indicated by scenario, tactics, and other factors, in assessing battlefield utility, which is the realm of level 4 analysis.

The damage and capability states discussed previously may intrinsically vary with time after attack. For example, should an attack result in a small perforation of a fuel tank, the capability to move may not be lost until long after the attack. As with environmental factors, the importance of a delayed mobility loss depends on the scenario being analyzed. If the scenario involves extensive movement to contact, the fuel tank leak might prevent the system from reaching the battle and thus constitute a kill. (For a discussion on kill criteria, see Chapter 3.) On the other hand, in a firefight scenario, the hit might have no affect on the outcome. To be able to roll-up level 3 metrics as needed by the user, it is important that analyses take into account and preserve the time dependence of the results. Although developers are striving to address the issues of damage and failure over time for future

V/L software tools, most computer models now in use still assume that component damage and the resulting loss of system capability occur simultaneously and instantaneously; the analyst must therefore use best judgment as to the validity of such an assumption for a particular application.

The V/L Taxonomy and the MMF also serve as the basis for attempts to define a rigorous mathematics for the V/L process. The elements of the initial configuration, resulting physical damage, system capability, and tactical utility states have been strictly defined as discrete, observable entities. When assembled into an n-tuple, such a collection of elements may be seen to constitute a vector. If, at a given level, one were to consider all possible states and express each of them as a vector, the collection of all such possible vectors forms a set with promising properties. For example, given a few extensions to this concept, the set of states has the attributes of a vector space. Some thought has been given to defining meaningful metrics on such a space with the goal of rigorously addressing such questions as the closeness of two different states (Walbert 1994).

It is also important to consider the issue of probabilities. Probabilistic outputs play a vital role in many of the current scenario-playing simulations, as well as in combining several V/L results into a single number for comparisons and trade-offs. However, to this point, the methodology presented here has dealt only with discrete, observable entities. It is important, therefore, to associate these entities with probabilities. The process is as follows.

As emphasized previously, a single incident can have but one specific, discrete result. However, there are many factors at play in the interaction of a threat and target and in the subsequent response of the target. For the purpose of V/L analysis, these factors may be considered to be stochastic in nature. Many of these factors are included in the physical process algorithms that take a V/L analysis from level 1 to level 2 to level 3. Thus, if a particular encounter (specific initial configuration) were repeated several times, one might expect several different resulting physical states to result and, correspondingly, several different resulting capability and utility states. That is to say, any one attack will produce one and only one result; however, which of the many possible results will occur is a probabilistic function. To estimate this function, the V/L process is exercised many times for each initial configuration, and the resulting physical, capability, and utility states are recorded. Given that the physical processes have been well modeled, the probability of any particular result should be inferable from the frequencies of occurrence of each state. It is also important to note that these frequencies are end points; they play no subsequent role in the V/L analysis. For example, frequencies or averages of damage states are not used to compute some sort of average residual capability. Rather, capability states are stochastically realized, one at a time; it is from these states that frequencies and averages are found and probabilities inferred.

In the context of modern V/L analysis, the various models employed produce one output configuration for each input configuration. That is not to say that these models are deterministic; they are in fact stochastic. Repeated applications of a model for a given input will produce various output configurations. For example, consider the following:

1) In level 1, an actual firing of a weapon has a dispersion pattern from which the impact on the target is a statistical sample. When modeling this, each point in space 1 of level 1 can be considered to be a sample from that distribution.

2) In level 2, an actual perforation of armor produces a cloud of fragments of random number, size, and velocity. When modeling this, each replication of the model must use samples from those distributions.

3) In level 3, the result of partial power loss in a tank can be analyzed by a mobility model, each replication of which might predict a failed or a successful attempt to negotiate a muddy field, depending on random factors associated with the mobility process.

Understandably, users of the data (e.g., tacticians and system engineers) are usually not interested in the point values that result from a V/L analysis. Rather, they are interested in some of the summary statistics of the points. These statistics typically come from repeated (stochastic) application of the mappings and aggregations of the resulting points.

System engineers might be interested in level 2 data but not in the specific n-tuples that represent which components are killed or not killed in a particular replication. More likely, they would prefer n-tuples that represent the probabilities that the components are killed as a result of some set of particularly significant encounters.

Likewise, tacticians might be interested in level 3 data, but not in the specific n-tuples that represent a particular set of performance degradations resulting from a particular replication. Again, they would most likely prefer a set of probability distributions for the capabilities that result from some set of encounters (e.g., the probability that a truck is destroyed by an artillery round for various distances between the functioning round and the truck).

Finally, of course, the user may desire numbers that reflect more broad scenario-dependent issues. In this case, a full MMF analysis is required. Although this book is devoted to ballistic threats, the MMF and the V/L Taxonomy are equally applicable to nonballistic applications such as directed energy and chemical threats. In these cases, the mapping from level 1 to level 2 for directed energy (electromagnetic) threats might go something as below.

The intensity of the radiation incidents on the target would be specified in level 1. Next:

1) The process(es) by which the radiation can enter the target (particularly into critical areas) would then be evaluated. This may include rather normal paths such as antennas, optics, and cables and other paths such as joints, cracks, and nonmetallic materials.

2) How the radiation is distributed within the target (a form of cavity analysis to determine peaks and nodes) would then be analyzed. This determines which components are threatened. The geometric description of the target must be adequate to allow such an analysis.

3) Based on a criticality analysis (which is described in Sec. 2.III), damage to components (e.g., burnout) would be determined. Note that components may be, for example, an element in a chip, a chip itself, a circuit board, or an assembly, depending on the granularity of the analysis.

4) Finally, the resulting dysfunction of the component would be assessed. To what level does it perform compared to critical specifications?

II. Initial Representation and Damage Mechanisms

This section describes various mechanisms that constitute the first part of the $O_{1,2}$ mapping from level 1 to level 2. It is devoted to the physical processes by which the damage mechanism enters the target, traverses the target, and creates physical damage to components. Greater detail on these topics is found in the appendices.

When a target is subjected to ballistic attack, several mechanisms may work to inflict damage. To evaluate the vulnerability of a target, the analyst must address the effects of each mechanism individually as well as the synergistic effects of certain mechanisms in combination.

The primary damage mechanisms that come into play in ballistic attack are as follows:

Ballistic penetration Includes the effects of KE penetrators (including both projectiles and fragments), SC, and explosively formed penetrators (EFPs).

Blast The compression wave in free air produced by a detonation.

Ballistic shock Vibration in the structure of the target produced by a detonation or an impact on the target.

Vaporifics The tendency of high-velocity, finely divided metal particles, which are produced by ballistic penetration, to ignite in air and release energy. The effects of noxious gases produced by burning of various materials in air are also considered.

Directed energy and chemical effects The effects produced by lasers, high-powered microwaves, etc., and the interaction of threat-delivered chemical compounds with materiel and/or personnel.

Synergistic effects The augmented effects produced by a combination of mechanisms, which often are greater than the sum of the effects taken individually.

Secondary effects Effects such as fires, particularly in fuels, lubricants, and ammunition.

When it comes to the role these effects play in V/L analysis, much effort has been expended to understand the various physical phenomena involved and derive mathematical models to describe them. Considerable progress is also being made improving methodologies for analyzing ignition and sustainment of fuel fires, internal and external blast damage to structures, ballistic shock damage to components, and ammunition reactions. Although models currently available tend to be limited in the variable range wherein acceptable predictions can be made, the state-of-the-art is constantly improving.

In the broad sense of the term, ballistic *penetration* refers to any instance of a moving rigid or fluid body intruding into a second body and thereby creating a cavity in it. When the cavity in the second body extends completely through, the cavity and the event are termed a *perforation*. Munitions that carry penetrators are categorized into two basic classes, based on when and how the penetrators are formed. The first (and older) category is the KE munition. For this category, the penetrator acquires its final shape during its fabrication and acquires velocity by means of a propulsion system. For the second category, the chemical energy (CE) munition contains an explosive that detonates to deform a metal liner, which becomes the penetrator. Among these are tank-fired HEAT munitions and missile warheads producing shaped-charge jets (SCJs) and warheads producing EFPs, formerly called self-forging fragments (SFFs). Further details on these various munitions can be found in texts such as "Design of Projectiles for Terminal

Ballistic Effects" and *Fundamentals of Shaped Charges* (U.S. Department of Defense 1997; Walters and Zukas 1989).

There are also fragments that can penetrate the target. These can be generated by an HE warhead (which is discussed in Sec. 2.III.C) or result from another penetrator's "spalling" pieces of material off a back surface of the target. These spall fragments are components of behind-armor debris (BAD).

Ballistic penetration can degrade a target in several ways. If the object receiving the impact is a structural element, it may be severed, or it may be weakened to the extent that it will fail under load. If a functioning component of a target is hit, mechanical parts may be cut, electrical circuits may be broken, or fluid containers may be perforated.

The analytical process of evaluating the vulnerability of an entire target to a particular munition involves selection of a large number of hypothetical shotlines passing through the target. It is noted which components of the target lie along each shotline. This process identifies the set of components that are subject to being hit by the penetrator. The process must also estimate the level of damage that would be inflicted on each component. The other key phase uses analytical penetration models to make the following estimations: 1) whether or not the component would be perforated (which depends, of course, on whether all previous components along the shotline have been perforated); 2) the remaining penetrating capability if the component is perforated; and 3) the size of the hole in the component.

A. Kinetic Energy

1. Projectiles

The analysis of the penetration of KE projectiles is a complex problem in continuum mechanics, especially against targets with complex geometry. However, three rules of thumb are useful: 1) penetration capability is directly related to the penetrator's length; 2) penetration capability increases with the penetrator's density; and 3) subsequent penetration capability is reduced if a penetrator breaks up; therefore, toughness is an important property for penetrators.

The emergence of the tank during World War I and the continuous development of armor since then have stimulated the growth and evolution of the KE penetrator. First in the evolution of KE rounds fired from tank guns was the full-bore shot, so called because its major body diameter was essentially that of the cannon bore. The full-bore shot, referred to as AP, had a penetrator with an ogival (pointed-arch-shaped) nose. The steel at the nose was hardened so that it would have sufficient strength to withstand the stresses caused by impact and penetration. In addition, while the nose was intended to be hard enough to resist deformation, it also had to be resistant to cracking or shattering (which can happen with excessively hard, and thus brittle, steels). In addition, the penetrator body was expected to be of sufficient strength to withstand bending stresses.

Figure 2.4 shows that a typical AP shot projectile is equipped with rotating bands and a tracer. A rotating band is used to 1) impart spin to the projectile by engaging the rifled groove of the gun tube, and 2) serve as an obturator, thus

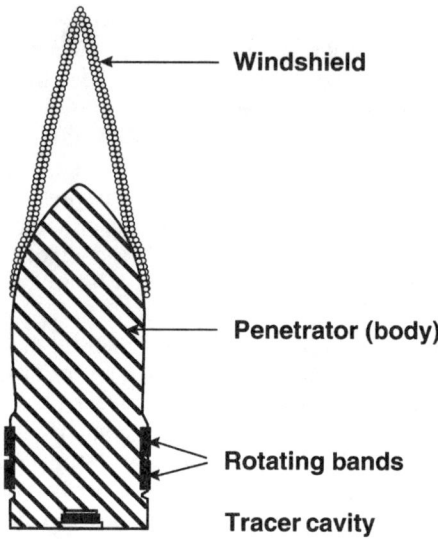

Fig. 2.4 Projectile T182, 105-mm AP (U.S., 1951) (Gillich et al. 1977).

inhibiting propulsion gases from blowing by the base of the projectile. The tracer produces a visual trail, which allows the gunner to modify his aim for a follow-up shot.

An advance that allowed the defeat of more massive and toughened armor without the need for differential hardening of the projectile body is the APC design shown in Fig. 2.5, which illustrates the application of a hardened APC in front of the main penetrator body.

Unfortunately (at least in terms of lethality), the increase in weight that came with additional length in newer projectile designs also resulted in a substantial reduction in muzzle velocity. Moreover, increased drag associated with the large diameter of the full-bore penetrator caused a substantial reduction in velocity during flight, which in turn caused an increased time of flight and decreased impact velocity at long range. The result was a loss of accuracy and armor perforation capability.

To respond to this situation and obtain higher muzzle velocities from existing guns, a family of ammunition having lighter payloads was developed. These HVAP rounds achieve maximum velocities of about 1500 m/s. To retain the in-bore advantage of a large diameter without the cost in drag paid by the full-bore penetrators, designers developed the armor-piercing discarding-sabot (APDS) round, which has a tungsten carbide core, either capped or uncapped, placed inside steel or light-alloy sheaths. This subcaliber penetrator is in turn placed inside a full-caliber carrier, which is called a sabot.

The sabot (a French term for "wooden shoe") is designed to attach to the subcaliber projectile and travel with it through the gun bore, thus imparting velocity and, in some cases, spin to the round before separating from it upon emergence

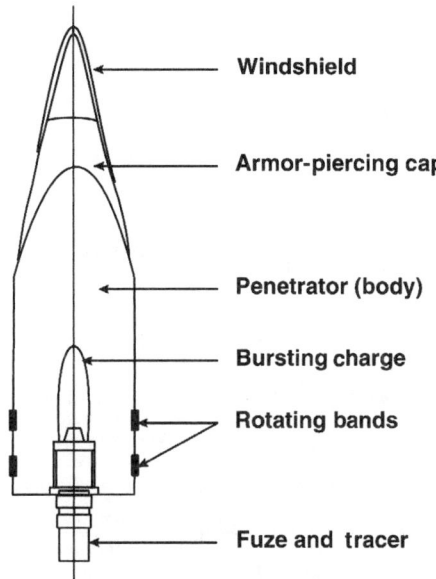

Fig. 2.5 Projectile, BR-412D, 100-mm APC (USSR, 1950s).

from the muzzle. To do this, the bourrelet (a French term for "pad") and the rotating bands are precisely machined to be slightly wider than the body of the sabot to fit against the walls (or the lands and grooves, in the case of a rifled tube) of the gun bore (see Fig. 2.6). This keeps the penetrator nose centered while traveling through the bore and prevents gas blow-by at the rear of the projectile. Sabots are usually made of some lightweight material such as aluminum or magnesium-zirconium alloy to minimize the so-called parasitic mass. Thus, when the sabot separates, the penetrator retains much of its momentum. The M392 projectile shown in Fig. 2.4 illustrates the high-velocity, armor-piercing, discarding sabot (HVAPDS) design.

Advances in armor technology in the 1970s created the need for longer penetrator cores and better penetration capability. The result was the development of the armor-piercing, fin-stabilized, discarding-sabot (APFSDS) round (shown in Fig. 2.7). The penetrator cores of APFSDS rounds typically have length-to-diameter ratios of 20 and beyond, and are made of high-density materials that are alloys of depleted uranium (DU) or tungsten.

Guns of a bore diameter of about 40 mm or less are considered to be medium or small caliber. Typical weapons include 0.30-cal., 5.56-mm, and 7.62-mm rifles; 0.30-cal. and 0.50-cal. machine guns; 25-mm tank autocannons; and 20-mm and 30-mm aircraft cannons. The development of small-caliber AP projectiles for such weapons paralleled development of tank-gun KE rounds. Beginning with full-bore steel shots, the development of small-caliber AP rounds has advanced to the use of high-density penetrator cores with HVAP and APDS

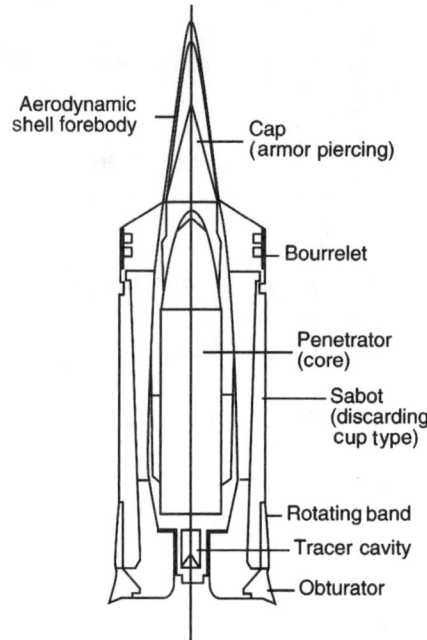

Fig. 2.6 Projectile M392A2, 105-mm APDS (U.S., 1957) (Gillich et al. 1977).

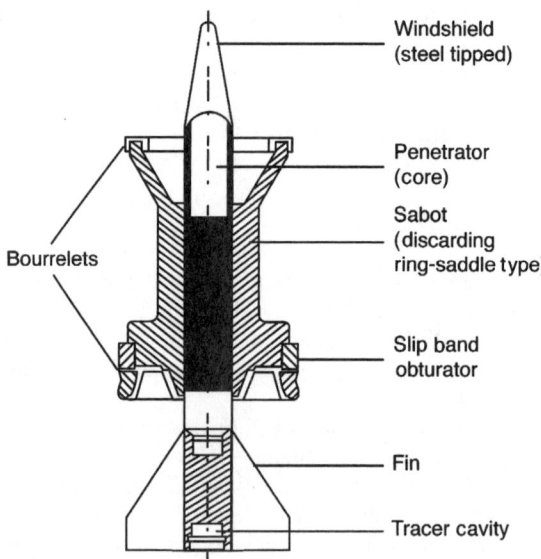

Fig. 2.7 Projectile M774, 105-mm APFSDS-T (U.S., 1980) (Gillich et al. 1977).

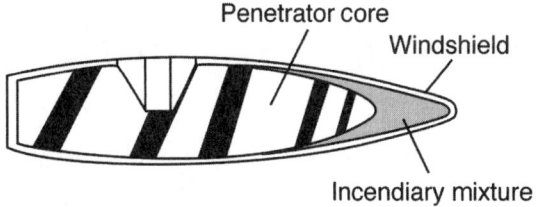

Fig. 2.8   Bullet, M8, 0.50-cal. API (U.S., 1940s) (U.S. Army Materiel Command 1962).

technologies in fielded systems. For the 30-mm cannon, the Air Force GAU-8 projectile uses a DU core in a nondiscarding aluminum carrier. The BFV fires APDS, APFSDS, and high-explosive incendiary-tracer (HEI-T) rounds from its 25-mm gun.

Related to the small-caliber AP bullet is the armor-piercing incendiary (API) round. On this projectile, a windshield protrudes significantly ahead of the penetrator core and incendiary mixture fills the space between, as illustrated in Fig. 2.8. The incendiary mixture is intended to ignite the fuel supply on board vehicles or aircraft. Fuel fires do not generally occur if the incendiary projectile functions only within the fuel cell and below the fuel level. Fire is far more probable if the projectile functions on the skin or armor of the target, close to the fuel cell, and then the penetrator core perforates the cell. In that case, perforation of the cell sprays fuel into the burning incendiary mixture outside the cell.

2. Fragments

In addition to the projectiles mentioned previously, fragments represent yet another form of KE ballistic penetrator. The general features of compact-fragment, high-velocity, multiple-plate penetration are shown in Fig. 2.9. Note that the fragment loses some of its mass near the impact side of the first plate because of shock-erosion and extrusion-shear mass loss mechanisms (and/or lateral erosion at high impact obliquities). The remainder of the fragment mass and most of the plate plug removed by the penetration appear behind the plate. Because momentum and energy are conserved, the residual velocity of the center of mass of the combined plate/penetrator system (including any particles sprayed out the front side of the plate) is less than the impact velocity.

For low-impact velocities, the residual fragment and plate plug will remain intact. For impact velocities above the penetrator or plate plug fracture threshold velocities, the material behind the plate will form a debris cloud, as shown in Fig. 2.9. The numbers of penetrator and plate debris particles in the cloud increase with increasing impact speed. For example, steel cubes impacting aluminum plates at speeds near 4000 ft/s fracture into 3 to 5 particles, but at speeds near 8000 ft/s, they will produce about 100 debris particles (not counting tiny dust-like particles) (Yatteau et al. 1994). In one series of experiments involving steel spheres striking copper plates at around 13,000 ft/s, researchers counted approximately

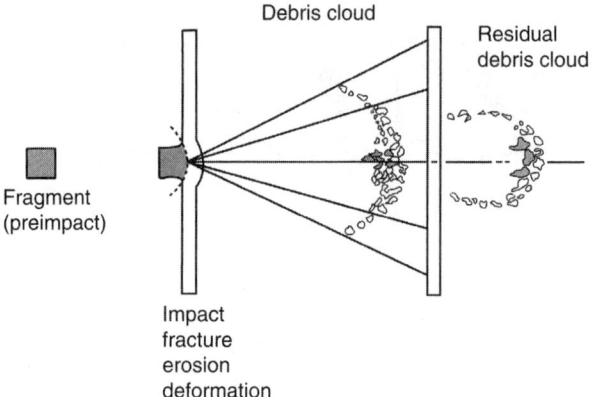

Fig. 2.9 Characteristics of high-velocity, multiple-plate penetration.

800 individual steel and copper particles and, using a sieve analysis of the smaller particles, estimated an additional 18,000 particles (Yatteau et al. 1995).

The shape of the debris cloud reflects the spatial variations in particle velocity and depends on fragment material and shape, impact orientation, velocity, obliquity, and plate material and thickness. For compact steel fragments perforating steel or aluminum plates at impact speeds up to about 16,000 ft/s, the debris clouds are nearly ellipsoidal in shape, with most of the debris particles residing on or near the leading surface of the hollow, expanding cloud (Dickinson et al. 1987). Penetrator particles have been observed to extend back into the debris clouds for aluminum and copper spheres and cylinders impacting thin aluminum plates at between 11,000 and 21,000 ft/s (Piekutowski 1990). Behind the plate, debris particles disperse radially, and the penetrator particles' trajectories usually fall within a tighter cone than those of the plate particles.

In general, some of the particles will also perforate the next plate, and others will be stopped. The perforating particles can lose mass and velocity and will drive new plate material into the debris cloud. Thus, the debris cloud behind the second plate will generally contain fragments of the penetrator and of both plates. The debris cloud penetration process is repeated until each particle has either perforated all later plates or been stopped.

Figure 2.10 illustrates typical damage to a plate caused by the impact of a debris cloud. Such damage includes craters and holes produced by nonperforating and perforating particles, respectively. Depending on the number, mass, and velocity of the incident fragments, there can also be an enlarged central hole where the plate material is completely removed because of blowout, petaling, or plugging of the plate. For hypervelocity impacts, there may also be material spalled from the rear of the plate even by nonperforating particles. [At extremely high projectile impact velocities (typically 4000–10,000 m/s), the mechanics of KE projectile penetration change. At these speeds, the impact force of the projectile so overwhelms the strength of the target that the target behaves like a fluid. Penetrator

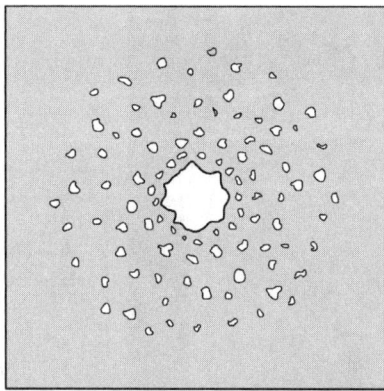

Fig. 2.10 Typical damage to a plate impacted by a high-velocity debris cloud.

and target densities are the only material parameters that have much influence on penetration. For more details on penetration mechanics in these velocity regimes, see MIL-HDBK-1226(AR) (U.S. Department of Defense 1997).] The plate will also be plastically deformed (dished) to an extent dependent on the magnitude and distribution of the impulse so that the particles' impacts transmit to the plate. For a more detailed discussion (and bibliography) on the penetration characteristics and modeling of fragmenting munitions, see Appendix A.

B. Shaped Charges

SCs, sometimes called lined-cavity charges, are commonly used in weapon systems intended to attack armored vehicles or heavily fortified positions. They are especially suited for these roles because they offer greater penetrating capability for a given warhead weight than other explosive devices, and they do not rely for their penetration capability on the KE with which they are delivered. SC warheads may be launched from cannons or may be delivered by guided missiles, unguided rockets, or mines.

As shown in the cutaway in Fig. 2.11, an SC is a hollow charge (i.e., a cylinder of explosive with a cavity in one end and a detonator at the opposite end), with the cavity assuming almost any geometric shape that is symmetrical about the centerline of the charge; the shape is usually a hemisphere or a cone. When the explosive charge is initiated, the cavity tends to focus the gaseous detonation products, thereby creating an intense localized force. When directed against a metal plate, for example, this concentrated force is capable of creating a deeper crater in the plate than a cylinder of explosive without a cavity. Moreover, when the cavity is lined, the penetration capability is even greater.

The increase in penetration resulting from the lined cavity charge is because of the formation of a jet when the explosive charge is detonated. Upon initiation of the charge, a spherical wave propagates outward from the point of initiation. This high-pressure shock wave moves extremely fast, typically around 8 km/s. When

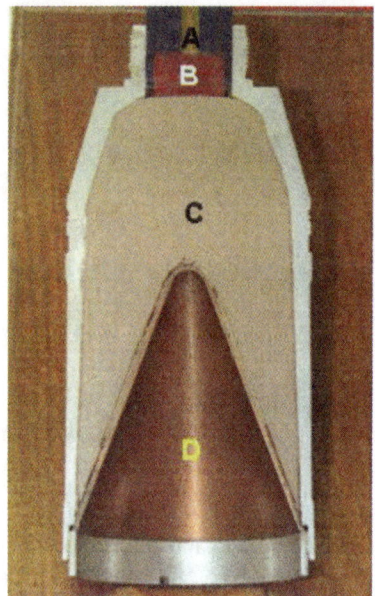

Fig. 2.11 SC cutaway.

the detonation front reaches the liner, the liner is subjected to the intense pressure of the front and begins to collapse. As the process of collapse progresses, the material of the liner collides on the axis of symmetry. This collision results in liner material under tremendous pressure being extruded at extremely high velocity along the axis of symmetry. This extruded material is called the jet.

Because there is more explosive energy available in the region of the liner apex than in the base region, the tip of the jet travels faster than the tail. This velocity gradient results in the jet stretching as it travels until it eventually breaks into a stream of particles. As is shown later, the stretching and eventual breaking of the jet ultimately limit the distance over which the jet maintains effective penetrating capability. SC warhead penetrating capability also decreases if jet formation occurs too close to the armor surface. SC warheads exhibit their maximum penetration capability when they are detonated at a specific distance from the armor surface. This distance is called the optimal standoff distance.

Analytical models capable of predicting the penetration of the jets from SC into a variety of target materials are extremely valuable in vulnerability analysis. Many disciplines are involved in deriving models because SC may be used to penetrate various metallic armors and structural materials, as well as nonmetallic materials such as cement, masonry, wood, and stone. Models also must address targets consisting of a single layer of monolithic material or several layers of different materials, including air or liquids. These models provide fast analytical predictions where the time and resources available prohibit analyses based on large hydrocode computer solutions or the direct measurement of penetration through field experiments.

C. Explosively Formed Penetrators

An EFP warhead consists of an explosive charge with a metallic lined cavity. The typical EFP warhead is a right circular cylinder (RCC) of explosive with an initiator on the axis at one end and a shallow cavity lined with thin metal at the opposite end. Iron, copper, and several high-density materials have been used as liner materials. Upon detonation of the explosive, the liner material forms into a high-velocity projectile with the ability to penetrate targets often at large distances.

Although some munitions could be classified as belonging to both the SC and EFP families, there is a distinction that can be made between EFP and SC warheads. SC warheads typically produce both an SCJ and a slug. The slug consists of residual-liner material, which is sacrificed to produce the high-velocity jet that penetrates the armor. The slug separates from the jet and contributes nothing of significance to the munition's performance. By contrast, an EFP warhead does not sacrifice liner material in producing the EFP. Although EFPs are produced in a variety of shapes, most of the liner is used to form the projectile, and no significant part separates initially from the EFP. However, the distinction between EFPs and SCJs is not definite. Many EFP munitions produce long, stretching projectiles that resemble SCJs.

The basic components of an EFP warhead are illustrated in Fig. 2.12. They consist of the explosive charge with a shallow cavity containing the liner (copper or other metal) at one end and an initiator of some type to detonate the explosive charge at the opposite end. The explosive is typically confined in a metal case that provides structural integrity. The thickness and type of casing material are dependent on the forces to which the warhead will be subjected in the intended system. For missile applications, the forces are relatively small, and the casing usually consists of a thin shell of aluminum. For gun-launched applications, the forces are

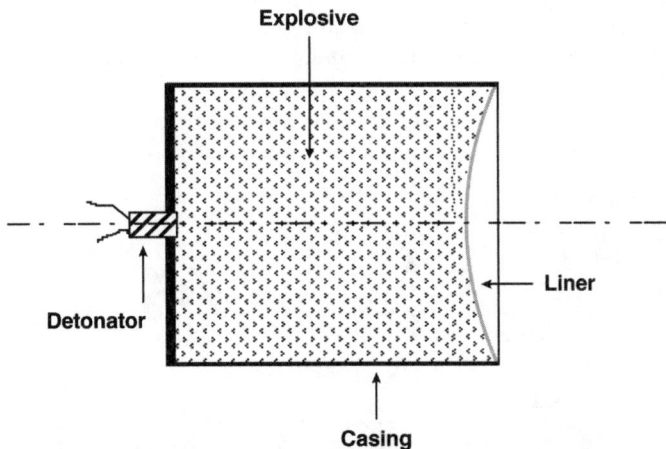

Fig. 2.12   Typical EFP warhead configuration.

substantially larger, and a thick steel case is required. The case also provides confinement for the explosive charge and thereby increases the magnitude of force exerted on the liner during the formation process.

Upon explosive initiation, the detonation wave travels through the charge and encounters the liner. The high pressure of the detonation products causes the liner to accelerate axially and collapse radially to form a projectile of a specific shape. The final shape of the projectile is dependent on the initial contour of the liner, and varying the liner contour can create many different projectile shapes. Typically, the design of the liner contour is an involved process, incorporating the use of a suitable computer code in addition to experimental testing.

For a given charge diameter (CD), the length of even the most elongated EFP is approximately one-fifth to one-third that of an SCJ at its optimum standoff. In addition, speeds of EFPs are less than those of SCJs. Consequently, EFP warheads produce penetrations substantially less than SCJ warheads of the same diameter. However, if an EFP achieves perforation into the interior of an armored vehicle, a larger perforation cross section is observed for the EFP than for an SCJ. The larger perforation results in more potentially lethal BAD.

The overall performance of an EFP is not only related to its material type and length but also to its geometry. EFPs can have a variety of shapes, from that of a sphere to a long rod. The geometry of the penetrator for a given application is determined by the warhead designer and is primarily a function of the target type and thickness and standoff. Figure 2.13 shows the collapse sequence for one design type of EFP liner design.

It is convenient to categorize various EFP liner designs according to whether their penetrators are compact penetrators or long-rod penetrators, and to further subdivide the long-rod category into multi-piece stretching rods, one-piece rods, and aerostable rods. Figure 2.14 presents typical penetration performance as a function of standoff distance for each of the four categories. The figure illustrates some of the tradeoffs associated with EFP warhead design. Multi-piece stretching rods maximize penetration capability, but only over a narrow standoff range. Aerostable rods and compact penetrators have less penetrating capability, but they maintain their penetrating capability across a wide standoff range; and, as mentioned previously, their larger diameters can produce more potentially lethal BAD.

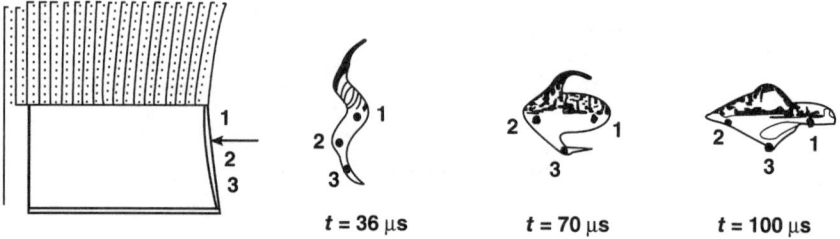

Fig. 2.13 The collapse sequence of a W-fold EFP design (Bender and Carleone 1990). Liner shape 36, 70, and 100 μs after warhead detonation.

VULNERABILITY/LETHALITY ANALYSIS PROCESS

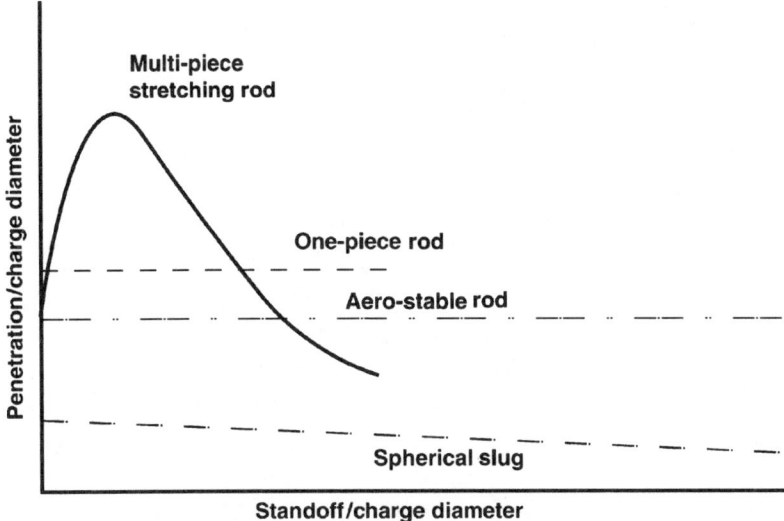

Fig. 2.14 Relative penetration capability for four types of EFPs.

D. Behind-Armor Debris

In analysis of penetrating threats, penetration or perforation is not the sole criterion for target defeat. The quantity and spatial distribution of BAD are also critical (Shnidman 1988).

When a munition perforates armor, a spray of potentially lethal fragments is formed behind the armor. These fragments, shown schematically in Fig. 2.15, consist of broken pieces of penetrator and armor material ejected by the penetration process and/or shock wave release. For high penetrator velocities, the shock wave releases a ring of fragmented material called the spall ring. Many of the fragments are large enough and are moving fast enough to be extremely dangerous to personnel and highly damaging to vehicle components. The importance of this debris as a damage mechanism is increased by the fact that it spreads away from the residual penetrator shotline, thereby potentially interacting with a large percentage of the vehicle's vulnerable contents.

BAD is essentially universal; that is, it arises for all types of penetrating munitions, including SC and KE penetrators and EFPs, and for all types of armor. It can also arise from nonpenetrating explosive warheads hitting monolithic armors.

A flash x-ray radiograph of an SCJ just after it has perforated armor is shown in Fig. 2.16. As can be seen, a cloud of behind-armor fragments accompanies the jet. Corresponding but larger fragments are produced when armor is defeated with EFPs and KE penetrators.

These fragments consist of broken pieces of residual penetrator and armor material ejected by the penetration process. The ejected armor usually forms from two specific locations within the target. The first location is the throat or channel of the penetration hole through the target that is produced along the path of the

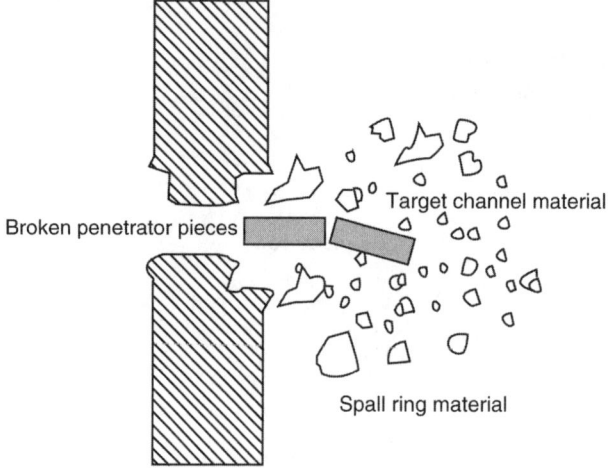

Fig. 2.15   Typical spall components.

penetrator. The second location is the rear surface of the armor, where the spall ring is released.

A common misconception is that lethality after a perforation is solely determined by the largest part of the residual penetrator. This large piece of penetrator can either be recovered from the event or viewed from radiographs, and specific information can be calculated in terms of residual mass and velocity. But equally important are the large numbers of small penetrator pieces that cannot be delineated from the ejected armor material within the debris cloud. In fact, the combination of the "small pieces" of residual penetrator and the fragments formed from the ejected armor material make an extremely lethal environment.

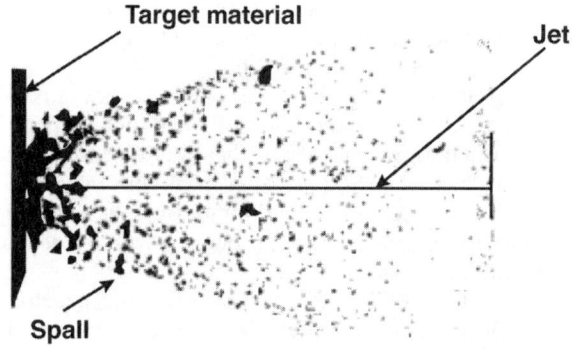

Fig. 2.16   Radiograph of an SCJ.

Technical advancements in armor and penetrator design continually challenge the understanding and prediction of the debris cloud formation phenomenon. In terms of simplicity, the easiest condition for which to predict or explain the outcome of a debris event is a well-known KE penetrator or SCJ perforating a single-element homogeneous armor at normal obliquity. For this type of event, the penetration process is well understood for many of the standard armor/penetrator materials. Typical material combinations are tungsten and copper against rolled homogeneous armor (RHA) and standard aluminum alloys (e.g., 7039, 5083, and 2024). Armor designers are now incorporating new materials in their designs, such as titanium, ceramics, and, in some cases, hybrids made of glass. The problem is further compounded when these materials are combined as multi-element solutions. The characteristics of the debris cloud are further complicated when these elements are fired upon at obliquity. Passive and reactive solutions to the base armor recipe also exacerbate the problem, as do the effects of spall-suppressing liners.

While the armor designers have been advancing the design of more survivable vehicles, the ballisticians have been designing more lethal penetrators. Penetrators are made of many different materials. KE penetrators are often made of tungsten or other dense materials, SCJs are made of copper, and EFPs are often made of tantalum or copper. In addition to material considerations, the geometries of these penetrators are constantly being modified. For example, the standard long-rod KE penetrator can be seen in the form of a series of segmented rods, and, through further advancements in detonation wave technology, EFPs can be shaped into many different forms. Even when small, all of these advancements, both on the armor side and the penetrator side, can have significant effects on the characteristics of the BAD cloud.

As can be seen, many complex physical phenomena contribute to the formation of BAD, and our understanding of the processes involved (e.g., fragmentation at high strain rates) continues to be incomplete. This lack of understanding has, moreover, precluded the complete computer modeling of first principles of the resulting behind-armor fragment properties. This is not to say that continuum-mechanics computer calculations of the behind-armor phenomena are worthless. Indeed, they can provide useful insights as to the overall extent of the fragment cloud and estimates of the range of fragment velocities. Nevertheless, field testing is still required, especially for complicated armors or munitions with new behaviors.

The analytical process of evaluating the vulnerability of a target to a particular munition involves selection of a large number of postulated impact locations on the target. An estimate is made of the extent of damage that would occur if the projectile impacted at each of the postulated points. To estimate the damage that would be inflicted by BAD, certain characteristics of the spray must be predicted. Because various armors may be encountered at various impact obliquities and striking velocities, it is economically infeasible to acquire all the needed information experimentally. Thus, recourse to mathematical modeling is necessary. To ensure the completeness of the mathematical process, a suitable design matrix must be developed to obtain adequate data of interest for a particular threat–target interaction.

When formulating a BAD program (no pun intended), several factors must be considered. Although every program is based on a unique munition–target interaction, the underlying issues regarding data collection are relatively constant: what is the target vehicle, what munitions are involved, and what are the funding and time constraints. As these questions are answered, a suitable matrix can be formulated for development of BAD models. This process considers target material types, obliquities, and conditions over which the munition will clearly defeat the target (i.e., overmatching conditions). Sometimes, as efforts are made to collect as much representative data as possible for the target in question, the matrix becomes extremely large. Unfortunately, in some cases the complete matrix cannot be funded, nor can it be executed under specific time requirements. Specific shots must then be systematically eliminated from the total matrix without jeopardizing model accuracy.

Repetition over specific shot conditions has proven to be necessary for model development; three shots at each condition is the typical number, due mainly to resource limitations. However, the question arises: how much repetition is necessary and reasonable? A first response is that more is always better, but there are the constraining issues. The BAD environment has considerable variability, even for nearly identical shot-to-shot conditions. Therefore, the strategy is to develop a matrix that provides enough repetition to understand the phenomena, thus allowing the development of BAD models with a confident degree of accuracy. A partial factorial matrix allows selection of specific overmatching shot conditions, whereas limited interpolation of the data allows development of these BAD models. For a more detailed discussion on BAD characterization and BAD model development, see Appendix B.

E. Blast

To explain the HE blast damage mechanism, we first must define a blast wave. A blast wave is a pressure wave generated in the air by the rapid release of energy stored in some source into the surrounding medium. Many sources of stored energy can give rise to such pressure waves. The stored energy in a compressed gas or vapor (either hot or cold) can be such a source. The failure of a high-pressure gas storage vessel or steam boiler and the muzzle blast from a gun are examples of explosions that give rise to blast waves in air. The more usual sources of energy for explosions and generation of blast waves are, however, either chemical or nuclear materials, which are capable of violent reactions when properly initiated. Here we confine our interest to blast waves generated by the detonation of HE charges (i.e., CE sources).

When energy from a source is rapidly released into the surrounding medium, a finite pressure wave of compressible gas propagates through the medium. If the front of the disturbance steepens as it propagates forward, the wave is termed a shock wave. The shock-wave front propagates at supersonic speed as it travels through the undisturbed air. The shock-wave front constitutes a jump discontinuity in the properties of the air behind the front. For our purposes, the term blast wave is taken to mean a shock wave generated by the detonation of an HE charge in the ambient air.

If we assume that an HE-charge explosion occurred in air, that it is spherically symmetric, and that the atmosphere is homogeneous, it follows that the characteristics of the air behind the shock front are functions of only the distance from the source of the disturbance and the time. If an ideal pressure transducer, which offers no resistance to the flow behind the shock front, were located a short distance from the source of the HE explosion, this transducer would record the variations of the pressure behind the shock wave as it sweeps over the transducer. The record produced by such a transducer would resemble that shown in Fig. 2.17. At the arrival time, t_a, the pressure would rise quite abruptly to a peak value, then decay to ambient, drop to a partial vacuum, and eventually return to ambient. The portion of the time–history above initial ambient pressure is called the positive phase; the portion below initial ambient pressure is called the negative phase.

The texts *Explosions in Air, Part One* (U.S. Army Materiel Command 1974) and *Explosions in Air* (Baker 1973) provide excellent treatments of the phenomenology of explosions and blast waves. In addition, as previously mentioned, "Design of Projectiles for Terminal Ballistic Effects" (U.S. Department of Defense 1997) and *Fundamentals of Shaped Charges* (Walters and Zukas 1989) provide detailed descriptions of SC design and performance.

F. Ballistic Shock

The impact of a high-velocity projectile onto a target has considerable damage-producing potential even if penetration does not occur. Consider a nominal KE penetrator with a mass of 20 kg and speed of 2 km/s. Upon impact, the kinetic energy (40 MJ) of this projectile must be transferred to the target. Although much of this energy is used in changes of state and is dissipated as heat, a significant amount of energy goes into transient and permanent deformation of the impacted surface. The transient deformation is propagated to adjacent surfaces and, potentially, throughout the entire target until it is ultimately dissipated through various processes, some of which may result in further, distributed target damage. The broad area of target deformation because of ballistic impact is called ballistic shock. For a given threat–target combination, ballistic shock is generally more severe for nonperforating shots

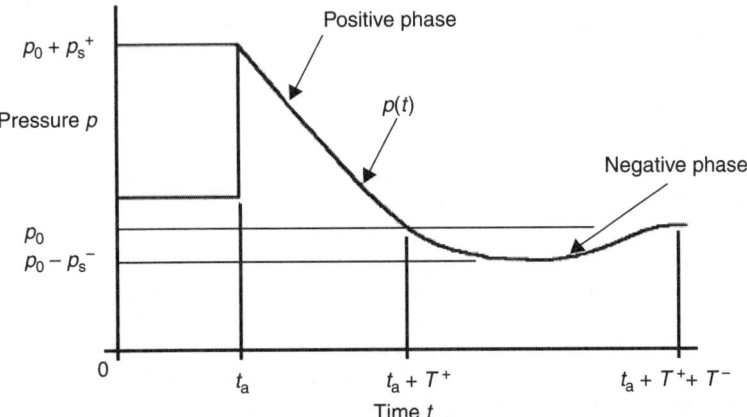

Fig. 2.17   Ideal blast wave.

because more KE is deposited in the structure by nonperforating shots. However, shock from perforating shots can still be significant.

The dynamic deformation of an impacted target is extremely complex. No currently existing models can determine the entire dynamic response of a complete target. As a result, ballistic shock analysis considers target response in the limit of low frequencies and in the limit of high frequencies. In common practice, a frequency of 1 kHz is taken as the boundary between the high- and low-frequency ranges. Experimental measurements of the accelerations of points on an impacted surface can be decomposed by frequency. From such a decomposition, it is possible to estimate the amount of energy in the low-frequency range and in the high-frequency range. The energy in each of these frequency ranges gives the starting point for analysis of ballistic shock in these corresponding frequency limits.

The dominant mechanism in the low-frequency regime is structural damage. Structural damage requires the insertion of relatively large amounts of energy, which translates into the work of permanent deformation or rupture of panels, supports, and edges (e.g., weld joints).

Analysis of the response of targets to the low-frequency portion of the ballistic-shock spectrum is similar to that employed in the analysis of blast threats. In the area of highly detailed analysis, finite-element methods have had some success for simple structures. Variants of these methods, including modal superposition, can predict the first few, lowest-frequency, vibration modes of simple structures. Beyond this level, accurate analyses cannot be achieved. This is because higher vibrational modes become increasingly sensitive to finer target details. Further, modeling of material failure also becomes problematic. Although techniques, such as the von Mises criterion, can model some material failure, these techniques are not adequately developed to accurately predict material failure at the high strain rates involved in ballistic shock.

In the frequency regime extending above 1 kHz, different phenomenology dominates. Although this regime contains rather little of the deposited energy, this energy is in the form of shock waves in the impacted surface. Such shock waves propagate with relatively little loss in many armor and structural materials. As a result, energy in this frequency regime may reach fragile components, such as electronics and optics, to which shock-propagation paths exist. Analysis of shock damage divides into the following three steps: 1) production of the initial shock (source term); 2) propagation of the shock to various components within the target; and 3) prediction of component failure.

Prediction of a detailed source term from first principles is not currently possible. However, sufficient experimental data exist to allow some semi-empirical scaling for many impactor–target combinations. Figure 2.18 presents a typical time trace of acceleration produced by the passage of a shock wave as it emanates from the point of impact.

Detailed propagation of shock waves throughout a target is also an intractable analysis for all but the simplest targets. Because of the propagation efficiency of waves of these frequencies in prevalent materials, the waves undergo multiple reflections. This results in extremely complex loading histories because of multiple reinforcements and cancellations.

Ballistic shock damage may be caused by a variety of threats. For thick-skinned armored vehicles, the classes of rounds that cause ballistic-shock damage are gun-fired anti-armor projectiles (both KE and HE), ATGM, artillery, and mines.

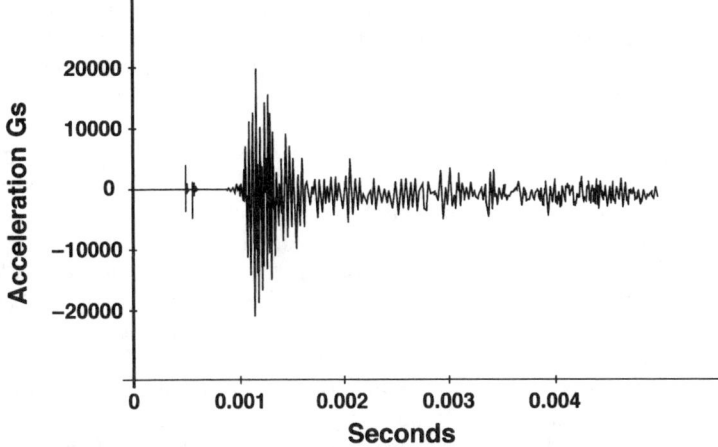

Fig. 2.18 Typical trace of a shock wave.

Each of these threat classes can damage both the vehicle structure and the electrical/optical/hydraulic units within the structure. Gun-fired projectiles cause extremely short duration impulse loading, which generates an intense shock response in the vicinity of the impact. This type of loading results in fracture of bolts and mounting screws for internally mounted units near the impact. A significant amount of energy can also be propagated far from the impact, causing damage to reference gyroscopes, gun resolvers, position-navigation subsystems, and fire-suppression sensors. Some gun-fired projectiles put enough energy into the loading, not only to perforate the structure, but also to deform the structure sufficiently to cause loss of function (LoF) in other moving parts of the vehicle because of mechanical interference. Another type of failure mechanism that can be exploited by these short-duration, high-energy loads is any manufacturing defect in a metal casting. This type of loading uses the defect to start a fracture that propagates and weakens the structure.

The ATGM offers a two-fold damage mechanism. The first is the large SCJ, which is a good perforating agent. The second mechanism is blast-induced shock caused by the high-explosive detonation of the warhead. The blast from the detonation causes deformation of any nearby flat surfaces and propagates shock throughout the structure, resulting in damage to any hard-mounted electrical/optical/hydraulic unit in the immediate area.

The artillery threat also produces damage by two mechanisms: blast shock (as with the ATGM) and fragments. Damage from each of these mechanisms can be superimposed to estimate the total damage to the structure. The blast effect often extracts the exterior window of sighting devices and, when of sufficient intensity, generates a repetitive impulsive load on the structure. This repetitive load causes optical and electronic components to become dislodged and nonfunctional. Typical damage includes cracked lenses, displaced optical prisms, distorted lens-mounting brackets, unseated circuit cards in the card cage, detached surface-mounted components on circuit cards, periscopes displaced from cupolas, internally damaged electronic sensors, and detached hydraulic-subsystem joints. The blast also generates shock in the structure that causes damage to delicate items such as the night-sight cooling mechanism, contact-mounted resolvers, and gyroscopes. The second damage mechanism involves high-speed fragments. These fragments are often of low mass and do not carry enough energy individually to cause shock damage to thick-skinned armored vehicles. These fragments may, however, cause damage to medium- to thin-skinned vehicles by generating intense point loads. The shock effect in this case would most probably be limited to the near impact area, and only units mounted directly behind the impact area would suffer damage. This is the same effect that occurs with small SC bomblets. The energy of the small bomblet is almost entirely consumed in forming the penetrating jet, and although the jet perforates as intended, the shock generated in the structure by the bomblet is limited to the local area around the perforation. In this case, the effect is a perforation with little, if any, shock-generated peripheral damage.

Mines constitute one of the major threats in ballistic-shock generation. As a threat class, mines cause not only possible overturning of the vehicle, structural deformation, and LoF, but they also generate large amounts of high-energy shock that propagates throughout the structure. Any hard-mounted unit in the area of the mine detonation is subjected to ballistic shock. This shock may or may not damage the unit, but most susceptible to damage are hard-mounted items such as hydraulic/electric slip rings, hydraulic couplings, hydraulic reservoirs, position-navigation units, and any electronics within a unit in direct contact with the excited surface. As with the high-energy inputs described previously, mines can generate sufficient energy to cause ballistic full-penetration welds to fracture, fire-suppression subsystem sensors to fail, hatch-locking and lifting mechanisms to jam, and bolt heads to shear.

When examining the ballistic-shock-caused damaging effects on thin-skinned vehicles, some effects seem to be scalable. Field tests have generated data that suggest that a small KE projectile impact on a thin- to medium-skinned structure generates shock levels comparable with large-caliber projectiles impacting on thick structures. Therefore, at first look, it seems that damaging shock can be generated even from small KE projectile impacts, given that the structural thickness is in the order of less than twice the projectile diameter.

Ballistic-shock damage as a far-field effect is often treated as secondary to a large hole in the structure. Granted, a large hole in the structure will almost certainly cause problems in this age where combat vehicles contain densely packed electronics, complex optics, and sensitive measurement sensors; however, shock damage, while not always physically evident, may be the major cause of loss of combat function.

In summary, typically observed ballistic-shock damage consists of the following: 1) damage to optics; 2) damage to mounting brackets for electronic/hydraulic/optical units; 3) distortion of basic structure; 4) damage to sensors in contact with the excited surface; 5) damage to slip rings and hydraulic reservoirs; 6) damage to cooling mechanisms for night sights; and 7) damage to display unit elements.

G. Vaporifics and Noxious Gases

It has long been recognized that aluminum armors, when penetrated by projectiles or SC warheads, produce an environment behind the armor plate that is different from that seen with steel. These differences include spalling characteristics, the so-called *vaporific effect*, and the potential of producing noxious gases in amounts that may be hazardous to personnel in an enclosed volume, such as an armored fighting vehicle (AFV).

In the case of an armored ground combat vehicle, the vaporific effect associated with aluminum armor is the result of aluminum spall particles/fragments being heated and injected into a volume of air by the penetration process. The favorable environment for combustion concomitant to the high velocity (6000–8000 ft/s) of the spall particles and the heat generated by the penetration and impact processes leads to a fuel–air event. The high-speed pictures contained on the right-hand side of Fig. 2.19 illustrate this phenomenon. This plate contains high-speed pictures of a tube-launched, optically tracked, wire-guided (TOW) SC warhead being fired into an armored test fixture having the internal volume of a typical AFV. The series of pictures on the left shows the behind-armor environment when the TOW penetrates 16 mm of the RHA on both the front and rear walls. The pictures on the right show the behind-armor environment for penetration of 45 mm of 5083 aluminum armor on the front and rear walls of the

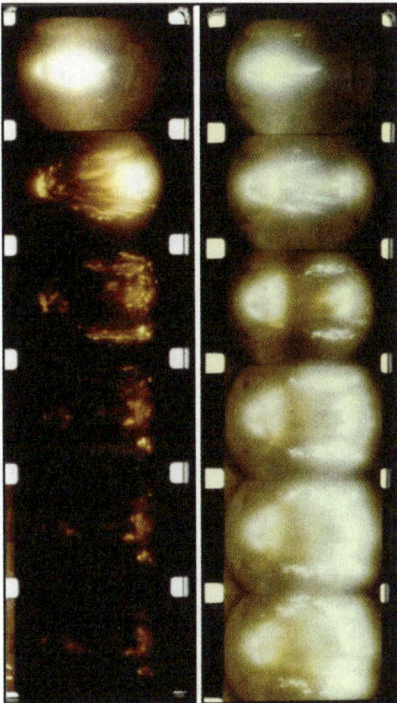

Fig. 2.19 Behind-armor environment for RHA (left) and aluminum (right) plate penetration.

compartment. The RHA represents protection against some small arms AP projectiles that is roughly equivalent to the aluminum.

This pyrophoric reaction produces increased pressures and temperatures as well as potentially hazardous combustion products, which include nitric oxide (NO), nitrogen dioxide (NO_2), and carbon monoxide (CO). The severity of this behind-armor environment increases as the size of the attacking warhead increases. The important question, however, is whether these additional effects of pressure, temperature, and toxic gases are of sufficient magnitude to incapacitate personnel in the vehicle when the attacking munition is of a size such that it does not cause a catastrophic failure of the armor plate. In those situations where the threat is large enough to produce catastrophic damage on impact, the enhanced temperatures and pressures, along with toxic fumes, will likely be superfluous because the target will have been killed by the other effects of the weapon, such as the main penetrator, spall, blast, and fuel and/or ammunition explosion or fire.

Hazards of these behind-armor effects (BAE) associated with aluminum armors were considered in U.S. and British studies (Kennedy 1973; Zeller 1962; Platt 1973). Data from these and other studies, coupled with various news papers characterizing the BFV as a "death trap" because it is an aluminum armored vehicle, resulted in a request from the Department of the Army for the U.S. Army BRL to conduct an investigation of BAE associated with aluminum armor. BRL's extensive phenomenology investigation of the vaporific effect in 1984 served to verify the findings of earlier investigators; namely, that the BAE associated with aluminum armors are enhanced as compared to that observed behind steel armor. This enhancement of both blast overpressure and free-air temperature varies with the size of the attacking munition. The key, however, is the consequence with respect to producing crew/passenger casualties or damage to the vehicle (Schmidt et al. 1986; Zeller 1962; Platt 1973).

H. Secondary Effects

Secondary effects are those phenomena caused indirectly by the attacking threat, such as fire, electromagnetic pulse effects on electronics, and some software failures. Except for threats such as flamethrowers and thermobaric munitions, fire is a damage mechanism brought about by a combination of conditions resulting from physical attributes of the threat and the target. Electronic failures also can occur because of secondary effects associated with power surges or short circuits caused by damage elsewhere in the target system. Software failures can occur for a variety of reasons, but improper handling of interrupts caused by failure of dysfunctional components is a major concern in modern combat systems. This type of software failure is a secondary effect, which, in extreme cases, may render an entire platform inoperable from loss of a relatively minor component. Although systematic treatment of many secondary effects is beyond the scope of this text, there is an important body of knowledge on the initiation and spread of fire to which the remainder of this section is devoted. The interested reader is referred to Dehn (1979), Wright and Slack (1980), Repa, and MIL-HDBK-684 (U.S. Department of the Army 1995) for a description of models describing portions of fire-related phenomena.

There are two distinct areas of interest as far as vehicle fuel fires are concerned. The first is injury to personnel. The Walter Reed Army Institute of Research (WRAIR) has provided criteria to assess burn injuries to personnel in vehicles (Ripple and Mundie 1989). One approach is to use free-air temperature at each crew-member location. A second approach is to use heat-flux calorimetry.

The second area of interest is fire damage to the vehicle itself. The questions that need to be answered are what is damaged by a fire, and how badly is it damaged? Testing and data collection in this area are still ongoing, and data from the exposure of vehicle components to controlled fires or heat fluxes are not available at the time of this writing. Nonetheless, the following rules of thumb are useful to guide V/L analyses of fuel fires in ground combat vehicles:

1) Combat-induced hydrocarbon fires in crew compartments must be extinguished within 250 ms to prevent second-degree or worse burn injuries on exposed skin.

2) Liquid hydrocarbons that are far below their flash points will burn as a spray. If an overmatching weapon perforates a fuel cell or breaks open a hydraulic-fluid line, the resulting spray will burn in air when an adequate ignition source is present, even if the hydrocarbon is 100° F below the flash point (Finnerty et al. 1985).

3) The probability of a sustained fire, given a perforation of a fuel cell, approaches 1.0 at the flash point of the fuel, given an appropriate ignition source and ideal initial conditions (Dehn 1979; Finnerty et al. 1985; Finnerty 1987).

4) The probability of a sustained fire, given a perforation of a fuel cell, decreases the lower the fuel temperature is below its flash point (Dehn 1979; Finnerty 1987).

5) If the fuel temperature is set at a certain value, there will be a smaller probability of sustained fire for diesel fuel numbers 2 and JP-5 than for diesel fuel numbers 1 and JP-8. The probability of sustained fire will be highest for gasoline and JP-4. However, lacking data or use of a sophisticated fire model, the analyst may not be able to distinguish between the probabilities of a sustained fire for diesel fuel number 2 and JP-8.

6) Measures such as jacketed fuel cells, powder packs, and spall curtains can significantly lower the probability of a sustained fire, given a perforation of a fuel cell (Finnerty and Dehn 1994).

I. Synergistic Effects

The term *synergism* refers to those phenomena involving the combination of more than one damage mechanism in which the result cannot be deduced from the results of the damage mechanisms computed independently and then simply added. For example, a particular communications shelter might be able to withstand a given blast wave and have a low probability of kill P_K from a certain fragment spray. However, if the fragment spray weakens the structure such that a subsequent exposure to the given blast wave destroys the shelter, the combination is said to be synergistic: the result is not derivable as a summation of the two damage mechanisms independently. The most common combination considered in V/L analysis of surface targets is the combination of blast and fragments.

Synergistic interactions are thought of as falling into two classes: 1) preconditioning, and 2) interactive. The previous example of the communications shelter illustrates synergism via preconditioning. Other examples include the following: 1) the presence and resultant ignition of fuel in a dry bay caused by a prior fragment penetration of a fuel tank; 2) the concave reshaping of a surface by prior fragments, thus increasing the coupling of a subsequent blast wave; 3) the creation of large holes in a surface by prior fragments, thus decreasing the coupling of a subsequent blast wave; and 4) the heating of a surface by radiative or conductive thermal loading, thus weakening its blast tolerance.

As seen, preconditioning accounts for the fact that a target, once exposed to a damage mechanism, is different and thus may not present the same resistance to a subsequent damage mechanism. Axiomatically, the timing of the exposures is important. Whereas a subsequent fragment may more easily penetrate a structure that is first impinged by blast, clearly the mechanism and expected degree of synergistic response is different if the fragment arrives first. Although one normally thinks of synergism as a damage multiplier, it is possible (as suggested by the previous example of surface holes decreasing the coupling of a blast wave) to find less damage than would be expected by a simple sum of independent damages.

It is also possible that the simultaneous presence of multiple damage mechanisms can produce a combined, hybrid mechanism that is itself different from the simple sum of the mechanisms. This is termed interactive synergism. For example, the pressure–time trace of the impingement of a pure blast wave on a flat surface shows a rise as the wave builds up and reflects, followed by a decrease as the high-pressure air spills off the edges of the impinged surface. However, in the presence of a sufficient density of fragments, this spill-off is impeded. The result is a more complete coupling of the blast-wave energy to the surface, a result that cannot be predicted from the blast and fragment impingements separately.

To at least the first order, it may be possible to model preconditioning synergism through the time-phased application of damage models that capture the intermediate states of the target. However, it is clear that interactive synergism will require the simultaneous modeling of blast, fragments, and target response. Commonly, phenomena of such complexity are not modeled explicitly in V/L analysis codes. Rather, high-resolution models are built, tested, and run off-line to produce results that can then be fitted with semi-empirical algorithms for insertion into V/L codes. Unfortunately, in the case of blast-fragment synergism, even the high-resolution models currently are incapable of such complex analysis.

The most common high-resolution, numerical analysis techniques for structural response are the finite-element codes, either Eulerian (grids fixed in space) or Lagrangian (grids fixed to the elements). A finite-element analysis of a penetrator against a target would likely use a Lagrangian grid of elements to represent the target and another grid to represent the penetrator. Upon impact, Newtonian physics is applied to the interacting elements, each of which is connected to its nearest neighbors by element interaction functions that model the properties of the materials from which the target and penetrator are made. On the other hand, an analysis of a blast wave against a target would likely use a hydrocode (Eulerian formulation) of the propagation of the blast wave and its subsequent interaction with the target. From this analysis, load functions are extracted to apply to a finite-element analysis of the target response. In advanced analyses, the hydrocode and finite-element analyses may be coupled such that the hydrocode grid deforms in the neighborhood of the target as indicated by the finite-element analysis.

It has not been possible, however, to include both blast and penetrators in the same analysis technique. Blast waves exhibit such extreme distortion that Lagrangian-grid representation is prohibitive, whereas the use of Eulerian grids is unmanageable for the highly interactive elements of a solid body. Recently, significant success has been had with the so-called gridless Lagrangian techniques. In these techniques, the interacting material is quantized into elements as in conventional Lagrangian formulations. However, each element is considered to be an independent entity, tied to its neighbors by force functions but not by a computational grid. As a result, no computational difficulties are introduced by an element that is forced out of its original neighborhood. Because it is not compelled by an artificial grid to interact (only) with its original nearest neighbors, it is free to be expelled and subsequently interact with other elements. The cost, of course, is the daunting need to simultaneously track the location and relationship of the myriad of independent elements. Although modern computational power is beginning to make realistic gridless Lagrangian calculations possible, such analyses generally require simplifications and shortcuts, as well as relatively coarse elements, to run in reasonable times.

One of the popular gridless Lagrangian techniques is called smooth particle hydrodynamics (SPH) (Libersky et al. 1993). In an SPH formulation, the shortcut is the construction of force fields. Each element contributes to a composite field (x, y, z) through a smooth function of mass and other properties. Subsequently, the motion of each element is calculated based on the influence of the composite field upon it. It is this use of a field in a fixed (x, y, z) space that lends the term *hydrodynamics* to the technique. This process is time-step-iterated to model an interaction event. By using the field, the analysis code need only make two passes through the elements per time-step: one to calculate the field and one to calculate the element responses. Thus, the number of calculations is on the order of $2N$, where N is the number of elements. This is in contrast to the $N(N-1)$ possible calculations to be made had the element-to-element calculations been made.

Applications of SPH to the problem of combined blast-fragment impingement have thus far not met with much success. In the formulation of the composite field, the densities of the interacting elements appear in the denominators of interaction functions. The wide disparity of the densities of fragments and shocked air results in numerical instabilities that have so far proven intractable.

Thus, the modeling of synergistic interactions between damage mechanisms in general, and blast-fragment synergism in particular, has not been solved. Rules of thumb (such as damage multipliers) have been proposed, based on experimental observations. However, their use has fallen out of favor, the community preferring to err knowingly on the side of underestimating damage, rather than to introduce another unknown uncertainty in the form of a multiplier.

Until now, there has been little interest in synergistic effects, partly because they often are small compared with primary effects and partly because of the present uncertainties in estimating them. However, as weapons become increasingly accurate and warheads are tailored to specific applications, the likelihood of events in which multiple environments are present will increase. Thus, increased emphasis on the modeling of synergistic effects is anticipated.

III. Induced Dysfunction of Components

In this part of the $O_{1,2}$, mapping, we are concerned with the ability of threat mechanisms to damage components to the point that they are unable to perform up to (critical) specifications. Clear definition of the terms *component, critical component*, and *criticality* are needed to frame the discussion.

The definition of a component is largely contextual. Depending on the granularity at which the analysis is done and the level of knowledge of the system and its function, an entity defined as a component in one study may be a subsystem in another. For example, an electronic box may be a component in one context while its internal circuit boards, transformers, and other parts may be denoted as components in another context. At a deeper level, the integrated-circuit chips and individual wires could be considered as components, and so on. Except for properly accounting for shielding and masking of other components, it is typically unnecessary to determine whether noncritical components have been damaged or to analyze the effects of such damage on continued operation.

Critical components are those components required to achieve specific system capabilities related to mission/scenario capability. Further, the criticality of a component depends in great measure on the robustness or redundancy of the system, such as results from parallelism in design. For example, a turbocharger is a critical component for a system such as a tank during a combat engagement, when adequate power is required to provide mobility capabilities and there is no parallel source of boost for the engine. A headlight, on the other hand, is generally a noncritical component in a combat scenario because headlights are seldom required in combat. Note, however, that components that are noncritical in terms of combat capability may be designated as critical for analyses in which maintenance is evaluated. Thus, the headlights may be critical in an interdiction scenario in which a unit not engaged in battle is moving long distances at relatively high speeds at night. Clearly, therefore, the term *critical* is mission/scenario dependent.

A criticality analysis determines which components are critical and how failure affects the operation of subsystems of which they are a part. A criticality analysis consists of three interrelated parts:

Functional analysis   How a subsystem and its constituent components operate and the interdependencies between components and subsystems.
Mission analysis   What components are necessary for subsystem functions required to perform various combat missions.
Failure Modes and Effects Analysis (FMEA)   For damage mechanisms of interest, how component function may be degraded and the effect on subsystem function and system capability.

Criticality analyses are conducted on aircraft as well as ground mobile systems. However, by tradition, and based on analyst prerogative, terminology is often somewhat different between the analyst communities addressing the two classes of targets. For example, the modeling is often referred to as criticality analysis for ground mobile systems and as Failure Mode, Effects, and Criticality Analysis (FMECA) or FMEA for air systems.

Regardless of the label, the fundamental analytical processes are much the same. In fact, the process was originally developed to support reliability, availability, and maintainability (RAM) studies. Criticality analyses in vulnerability/lethality studies are a simplified adaptation of the RAM process. The basic procedure for conducting a criticality analysis is well documented in MIL-STD-1629A (U.S. Department of Defense 1977). The process as applied to aircraft is described in detail in *The Fundamentals of Aircraft Combat Survivability Analysis and Design* (Ball 2003).

To preserve consistency throughout the V/L analysis, the criticality analysis should be conducted before, or at least concurrent with, the development of the geometric target description. Although sometimes the target description is developed first, doing so often leads to improper definition of components, omission of critical components, inclusion of unnecessary noncritical components, and other problems that affect results.

One of the most important products derived from the criticality analysis is the collection of fault trees (sometimes referred to as deactivation diagrams), which graphically show system functionality, component criticality, and interactions among components and subsystems. Figure 2.20 is an example of a typical fault-tree diagram developed for an armored vehicle. There are, obviously, many such trees needed to describe the vehicle fully.

A. Mechanical and Electrical Component Failures

One of the fundamental inputs required during the vulnerability analysis process is a determination of whether a specified threat damage mechanism is capable of causing dysfunction of critical mechanical and electrical components. For purposes of this discussion, mechanical and electrical components include items such as fluid containers, control rods, shafts, gear boxes, electrical boxes, sighting devices, engines, suspension parts, hoses, electrical wires, armaments, filters, pumps, radios, intercoms, and electric motors.

The probability of component dysfunction given a hit ($P_{CD|H}$) is the modern metric commonly used to quantify the dysfunction; it allows the notion that a component does not necessarily have to be "killed" to have a degrading effect on system or subsystem capability or performance (Deitz et al. 1990). In the past, critical component damage was cast in terms of a kill of the component, resulting in LoF of the system being analyzed. The term applied to this metric was probability of kill given a hit $P_{K|H}$. {Note that, within the V/L community, the typographical notation for expressions such as probability of kill (P_K), probability of kill given a hit ($P_{K|H}$), probability of kill given damage ($P_{K|D}$), probability of component dysfunction given a hit ($P_{CD|H}$), probability of incapacitation given a hit ($P_{I|H}$), and probability of survival (P_S) has varied throughout the years and across work disciplines. In some cases, variables are enclosed in parentheses instead of being subscripted, slashes (/) are used instead of vertical lines (|), and different schemes are applied for uppercasing and lowercasing letters [e.g., to distinguish between damage at the system (uppercase) and component (lowercase) levels].}

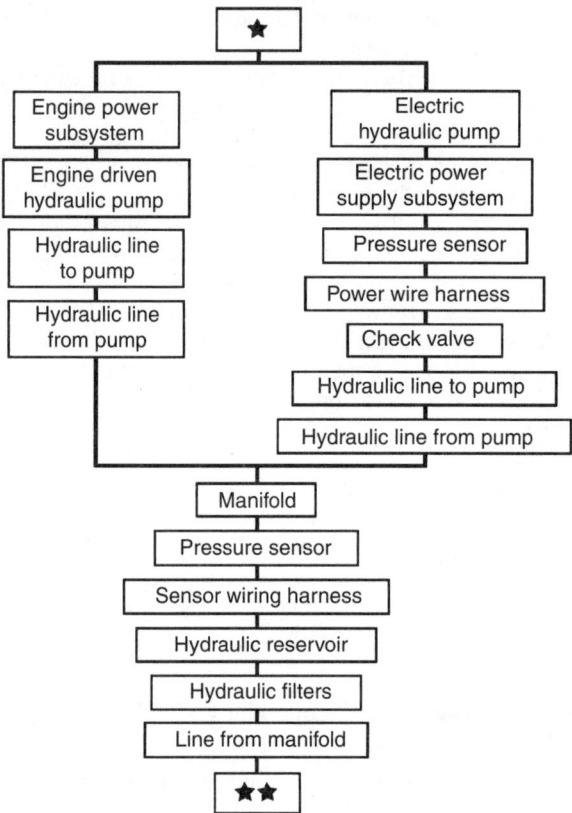

Fig. 2.20 Fault tree for the hydraulic power supply subsystem of an armored vehicle.

The analysis of any single component involves a number of factors, including how the component is used, what operating loads are placed on the component, and which materials are used in the construction of the component. After this information is gathered, the vulnerability analyst must determine the failure mode(s) of the component, the type and amount of damage required to cause failure or dysfunction, and the damage mechanism(s) capable of causing such damage. Adding to the complexity of the task is the fact that this type of analysis must be conducted for virtually every critical component in a vehicle because generally no two components operate under the same conditions, serve the same function, or are constructed of the same material in the same quantities.

Penetrators capable of causing component dysfunction include bullets (projectiles), EFPs, SCJs, and fragments. This discussion focuses on fragments, as the other penetrators are generally considered to be capable of causing complete component dysfunction or can be addressed in a similar manner as fragments.

The traditional method used for deriving component $P_{K|H}$'s and now $P_{CD|H}$'s was first documented by Kruse and Brizzolara (1971). The basic principle is to estimate the ratio of the total critical damage area to the total presented area of the component. This ratio can be estimated for a particular angle of attack or averaged across all possible angles.

The $P_{CD|H}$ function thus described provides the probability that a single fragment is capable of producing sufficient damage to render a particular component nonfunctional at a prescribed level of capability. Historically, V/L codes analyzed the effects of fragment impacts on a target one at a time. If a particular fragment striking a critical component did not meet the penetration requirement, as provided in the $P_{CD|H}$, it was discarded as a nonlethal fragment and the V/L code would move on to analyze any subsequent fragments impacting the component. It has always been obvious that this practice was flawed for certain types of components such as fluid containers. If a single fragment, for example, does not create a hole large enough to drain the fluid out of a container in a prescribed time limit, then perhaps two, three, or more fragments could surpass the hole size criterion. For fluid containers and tires, modules that keep track of multiple fragments have been incorporated directly into V/L (Grote et al. 1996).

Currently, there is limited blast methodology in place for mobile ground target components. Blast has been included in V/L analyses but only in terms of total system function and with little to no mapping to the component level. Similarly, current V/L codes do not have the capability for adequately determining component dysfunction because of ballistic shock, nor is there a suitable component failure criterion.

Where ballistic effects are concerned, much of the analysis of electrical components is similar to that of mechanical components as they are subjected to banging and shaking. The analysis of the subsequent performance of the electrical components is an area for electrical engineering rather than mechanical engineering, but the overall logical process is similar. [These are also important for consideration of DE weapons, where one might be concerned with the effect of various levels of incident radiation or surges on the lines on a chip, a circuit board, or a major component (e.g., a radio), and what any such damage implies for the operation of the component or assembly that contains the component.]

B. Damage to Fuel System Components

The 20th century saw the introduction and widespread use of tanks and aircraft in battle. It can be said that the history of tanks and aircraft in combat is the history of burning tanks and aircraft. Modern ground and air combat craft must carry large quantities of combustible fuel in order to fulfill their missions. This fuel presents a serious hazard for the occupants of these weapon platforms, as well as for the systems themselves. An attacking weapon needs only to break open a fuel container and act as a trigger to ignite the fuel. The platform is then destroyed by the heat produced by combustion of its own fuel.

In World War II, the U.S. Army and its enemies used gasoline-powered ground combat vehicles and aircraft. Even though some airplanes had self-sealing gasoline tanks, most air and ground vehicles burned when the fuel systems were perforated in combat. The Army had no effective fire-extinguishing systems in these combat vehicles, and it was accepted that a hit on the fuel (gasoline) system would cause a catastrophic loss.

In the Korean conflict, the attention of the U.S. Army was drawn to diesel fuel because it was used in the Soviet tanks that were employed by the North Koreans. Test firings at Aberdeen Proving Ground showed that this fuel did not always undergo sustained combustion when struck by a weapon. Further tests (summer vs winter) demonstrated that warm diesel fuel, when attacked, was more likely to undergo sustained combustion than cold diesel fuel (Zeller 1952).

With the emergence and widespread use of jet engines, Army helicopters were converted to use JP-4 fuel. This fuel has a low flash point and is similar to gasoline in its response to hits in combat. In addition, in more recent years, the concept of one fuel for the battlefield has emerged, and it has been suggested that this fuel be JP-8 since both diesel engines and gas turbine (jet) engines perform satisfactorily on JP-8. The flashpoint of JP-8 is below that of diesel fuel but is above the flashpoints of gasoline and JP-4. Based on volatility and flashpoint, it is likely that the JP-8 response to hits in combat will be somewhere between the responses of diesel fuel and gasoline.

Since the early 1950s, the Army has conducted firings at diesel-fuel loaded vehicles and containers. Aluminum and steel vehicles and containers, as well as surrogates, have been used. Fuel at temperatures from just above 0°C to over 85°C has been used. It is clear that at higher temperatures, sustained fuel fires are more likely, given a hit by an SCJ (Hanna and Goodwin 1955; Beichler 1956; Skinner 1959).

Data have been collected on over 400 SCJ and KE round firings at diesel fuel containers, inside and outside vehicles. Curves have been constructed to represent the probability of sustained fire given a hit on the fuel cell of a steel vehicle and on the fuel cell of an aluminum vehicle as a function of fuel temperature (Finnerty 1987; Groves 1986).

The question of whether or not a fire is initiated when a weapon strikes a fuel container in a combat vehicle has traditionally been answered using simplistic look-up tables or by relying on the judgment of personnel experienced in the area of combat vehicle fires. The simplistic approach says the probability of a sustained fire when a weapon perforates a fuel cell containing diesel fuel is 0.1. The probability of a sustained fire when gasoline is the fuel is 0.9. For fuels other than diesel or gasoline, the analyst must use best judgment. When experienced personnel estimate the probability of sustained fire when a fuel cell of a vehicle is perforated, parameters such as fuel type penetrating ability of the weapon, and the presence of an automatic fire-extinguishing system, are all considered (Finnerty 1987).

C. Damage to Ammunition

1. *Commonly Used Terminology*

As discussed in Chapter 3, reactions in stowed ammunition account for the majority of K-Kills in armored vehicles and can contribute significantly to the vulnerability of other weapon systems. Therefore, it is important that vulnerability analysts understand how ammunition responds to attack and how these reactions affect the systems that carry ammunition.

When considering ammunition damage, the following terminology is used:

Explosion Any rapid release of gas and heat that has a destructive effect.

Explosive   Any material capable of undergoing a rapid chemical reaction to produce an explosion. The word encompasses both HE and propellants (see definitions that follow).

Detonation   An exothermal chemical reaction that propagates through the explosive following a supersonic (with respect to the speed of sound in the explosive) shock wave. For most detonations, the propagation velocity of the detonation wave and the pressure behind the detonation front are determined by a thermodynamic condition called the Chapman–Jouguet (CJ) condition. The pressures and velocities associated with CJ detonations are extremely high (typically 6 to 10 km/s and several tens of giga pascals) and are much higher than for nondetonative explosions. Likewise, CJ detonations produce much higher fragment velocities than do nondetonative explosions. In addition, it is important to realize that not all explosions are detonations.

Low-velocity detonation (LVD)   A detonation where the propagation velocity is much below that predicted by the CJ condition, but still supersonic. Some literature refers to this condition as "low-order" detonation. Pressures and fragment velocities are also much smaller than for CJ detonations. LVDs are rare under conditions of ballistic impact, but they can occur in some materials.

Deflagration   Extremely rapid combustion, generally implying the burning of a substance with self-contained oxygen so that the reaction zone advances into the unreacted material at significantly less than the speed of sound. Deflagration is not accompanied by a shock wave.

Energetic material   Any material capable of undergoing a strong exothermic reaction by itself (without the necessity of an external supply of oxygen).

High explosive   An energetic material that detonates when used as intended. HEs are used in warheads to accelerate fragments, push metal, and produce blast. Although they detonate when used as intended, they can also undergo slow burns or nondetonative explosions.

Propellant   An energetic material that burns (to push a bullet or missile) when used as intended. However, many propellants can be made to detonate under some conditions.

Pyrotechnic   An energetic material that produces light or smoke when used as intended.

Shock   This word is used in different ways in different parts of the vulnerability community. In the energetic-materials community, the word is used in the fluid dynamics sense to denote a step discontinuity in pressure, density, and particle velocity that propagates supersonically through a material. Shocks are uniquely effective at initiating reactions in energetic materials.

Failure diameter   The minimum diameter at which any particular explosive will sustain a detonation. When the diameter of an energetic material is below its failure diameter, detonation cannot be sustained. The failure diameter is reduced by the presence of confinement. Also, when energetic materials are used in slab geometry, the critical thickness is typically about half of the critical diameter.

Initiation   The process by which reactions in an energetic material begin and grow to an explosion or detonation.

Sensitivity   The ease with which reactions are initiated in an energetic material. The relative sensitivity of a series of explosives can depend on what stimulus (e.g., shock, electrostatic discharge, heat) is being considered, but in general there is some correlation between these sensitivities so that an explosive that is extremely sensitive to one threat is likely, but not necessarily, to be quite sensitive to another threat.

Explosiveness The ease with which a reaction builds up in an energetic material after an ignition has occurred. This terminology was introduced by the British and is useful for distinguishing between materials that have similar sensitivities but that produce different levels of reaction after ignition. Brittle materials tend to have high explosiveness under ballistic-impact conditions because cracks generate paths for flame to spread.

2. Ammunition Components

Different ammunition components respond in different ways to ballistic attack. One or more of the following types of ammunition are typically found in ground combat systems.

Explosively loaded warheads These contain HEs that accelerate metal or produce strong blast effects when the warhead is used as intended. The HE is normally present as a solid charge with little porosity. Bulk-loaded warheads have a single explosive charge. Improved conventional munitions (ICMs) contain a multiplicity of smaller bomblets, each containing its own explosive charge. ICMs can present special vulnerability problems, partly because they usually contain rather sensitive explosives but also because of the possibility that armed but unexploded submunitions may be scattered as the result of a ballistic impact on an ICM round. Warhead explosives may differ markedly in sensitivity, but all are capable of being detonated by some ballistic threats, especially SCJs.

Solid-propellant rocket motors When used as intended, these burn in a highly controlled fashion to propel a missile. However, they also contain highly energetic materials that can produce destructive effects when subjected to ballistic impact. Like warhead explosives, solid rocket propellants usually have low porosity. However, unlike warheads, solid-propellant rocket motors have a hollow center core that can have a significant effect on their response to ballistic threats. Many solid rocket propellants have failure diameters that are greater than the motor diameter; consequently, the motors cannot detonate but may still burn violently. This is especially true of composite propellants based on combinations of ammonium perchlorate with a plastic binder (and sometimes aluminum). However, these propellants are smoky and leave trails that can disclose the user's position. To eliminate smoke trails and to achieve higher specific energies, missile designers frequently turn to higher-energy, detonable propellants (minimal-smoke propellants). These usually contain some of the same ingredients used in explosives (such as the nitramines RDX and HMX), and they respond like explosives in many ballistic impact tests.

Liquid-propellant rocket motors These are older systems that combine highly reactive oxidizers (e.g., liquid oxygen, red fuming nitric acid, nitrogen tetroxide, or chlorine trifluoride) with a high specific impulse fuel such as monomethylhydrazine or unsymmetrical dimethyl hydrazine in the motor at the time of combustion. Although hypergolic (they ignite on contact with no exterior source of ignition), they do not present a strong explosion threat unless both fuel and oxidizer are released simultaneously and mixed. However, they can release extremely corrosive and toxic vapors if the storage tanks are ruptured. Most of these systems are now out of military service, and thus they are not discussed here.

Solid gun propellants Gun-propulsion charges usually consist of a bed of loosely packed grains, each of which may be perforated with small holes. The resulting bed has large-scale porosity and a large surface area. In some cases, an array of cylindrical sticks is used in place of grains, but in all cases there is significant porosity and significant surface area, which facilitates flame spread and combustion. As a result of this physical structure, the vulnerability characteristics of solid gun propellants are quite different from those of solid explosives and solid rocket propellants. Gun-propulsion charges frequently have rather large failure diameters; as a result, they are not often detonated by ballistic threats. However, the porosity and high surface area in the bed facilitate flame spread and combustion so that ballistic impacts can cause violent burns.

Liquid gun propellants No liquid gun propellants are currently in service, but research on liquid-propellant guns continues. The propellants that have been investigated include those based on water solutions of hydroxyl ammonium nitrate and a fuel, and their vulnerability characteristics are vastly different from solid explosives, solid rocket propellants, or solid gun propellants.

Pyrotechnics There are a wide variety of these materials. They generally do not detonate, but they are easily ignited and may produce highly toxic gases during combustion.

Fuzes and boosters These components initiate detonation in warheads when used as intended. Fuzes contain highly sensitive energetic materials, but reactions in the fuze cannot propagate to the main charge until the fuze is armed. Therefore, impacts on fuzes are not normally important in vulnerability analysis. Booster explosives transfer the detonation from the fuze to the main charge. Booster explosives are more sensitive than main charge explosives, but they are usually small and buried deep inside the warhead. Consequently, they usually do not significantly influence vulnerability.

Primers Primers ignite gun propellants when used as intended. For medium- and large-caliber charges, they frequently consist of a tube containing a fast-burning propellant that runs down the center of the charge. There is some dispute in the literature about the importance of primers in determining vulnerability. Some data indicate reactions in a propellant bed are more violent when a primer is hit, but other data indicate that the primer makes little difference. For an extremely strong threat, such as an unconditioned SCJ, the primer will likely have little effect since the jet initiates a violent reaction in the propellant anyway. However, for lesser threats, impact on a primer may give a stronger reaction than an impact that misses the primer.

3. Initiation Mechanisms

The vulnerability analyst should also have a cursory understanding of the mechanisms that lead to explosions or detonations in energetic materials. The better-known mechanisms that lead to explosions or detonations in energetic materials are as follows:

Shock-to-detonation transition (SDT)   This is the process by which a shock wave builds up to a detonation. SDT is the most thoroughly studied initiation mechanism (Lee and Tarver 1980; McGlaun and Thompson 1990).

Deflagration-to-detonation transition (DDT)   This is the process by which burning in an energetic material produces a shock wave that then runs up to detonation.

XDT   The acronym XDT originally stood for "unknown detonation transition." It is now recognized that XDTs are events in which a ballistic impact damages a material (increasing its sensitivity) and then ignites the damaged material, all in one ballistic event (Finnegan et al. 1995).

Hot embedded fragments   Hot fragments can cause burning reactions in energetic materials if they become embedded in the material.

Bullet or fragment impact on solid rocket motors   Figure 2.21 illustrates the complex behavior that can occur in a rocket motor after impact by a bullet or fragment. Burning reactions are likely whenever the motor casing is perforated. At low impact velocities, these may be mild burns that do not produce blast or fragments. As impact velocity increases, these burns may become violent. This happens more easily in a rocket motor than in a solid explosive because the hollow center of the motor facilitates fracture of the propellant and flame spread. At still higher velocities, the reaction violence may diminish because the projectile penetrates though the motor and creates an exit hole that relieves confinement. At a still higher velocity, an SDT occurs. In some cases, XDT-generated detonations may also occur at velocities below the SDT threshold.

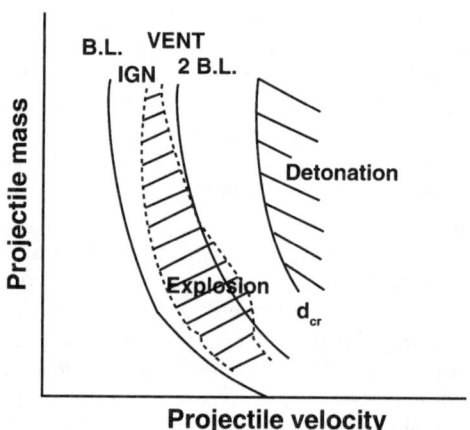

Fig. 2.21   Schematic graph of the response of rocket propellants to projectile impact (Boggs et al. 1987). For a given case and a given propellant, lines/areas change with other case, other propellant, temperature, and damage.

Bullet and projectile impact on warheads and other solid-explosive charges Most of what has been said about rocket motors also applies to warheads and solid explosives (Frey et al. 1979).

SCJ impact on rocket motors and solid explosives SCJs must be considered separately from other projectiles because they are capable of penetrating explosives supersonically (Chick et al. 1989).

Bullet and fragment impact on solid gun propellants Gun propellants normally do not detonate under bullet and fragment impact. There are exceptions to this rule, however, especially when fine-grained propellants are considered (Wise et al. 1980; Baker and Stegall 1997).

SCJ impact on solid gun propellants Some gun propellants will detonate in response to SCJ impact, but most do not, at least not at 20°C. However, although they do not detonate, they will normally react with considerable violence (Watson et al. 1991; Dispersio et al. 1965).

Liquid gun propellants As mentioned previously, none of these is currently in service. Liquid propellants based on water solutions of hydroxyl ammonium nitrate are extremely insensitive to shock and do not burn at atmospheric pressure in the neat (uncontaminated and bubble-free) condition. However, reactions akin to XDTs can occur when a ballistic impact induces cavitation and then ignites the cavitated material. A discussion of the vulnerability of these materials is contained in Frey et al. (1996).

Other mechanisms, not as well understood, are also possible. Heavily confined explosives often ignite under ballistic-impact conditions whenever the metal confinement is perforated. It has been postulated that a localized shear of the energetic material occurs if the metal case fails in a plugging fashion, and this localized shear may also ignite the explosive. But these types of events usually result in explosions, not detonations. The violence of these events is controlled by the degree to which the material fractures and generates surface area under ballistic impact and by the confinement.

D. Personnel Vulnerability and Soldier Survivability

The applications for data and models on the vulnerability of personnel to weapon effects are numerous. Virtually any weapon-effectiveness study, whether it is to evaluate weapon design improvements, establish the relative performance of competing munition candidates, or evaluate the effects of individual protection measures, will require an assessment of the weapon effects on the target's human elements.

Similarly, numerous requirements exist for human vulnerability data in connection with vehicle vulnerability and crew-survivability analyses. These analyses support, among other things, the soldier survivability portion of the manpower and personnel integration (MANPRINT) program, Live Fire and JLF test evaluations, concept-exploration studies, and basic phenomenology studies. All of these applications involve an evaluation of the effects of hostile environments on a soldier's ability to perform required functions, the soldier's mortality (or survivability), or both. Although the weapon developer's interest is focused on the operational consequences of weapon effects, there is also an increasing need within the V/L community to make statements regarding the medical significance of weapon-induced injuries (e.g., for Live Fire testing).

Additionally, ongoing soldier or combatant survivability research conducted at the ARL has resulted in the application of trauma severity scoring for use in the design, development, and evaluation of U.S. Army body armor against ballistic threats (Davis and Neades 2002b). In civilian sector medical research, trauma severity scoring systems provide a means to quantify severity, and these have widespread application. In general, trauma severity scoring systems can be characterized as anatomical-based, physiological-based, or a combination of anatomical- and physiological-based systems. Several examples of anatomical-based systems include the Injury Severity Score (ISS) (Baker et al. 1974; Baker and O'Neill 1976), New Injury Severity Score (NISS) (Osler et al. 1977), and the Anatomic Profile (AP) (Copes et al. 1990). Also, it has been proposed that contemporary medical trauma severity scoring systems have potential widespread application in other aspects related to military wound ballistics (Davis and Neades 2002a). Furthermore, a new and improved trauma severity score has been developed to characterize both civilian and military penetrating- and blunt-injured trauma patients (combat soldiers)—the Revised Injury Severity Score (RISS) (Davis 1998). The RISS score has application in combat casualty wound epidemiology, individual protection effectiveness evaluation (e.g., body armor vests, helmets, etc.), and soldier or personnel survivability analysis. The RISS severity score has been shown to have increased capability to characterize both penetrating and blunt injuries compared with the ISS, the NISS, and the AP (Davis 1998). (See also Secs. 2.III.D.1, 2.III.D.2, and 4.III.E.)

1. Common Personnel Vulnerability Definitions

Definitions of common terms within the personnel vulnerability community are as follows:

Incapacitation   The most common anti-personnel measure of effect is simply "the inability to perform, at a level required for combat effectiveness, the physical or mental tasks required in a particular combat role." Clearly then, whenever specifying a given incapacitation value, whether it is an expected value or actual probability of a specific level (e.g., 100 percent), one also needs to state the task or set of tasks (role) that the individual is expected to perform. In addition, because post-injury performance can also depend on how long it has been since the injury occurred, time after wounding is usually specified as well. For example, an expected probability of incapacitation given a hit (P_{IH}) using the assault 5-min criterion might be written as follows: $E_{IH\ A5} = 0.40$.

Insult   An external, munition-produced physical agent capable of producing an injury (examples: 1-g fragment, a 50-kPa pressure pulse, a 3-cal/cm^2 thermal dose).

Standard injury   The anatomical damage and/or physiological response produced by an insult. Standard injuries require a minimum insult dose (threshold) and may change in severity as a function of time after insult (e.g., 10-mm hole in thorax muscle tissue and associated weakness, lung petechia, edema).

Casualty Joint Publication 1-02 (U.S. Department of Defense 2005) defines a casualty as any person who is lost to the organization by reason of having been declared dead, wounded, injured, deceased, interned, captured, retained, missing, missing in action, beleaguered, besieged, or detained. For Army analysis purposes, a more precisely defined definition is used, and a further distinction can be made between medical and operational casualties.

Medical casualty An individual who has experienced an injury that requires medical intervention and usually evacuation from his/her unit so that medical treatment can be administered. (See also *Operational casualty* below.)

Operational casualty An impaired individual whose performance is less than that required for combat effectiveness (i.e., an incapacitated individual). It should be noted that operational casualty measures provide no indication of wound severity and therefore cannot be used to infer lethality or the need for any particular medical attention; an operational casualty may or may not also be a medical casualty. By the same token, a medical casualty is not necessarily an operational casualty since in many cases the injury will not preclude all activity. For example, a radio operator who receives a leg injury can probably continue to function effectively (at least for some time periods) in spite of the need for medical attention.

2. Blast Overpressure Injuries

Blast overpressure injuries are one of the secondary causes of predicted casualties in armored vehicles perforated by threat munitions. Primary blast overpressure injury is limited to the body's air-containing organs (i.e., the lungs, gastrointestinal tract, and ears) and occurs as a result of an incident pressure wave directly loading the body. The resultant loading is distributed over the entire body surface and depends on the body's orientation to the incident wave. The exposure conditions that result in primary blast injury have been roughly determined; however, the precise injury mechanisms are not clearly understood.

In contrast to blast waves generated in a free-field environment, the blast overpressure created inside of a defeated armored vehicle or other enclosure is characterized by a complex pressure wave. In these circumstances, the surfaces produce multiple reflections of the blast wave, resulting in a complex delivery of the blast energy that governs how the body responds. The blast wave starts out as an initial fast-rising wave emanating from the point of penetration. Other shocks produced by the passing of the primary penetrator and waves from multiple internal reflections are superimposed on the initial wave. The overpressures thus created may be followed after several milliseconds by a quasi-static pressure because of heating of the air in the enclosure's internal volume and the accumulation of combustion products. This quasi-static pressure depends on how rapidly venting occurs. Potentially, other blast events may occur from detonation of onboard explosive devices or vaporized fuel and/or hydraulic fluid.

The current method for predicting blast injuries and resulting incapacitation for blast exposures in enclosures is to conform the observed overpressure-time histories into a free-field waveform and then compare the peak and duration to Bowen's free-field injury curves for incident orientation. These curves, developed by the

Lovelace Foundation in the 1950s and 1960s, are based on extrapolations from effects of fast-rising, long-duration, exponentially decaying shock waves (i.e., nuclear blast), and they are widely accepted for free-field exposures. These free-field blast waves are typically defined in terms of peak pressure and duration of the positive phase (A-Duration). At the time of this writing, the body's interaction with complex blast waves has not yet been completely defined, and extensive data to support injury criteria for complex wave environments do not yet exist. In addition, data on short-duration (conventional) blast wave effects are limited. As a result, blast-injury assessments for exposure inside the reverberant space of a defeated armored vehicle must be related to the criteria developed for free-field situations. This is accomplished by conforming the observed overpressure-time histories into a free-field wave form by establishing an effective peak pressure (Ripple and Mundie 1989). This value is then compared, along with the positive phase duration, to injury curves established for incident exposure in a free-field situation.

The effective peak pressure technique involves the following steps. First, the total positive duration is determined, and an effective peak pressure is graphically interpolated. The total positive duration is likened to the A-Duration of a Friedlander blast wave in the free field. The effective peak pressure ignores the random pressure spikes that do not significantly contribute to the overall impulse of the waveform. Pressure pulses are not corrected for transducer orientation relative to the direction of travel of the blast wave. In an armored vehicle, gauge position and orientation relative to the recorded shock waves cannot be accurately determined, but the pressure reflections are already accounted for in the pressure trace itself. The effective peak pressure and duration are then compared with the Bowen pressure-duration injury curve for a prone body where the long axis of the body is parallel to the blast winds in a free-field blast wave environment. Because the effective peak pressure technique takes into account the total duration, quasi-static pressure is included in the injury prediction. Injury predictions with this technique have correlated well with injuries observed in studies involving complex blast waves from point-source explosions in enclosures and from anti-tank-round penetration of an armored combat vehicle.

To estimate the operational impact of sublethal blast injuries, in 1983 a panel of blast experts agreed on a relationship between incapacitation and the injury levels described by the Bowen curves. These estimates were based on experimentally observed physical damage from Friedlander blast waves. Accordingly, affected soldiers are expected to be one percent incapacitated at conditions that result in threshold injuries to the lung. Similarly, conditions that cause death in one percent of the exposed population are equated to 50 percent incapacitation. An exposure that is lethal for 50 percent is assumed to cause 99 percent incapacitation of the exposed population. Intermediate degrees of incapacitation are estimated assuming a log-normal probability distribution based on these three points. The injury predictions determined from application of the effective peak pressure technique are applied to this curve to predict levels of fractional incapacitation.

Data gaps exist in both the injury and performance-degradation portions of blast-overpressure assessments. The current effective peak pressure technique discussed previously is merely an approximation method with little actual data to back it up.

Specific areas for advancement that could lead to improved criteria include the following:

1) *Basic information that better defines the mechanism(s) of blast-overpressure injury for personnel in enclosed environments:* Studies have shown that the traditional mechanisms associated with free-field blast injury are not well correlated with complex-wave injuries. In armored vehicles and other enclosures, the surfaces produce multiple reflections of the blast wave, resulting in a complex delivery of the blast energy, which governs how the body responds.

2) *Data on the mitigating effects of clothing and equipment items on nonauditory injury:* Data indicate that the use of Kevlar body armor may actually increase the severity of lung injuries and that lung threshold injuries also occur at lower levels. However, the fragment protection provided by the Kevlar body armor is currently regarded as outweighing the enhanced blast-injury potential.

3) *Combined effects data for combinations of blast, penetrating wounds, thermal, and other injuries.*

4) *The synergistic effects of blast injuries in combination with other wounds:* The present method for modeling multiple injuries involves mathematically combining the effects of individual injuries ($P_{I/H}$'s) using the so-called *survivor rule*. This rule, which is written as

$$P_{\text{Total}} = 1 - \prod_{i=1}^{n}[1 - P_i]$$

requires one to assume that the events (injuries) are statistically independent, an assumption that intuitively seems wrong.

5) *Data on the relationship between injury and performance degradation:* Although they allow determination of the operational impact of sublethal blast injuries, the various levels of incapacitation that were equated to the Bowen injury curves at the 1983 meeting of blast experts are highly subjective in nature. No experimental data to date relate level of performance to level of blast injury. Knowledge of the relationship between blast-induced pathology and the resulting physiological consequences is critical if we are to accurately assess primary blast-induced performance degradation. Additionally, the incapacitating effects of hearing loss or pain from ear injury are not known and cannot be modeled because of a lack of data on pain and impairment of communication on normal soldier functions.

As a result of an ambitious multi-year modeling effort sponsored by the U.S. Army Medical Research and Materiel Command (MRMC), including the work conducted by the Department of Respiratory Research at WRAIR, major advances in the prediction of lung injuries from blast overpressure have been made in recent years. This work has produced a number of computer models that have been demonstrated to be useful in the assessment of both auditory and nonauditory hazards to personnel. Version 5.0 of the INJURY code models the probability of lung injury as a function of the normalized work done on the lung. This model has been favorably compared with experimental results collected as part of injury studies. In addition to calculating lung injury, the model also calculates tracheal injury as well as auditory limits.

E. Structures

The response of a target system to loads from the detonation of a nearby HE weapon can be quite complicated. The response usually entails nonlinear, plastic behavior of host materials, which can lead to tearing, fracture, or failure of the material. The response is also directly related to the applied loads, which tend to be extreme. Applied pressure loads are extremely high and of short duration. They also involve significant gradients or spatial changes in the load environment over the target surface. Fortunately, geometric nonlinearities usually do not come into play owing to the rapid rise of the detonation loads. The response periods of the target and target components are usually long compared with the duration of the threat environment. In other words, the target usually does not have time to deform or change from its initial geometry before the ending of the threat environment. In this instance, the threat environment is impulsive in nature, imparting an initial momentum or energy to the target components.

Structural-response analysis of target components to the applied threat environment can be performed at varying levels of sophistication. Examples of analytical methods in increasing order of sophistication include the following: energy-based analysis of single-degree-of-freedom approximations to component response (Bingham et al. 1996); numerical, time-marching analysis of single-degree-of-freedom approximations to component response (Biggs 1964); finite-element analysis, and smoothed-particle-hydrodynamic, numerical calculations (Libersky et al. 1993). Each of these methods must be performed in conjunction with some approximation to the threat environment, which can also take on varying levels of sophistication. However, increasing levels of sophistication usually come with corresponding increases in computer run time. The choice of analysis method usually depends on the availability of time and money, the applicability of the analysis models, and the criticality of the answers. If one can afford sophisticated analysis approaches, these are usually performed in conjunction with lower methods to approximate the answers or verify the reasonableness of the answers.

Target systems can usually be broken down into structural components. Response of these components can be approximated as modes, ranging from beam response to plate response to shell response, depending on the geometry and material makeup of the component as well as support conditions (intermediate or boundary). Many of these modes have been tabulated as equivalent single-degree-of-freedom systems in Biggs (1964).

A typical example of the effects against a real target can be found in Halsey and Roquemore (1987), which contains results of four Mk-82 bomb firings using radar vans as targets.

IV. Target Response

This section concerns itself with the MMF mapping ($O_{2,3}$) from level 2 to level 3, whose points represent degradations in performance associated with the damaged components (e.g., degradation in mobility, firepower, command).

A. Engineering Aspects

Not surprisingly, the mapping from MMF level 2 to level 3 (system capability state) is primarily an engineering problem. This mapping relates component dysfunction to system capabilities. Generally speaking, this is a multi-step process in which component dysfunction is mapped to subsystem LoF, which in turn is mapped to the system capability state(s). Further, a given component dysfunction may affect more than one subsystem, and there may also be subsystem-to-subsystem interrelationships to consider. For example, a battery may be a critical component in a tank fire control subsystem as well as in the communication subsystem. The battery must therefore be considered with the other components in each subsystem. Further, damage to either subsystem (e.g., short or open circuits) may well affect the operation of the other subsystem if electrical power becomes disrupted.

After component criticality has been established, it is then necessary to determine loss of component and subsystem function resulting from specific component physical damage. The tools for accomplishing this are broadly referred to as engineering models. These models may take a variety of forms and sophistication depending on the damage, component characteristics, and state of knowledge of modeling nonstandard component and subsystem operation. Implementation ranges from complex computer codes to simple empirical data curve fits. The purpose of this discussion is not to describe specific engineering models that are available, but rather to illustrate the nature of available models and suggest the variety of modeling constructs that are used in practice.

Ideally, closed-form algorithms relating critical parameters would be available. In certain special cases, such algorithms exist. For fluid containers, for example, leakage rates can be predicted given accurate characteristics of the hole. Note, however, that even then the closed-form algorithms address only part (leakage rate) of what is usually a complex engineering problem involving hole size, shape, and number, bomblet fragmentation patterns, and threat-target initial conditions.

To illustrate, consider the problem of analyzing bomblet fragment damage to truck tires and the resultant effect on the system mobility capability (Fig. 2.22).

Small bomblet fragments are capable of perforating truck tires, but the size of a single fragment hole is quite small, resulting in a slow leak that may take considerable time to deflate the tire. A single fragment hole can be analyzed quite easily given knowledge or assumptions about the characteristics of the hole, but usually a threat-target encounter results not only in several holes of various sizes in a given tire but in holes in several tires. The situation is further complicated by the fact that some truck designs have on-board tire inflation systems that can replenish leaking air on the go. Thus, while closed-form approximations of leakage rates for single and multiple tires were developed (Grote et al. 1996), the overall problem is still highly stochastic. Individual bomblet fragment characteristics (size, shape, orientation, and speed) vary, influencing not only the size of any given hole but also the likelihood of fragment perforation. The number of such holes, the number of tires impacted, tire inflation at the time of attack, and inflation system performance (which in turn depends on engine speed) are other stochastic variables that must be taken into account in the analysis. So, while closed-form algorithms can address part of the problem, the implementation of the algorithm and the availability of necessary input data determine the realism that can be simulated and the accuracy that can be achieved.

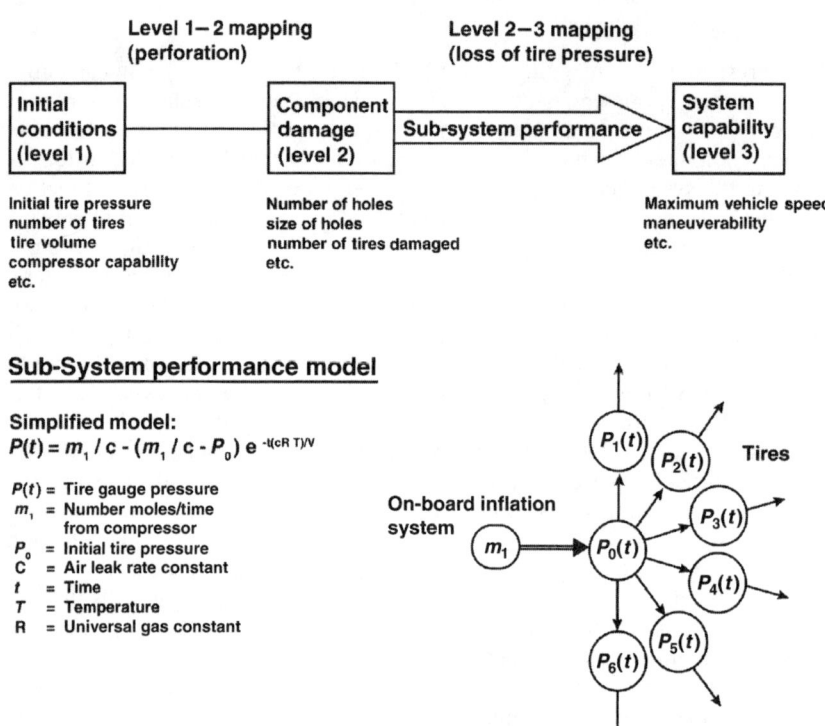

Fig. 2.22 Level 1 to levels 2 and 3 mapping of tire damage using closed-form algorithms.

As a second illustration, consider controlled damage experiments that were conducted on diesel engine variants with orifices installed in intake air components to create specific leakage rates (Mahaffey, to be published). Dynamometer tests were then conducted to determine engine power loss (level 2) followed by vehicle tests to determine mobility degradation (e.g., reduction in acceleration) (level 3). Figure 2.23 illustrates the process for the BFV. Again, although the data were collected for specific engines and vehicles, the results can be extrapolated to engines and vehicles of similar design.

Even as engineering models range from simple to complex and from empirical to analytical, their implementation is accomplished in many ways. At one extreme, algorithms are "hard wired" into V/L analysis codes for specific analyses. Many are exercised independently of the analysis codes and provide input data sets that are input to the codes. Perhaps the ideal situation is to implement them as modules or libraries, depending on the analysis code architecture, so that they are available and applicable for a wide range of analytical scenarios. In this regard, excellent progress has been made with blast overpressure damage analysis.

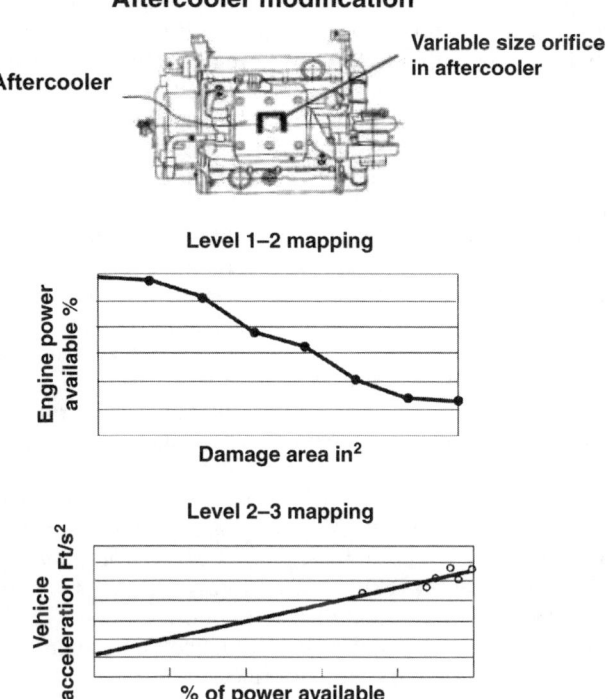

Fig. 2.23 Mapping from level 1 to level 2 and from level 2 to level 3 using empirical data.

B. Systems of Systems

In light of increasingly complex and networked military systems, it is increasingly necessary for the V/L analyst to consider a number of different ground combat systems acting in concert with one another as an ensemble to accomplish a task. For such cases, commonly termed SoS, survivability of individual platforms is not only a function of conventional platform vulnerability but also a function of joint actions of the ensemble. At the end of the 20th century and the beginning of the 21st century, changing world political conditions placed greater emphasis within the U.S. DOD on ground forces that are capable of rapid deployment. As a result, the burgeoning communications and electronics field is enabling networked operations to a level unimagined just a few decades ago. In addition, the need for more readily and rapidly deployable systems has driven ground combat platform weights steadily downward, reducing the level of conventional armor possible for such systems. Fortunately, advances in materials science have also been rapid; the advent of new materials and configurations for armoring ground combat systems, both actively and passively, has contributed to increased system and crew survivability despite the trend to lightweight systems.

A number of factors contribute to increased survivability of lightweight systems, including the use of networked fires for improved ensemble lethality at longer ranges. Networked surveillance; improved, timely situational awareness, especially through the use of unmanned ground and low-altitude aerial systems; software decision aides; and a general reduction in engagement duration all improve survivability and are changing the conduct of ground warfare. For the V/L analyst, this means that a greater emphasis is placed on reduced susceptibility to improve survivability more than ever before.

In conventional tanks, for example, one might consider that all functions, from detection of targets through engagement and battle damage assessment, are accomplished in a single spot, namely the turret. In an SoS, however, detection might be from one platform, tracking and acquisition from another, engagement from yet another, and battle damage assessment from still another. This configuration might be used to prevent the threat system from engaging a friendly system that took no part in the action. In this sense, the other platforms contributed to its survivability; the on-board survivability features played no role for this particular instance.

The method of fault trees discussed in Chapter 3 (Sec. 3.II) provides a ready means for analysis of SoS. Fault trees can be generated for functions, capabilities, or ensembles just as they are for individual systems. These fault trees include "components" from a number of different platforms. The major difference between ensemble fault trees and individual platform fault trees, however, is that for the ensemble the fault trees are dynamic; that is, the fault tree configuration changes over time as the same functions are performed by different systems during the course of a battlefield scenario. while each platform has a predetermined manner in which to perform its functions, an ensemble, or SoS, does not [see Walbert (2004b)].

It is also evident from the preceding discussion that assessing the vulnerability of an ensemble and its constituent platforms is highly mission- or scenario-dependent. Unlike conventional ground combat system survivability, SoS survivability analysis ties the analyst ever closer to MMF levels 4 and above. These concepts are also discussed in greater detail in Chapter 3 (Sec. 3.V).

V. Tactical Utility

Mapping from level 3 to level 4 (combat utility) is an operations research process. This mapping takes into account system performance requirements based on tactics, terrain, visibility, and other scenario-dependent factors, and it produces metrics indicative of remaining combat utility (i.e., usefulness of the system to the commander responsible for conducting a specific mission).

Traditional measures of effectiveness (MOEs) have focused on forces-based, materiel-centric measures such as loss-exchange-ratios, force-exchange-ratios required to achieve an objective, or the time required to complete an operation. In addition, most combat models require some form of grossly aggregated (e.g., over scenarios) V/L data as input. Historically, these have been such measures as the M-Kills, F-Kills, and K-Kills for armored vehicles. This process is suboptimal, however, because it is focused largely on the *materiel* (the physical means needed for successful military prosecution) without adequate consideration for, or linkage to, the *missions* (the end actions that must be accomplished to meet objectives). Thus, the MMF recommended practise is to focus on mission-centric, task-based MOEs. These MOE and standards are the codification of how planned/delivered task outcome affects mission success.

Ideally, the result of an analysis at level 3 (possibly aggregated to allow for the stochastic nature of the mappings) would be used as input to a combat model. That model would simulate the conflict between two forces with the system(s) in question being modeled as individual entities in one of the forces. After encountering opposing systems, the resulting conditions would be based on the data at level 3. Data would be accumulated for a variety of combat scenarios in some set of interest, and the combat worth of the system in question would be assessed on the basis of whether it helped its force to win, giving a level 4 result. Similar level 4 results would be compared for alternate systems or to show the value of a totally new concept.

At present, there are somewhat less complete definitions of combat utility. Often one wishes to compare two competing designs to establish whether one is superior to another in important characteristics. One of these might well be the vulnerability (or lethality) of the designs. In this case, the results of a level 3 analysis might be wrapped up by averaging over a variety of conditions, possibly derived from some form of scenario analysis.

Thus, the MMF enables DOD transformations, from concept through actual combat, with a framework to help the warfighter, engineer, and comptroller specify a common understanding of military operations, systems, and information, and provide quantitative mission assessment of alternative solutions. Moreover, the V/L analysis process is an integral part of this framework, not only in gaining an understanding of how one system might survive better than another, but also in encapsulating system characteristics to allow system development decisions to be made at various milestones in the acquisition process.

Chapter 3

Vulnerability Assessment and Measures of Effectiveness

I. Traditional Measures of Mission Effectiveness

THE basic principles upon which the traditional methods of vulnerability analysis are built were dictated, in part, by the applications of vulnerability evaluation results and the approaches to weapon system evaluation that were extant more than 40 years ago. Although the traditional methods have served well and are still used routinely, changing applications have suggested a reexamination of basic principles. Some modern alternatives to the traditional methods that are now being explored are discussed here. A historical perspective is also included to show how the traditional methods developed and how requirements are continuing to change.

A. Background

The foundations of the methods currently used to evaluate the vulnerability of military equipment were established in the years immediately following World War II. Even though military manpower was being reduced in that era, national policy required that a strong military force be maintained, and it was clear that more sophisticated weapons would be needed in the event of future conflicts. Accordingly, the Armed Services acted to increase scientific research in areas that might have military applications and to develop analytical capabilities that would be helpful in defining requirements and evaluating new or proposed weapon designs. The technology of vulnerability evaluation was one of several technologies with origins in that era.

Before 1948, the only attempts to study the vulnerability of ground-combat equipment were a limited number of firings against armored vehicles, conducted mostly during World War II. No efforts had been made toward derivation of analytical methods. The results of these tests were recorded as narrative descriptions of the observed mechanical damage. Although such experiments provided some qualitative insights into the effects of anti-armor munitions, no quantitative measures of the damage were obtained, and only subjective comparisons among projectiles or vehicles were possible.

In the latter part of 1948, the Department of the Army established a formal program to study the vulnerability of armored vehicles; a similar program had been established about a year earlier to study aircraft vulnerability. This program was intended to establish systematic experimental methods for studying the vulnerability of armored vehicles and, in the longer term, to derive analytical methods that would permit evaluations without actual testing. In later years, similar efforts were undertaken for various other targets, including trucks, missiles, missile launchers, buildings, and bunkers.

While the basic approaches to vulnerability analysis and testing of ground combat equipment developed in the late 1940s and early 1950s are still used today, there has been significant improvement in the implementation of these approaches and in the empirical database over the intervening years. Testing now stresses combat realism (live fire testing), stochasm in the target–threat interaction and attempts to understand and quantify it, and, probably most significantly, model development and validation.

B. Measures of Vulnerability

A fundamental problem in deriving a systematic approach to vulnerability assessment is the definition of appropriate terms in which vulnerability can be measured. If the problem is viewed from the perspective of experimental evaluation, the most rudimentary approach is to fire against a target and document the results as a list of target components that were damaged or destroyed. The problem facing the analyst is how to derive quantitative measures of damage from the lists of damaged components; this step is essential to useful comparisons between targets and between weapons. Conceptually, the problem would be the same if the component damage were predicted analytically rather than observed experimentally. Such analytical methods did not exist when the issue of how to measure vulnerability was first addressed; however, the long-term goal was to develop such methods, and it was thought to be essential that experimental and analytical methods be consistent. Expressed in MMF terms, the problem is the definition of level 4 and derivation of the tools for mapping from level 2 to level 4 (with or without passing through level 3).

The intent of any attack on military equipment is to prevent the equipment from performing its intended combat role. Consequently, the first rudimentary notion in the minds of the analysts who first addressed the question was that vulnerability should be measured in terms that in some way indicate how the ability of the target to perform its role is likely to be affected by a specified attack. Beyond this simple idea, it was not clear how to proceed, although one idea was to use a P_K. In other words, target performance should be defined as the probability that the target would be damaged to the extent that it could not function in combat. This idea was attractive because its meaning is clear intuitively, and, presumably, criteria for deciding whether a target was killed could be related to the combat missions of the target and to the physical condition of the target required to perform the mission.

Because of the diversity among ground targets in their combat roles and the way their mechanical structures respond to ballistic attack, no single measure of vulnerability was found to be appropriate for all targets. For the purpose of this discussion, it is convenient to divide all combat equipment into a few broad categories and to address each separately. Brief mention of aircraft targets is made here even though the focus of this book is ground combat equipment.

1. Armored Vehicles

The initial perceptions of the response of armored vehicles to ballistic attack were derived from the results of limited proving-ground tests of tanks. Although these tests were few in number, they were sufficient to yield an important insight: a broad range of mechanical damage can occur to degrade the vehicle to some extent but leave it capable of performing some useful combat functions. The accumulated experience in tank warfare in World War II confirmed these experimental observations by showing that tanks in combat experience various levels of damage and that they can continue to operate in a degraded fashion after being attacked.

These observations were thought to have a critical bearing on the selection of a measure of vulnerability; they showed that the results of ballistic attack on a tank cannot be characterized simply as either 1) no significant damage, or 2) destruction of the fighting capability of the vehicle. Rather, it was necessary to recognize a wide range of possible effects. On the basis of this observation, then, it was judged that P_K was not the most appropriate measure of the vulnerability of armored vehicles because it ignored possibly important differences among vehicles in their susceptibility to levels of damage that are significant but not completely incapacitating. Instead, it was judged more appropriate to devise a continuous scale on which the effects of ballistic attack could be quantified and to characterize vulnerability as the average value for a specified set of attack conditions.

2. The Concept of Combat Utility

Each type of armored vehicle has an assigned primary mission (or role) that it is expected to perform in combat. The capability of a vehicle to perform its mission is called the combat utility. To explain what is meant by the term *combat utility*, however, it is not necessary to define the missions of armored vehicles; rather, it is necessary to indicate what performance capabilities are required of the vehicle and crew.

Combat utility involves more than measurable performance capabilities of the vehicle, such as speed over a specified terrain or the accuracy of the armament. The capabilities of the crew to use the physical capabilities of the vehicle are also critical. An armored vehicle must perform as part of a unit. Thus, combat utility involves less tangible factors such as coordinating fire and maneuvering with other elements of the unit, selecting travel routes across difficult terrain, and detecting and reacting to enemy activity. Also, communication and surveillance capabilities are essential to the effective use of the firepower and mobility of a vehicle.

If any of these capabilities are lost or degraded due to damage to the vehicle or crew, then the combat utility of the vehicle is reduced. Thus, the concept of combat utility promised to provide an appropriate scale on which vulnerability could be measured. It would serve as the desired tie-in to battlefield capability, and it would recognize all levels of damage.

At the time these issues were first addressed, analysts were concerned primarily with developing systematic experimental techniques for evaluating armored-vehicle vulnerability; the task of developing analytical methods had not yet been addressed. Some means to translate observed physical damage into a numerical measure was needed. It was thought that the concept of combat utility might provide this method and, presumably, could later be used as a way to present analytical results.

Unfortunately, there was no basis on which to form a mathematical construct that would relate degradation of one or more capabilities to a quantitative measure of the overall loss of combat utility. Nevertheless, it was thought possible to evaluate the combat utility of a damaged vehicle, at least in a subjective fashion.

3. *The Standard Damage Assessment List*

The selected approach to evaluation of combat utility involved the development of a device called the Standard Damage Assessment List (SDAL). This list contains most of the individual components and certain mechanical and electrical subsystems critical to the operation of a vehicle, as well as each member of the crew.

For each entry in the list, two numerical values are given, one for the mobility function and one for the firepower function of the vehicle. The values, presented on a scale of 0 to 1, indicate the fractional degradation of combat utility (DCU) that would result from the destruction of each component or subsystem or incapacitation of each crew member.

Because combat utility is not a quantity that can be physically measured, the values presented in the list are necessarily judgmental; they represent the composite judgment of a panel composed of military officers and enlisted personnel experienced in operation of the particular type of vehicle under consideration.

The panel members consider each item on the list individually and judge how much the combat utility of the vehicle would be degraded if the component or subsystem were destroyed or the crew member were incapacitated. To arrive at the required values, panel members are instructed to consider all of the combat missions that the vehicle might perform and the expected frequency of occurrence of each mission. Of course, it was not expected that this could be accomplished explicitly, but, on the basis of the composite experience of the panel, it was possible to make a coarse categorization of mission types and to form an impression of how often each was likely to occur.

To assist the deliberations of the panel, the following kill criteria are used:

Mobility (M-Kill)   Loss of tactical mobility resulting from damage that cannot be repaired by the crew on the battlefield. A vehicle has sustained an M-Kill when it is incapable of executing controlled movement on the battlefield.

Firepower (F-Kill)   Loss of tactical firepower resulting from damage that cannot be repaired by the crew on the battlefield. A vehicle has sustained an F-Kill when it is incapable of directing controlled fire from its main armament.

Catastrophic (K-Kill) A vehicle has sustained a K-Kill when both an M-Kill and an F-Kill occur and the damage is judged not to be economical to repair.

These criteria serve to define the level of degradation that should be regarded as complete loss of utility for the SDAL process. For example, M-Kill does not mean complete loss of mobility; rather, it means failure to perform a combat mission due to degradation of the mobility function. The criteria are as originally worded in the armored vehicle kill criteria written sometime in the early 1950s. These definitions were later modified to specify that LoF must occur within 10 min after the attack to qualify as a kill. This modification was made because it was observed in tests that certain vehicle components could sustain ballistic damage and continue to function for a short time before failing. The definitions relate to direct-fire engagements, and such engagements typically last only about 10 min. Any degradation of function that occurs after that time is irrelevant to that engagement.

As a part of the process, the panel had to estimate how the loss of each item on the list affected the physical performance characteristics of the vehicle and crew as a fighting unit. This seemingly difficult task proved to be simple. In most cases, the panel members had experienced failure of the particular component under consideration owing to normal wear and understood how the vehicle performed without the component. Also, they had experienced occasions where it was necessary to operate vehicles with one or more crew members missing, and they understood how crew functions had to be apportioned among those remaining. In other cases, the analysts were able to cite tests in which vehicles were operated with certain components removed or destroyed to observe performance characteristics.

An SDAL constructed in this way applies to only one category of armored vehicles because the components present vary to some extent from one category to another. The first list developed applied to conventional turreted tanks. In later years, similar lists were developed for various categories of vehicles including armored personnel carriers, infantry fighting vehicles, reconnaissance vehicles, and armored self-propelled howitzers.

The original SDAL for conventional turreted tanks (minus actual numbers for the kill categories) is presented in Table 3.1 (in three parts), taken from MIL-HDBK-1226(AR) (U.S. Department of Defense 1997) and originally from the CARDE report Q-21 (Benjamin 1960). A more thorough discussion of the general problem of deriving SDALs is found in Zeller and Armendt (1987).

4. Application of the SDAL to Armored Vehicle Testing

The original approach to armored vehicle testing required that a panel of soldiers trained in operation of the vehicle be present for each test shot to assess the damage. The SDAL could not include all possible combinations of components that might be damaged; consequently, some judgments were required on each occasion. The assessors were expected to identify the vehicle components that were damaged or destroyed and to infer crew incapacitation from the damage to mannequins placed in the crew stations. They would then consider the aggregate damage and form a consensus as to loss of combat function.

Table 3.1 SDAL for tanks

Part A: Tank DCU when an interior component is destroyed			
Component	M[b]	F[b]	K[b]
Ammunition case main gun[a]			
Ammunition projectile, HEP, main gun[c]			
Ammunition projectile, KE, main gun			
Ammunition, machine gun			
Armament			
– Coaxial machine gun			
– Commander's machine gun			
– Main gun			
– Both machine guns			
– All other gun combinations			
– Recoil mechanism, main gun			
Battery			
Commander's vision device			
Driving controls			
Driver's periscopes			
Engine			
Elevating mechanism			
– Power only			
– Manual only			
– Both			
Fire control			
– Primary only			
– Alternate only			
– Both			
Intercommunication equipment			
– Complete			
– Commander's only			
– Gunner's only			
– Commander's and gunner's			
– Loader's only			
– Driver's only			
Rotary junction box			
Turret terminal box			
Bore evacuator			
Radio			
– Present battle, present engagement			
– Present battle, future engagement[c]			
Traversing mechanism:			
– Power only			
– Manual only			
– Both			

(continued)

VULNERABILITY ASSESSMENT AND MEASURES OF EFFECTIVENESS 71

Table 3.1 SDAL for tanks (continued)

Part B: Tank DCU when an exterior component is destroyed			
Component	M[b]	F[b]	K[b]
Buffer spring			
Idler wheel			
Road wheel (front end)			
– One			
– Two			
Road wheel (rear end)			
Road wheel (other)			
Shock absorbers			
Sprocket			
Track			
Track guide			
Track support roller			
Main gun tube			
Loss of fuel, one side			

Part C: Tank DCU when personnel are killed or rendered ineffective			
Personnel	M[b]	F[b]	K[b]
Commander only			
Gunner only			
Loader only			
Driver only			
Two crewmen rendered ineffective			
– Commander and gunner			
– Commander and loader			
– Commander and driver			
– Gunner and loader			
– Gunner and driver			
– Loader and driver			
Sole survivor			
– Commander			
– Gunner			
– Loader			
– Driver			

[a]Destruction will cause fire and/or explosion.
[b]For security purposes, actual numerical values have been deliberately omitted.
[c]This damage level may reasonably be assumed to cause loss of command control.

The SDAL was to serve as a guide to the assessors to ensure a degree of consistency in the assessments from one test program to the next. That is, the list provided values for individual components and a few combinations of components, but the assessors had to arrive at values for the particular set of components damaged in the test. Further, the SDAL was prepared for a certain category of vehicles, and in some instances a particular value might seem to be incorrect for the specific vehicle under test. In this event, the assessors were allowed to deviate from the list.

Later, it was decided to accept the SDAL as a standard rather than a guide and combine the values for the various damaged components mathematically instead of seeking the opinions of a panel of assessors. This approach provided essential consistency among test results, and a reproducible and more objective record of each assessment. Also, as discussed later, this approach satisfies a requirement of modern vulnerability models. These models predict physical damage analytically, and the resulting loss of combat function must be assessed in an unambiguous and computer-implementable way; thus, experimental and analytical evaluations should be similarly conducted.

In this approach, vulnerability analysts, often with the help of vehicle mechanics, inspect all internal and external components and check each subsystem for proper operation. The analyst then prepares a table of all damaged components and subsystems. By referring to the SDAL, fractional loss of combat function values are associated with each table entry, which is also given a value for each major vehicle function (mobility, firepower, etc.). An overall LoF value is then computed for each major function, as the expression

$$\text{DCU} = 1 - \prod_{i=1}^{n}(1 - \text{DCU}_i) \qquad (3.1)$$

where

DCU = expected degradation of combat utility resulting from the combined effects of damage to multiple subsystems
DCU_i = degradation of combat utility resulting from damage to subsystem i
i = subsystem index
n = number of subsystems of the vehicle

This formulation assumes that the various DCU values are independent. Before accepting the idea of a mathematical combination of the SDAL values, an analysis was conducted to learn whether this approach was consistent with the judgmental assessments. Values were calculated for each test shot in the data collection and compared with the judgmental values developed by the assessors. The comparison showed acceptable agreement and the computational approach was accepted.

By applying the process described previously, the effects of a ballistic attack against an armored vehicle can be measured in meaningful terms and expressed by a few numerical values. By performing similar tests with different vehicles or with different projectiles, quantitative comparisons could be made. It was then possible, for the first time, to address in a quantitative way such important questions as the following: 1) Which among competing projectile designs is more lethal? 2) Which among vehicle designs is less vulnerable to ballistic attack?

The ability to answer such questions, even with this slow experimental approach, was invaluable in making design and development decisions.

5. Some Observations Regarding the Measure of Vulnerability for Armored Vehicles

Whereas the discussion has thus far concentrated on experimental evaluation, analytical methods are based on the same measure and the observations that follow are applicable to both evaluation techniques.

The SDAL method provided a way to measure vulnerability, a way that has served well for many years, both in quantifying experimental damage and, later, as a foundation for analytical methods. Despite its long-term acceptance, however, it has certain fundamental weaknesses that should be noted. The scale upon which vulnerability is measured (i.e., the DCU) cannot be defined precisely; loss of combat utility can be evaluated only in the context of a particular mission, not the subjective manner of the SDAL. This characteristic has been the subject of criticism because the numerical values produced by this method cannot be interpreted in terms that have intuitive meaning. For example, if a value of 0.50 loss-of-mobility function is given, one might ask what this means in terms of measurable automotive characteristics (e.g., does it indicate a loss of speed, a loss of acceleration, a loss of traction, or some other physical characteristic?). In fact, the value cannot be interpreted in terms of any measurable characteristics. Rather, it is an index based partly on the susceptibility of the target to physical damage and partly on a subjective interpretation of the importance of the damage.

Nonetheless, the inherent subjectivity in the process may be justified, at least to some degree, by the fact that actual combat involves subjective decisions similar to those upon which the SDAL is based. When an armored vehicle is damaged in combat, the field commander on-site must decide whether the vehicle is useful for further action, and, if so, he must judge what combat functions the vehicle and crew can and cannot perform. He must also consider whether the damage can be repaired and whether the attendant time delay is acceptable in the particular tactical situation. In other words, the commander must judge the combat utility of the damaged vehicle and do so entirely on his own intuition. Thus, it may be argued that vulnerability is dependent, not only on susceptibility to physical damage, but also on the perceptions and judgment of the people who use the vehicles. The intent of the SDAL is to capture these immeasurable and intangible aspects of combat utility; of course, the SDAL tries to represent the average case rather than any specific set of circumstances that a field commander might face.

Additionally, although DCU has fundamental deficiencies as a measure of vulnerability, it was probably the best choice available at the time. The other possibility that was considered (and that has been used for certain other targets) was the previously discussed P_K, the probability that combat utility is reduced to the extent that the vehicle can perform no useful function. As mentioned, the concept of P_K is based on the assumption that a given degree of damage can have only one of two possible consequences: complete loss of utility or no loss of utility. From the perspective of some 50 years of testing, this "all-or-nothing" assumption does not seem appropriate for armored vehicles.

The development of combat simulation models beginning in the early 1960s raised new issues regarding the selection of a measure of vulnerability for armored vehicles. These models mathematically simulate combat between combined arms units and predict the equipment losses by each side. Such models can be used to compare the combat effectiveness of particular weapon systems in the context of a combat unit in action rather than in isolation. Thus, all of the performance characteristics of the subject weapon system are taken into account, as well as those of the other weapons on both sides of the conflict. Models of this sort have become extremely important in the weapons acquisition process by providing a basis for selection among competing systems.

Among the various inputs required by combat simulation models are estimates of the vulnerability of all of the weapon systems involved. Current models are designed so that vulnerability must be quantified in terms of P_K even though it has long been argued that this measure of vulnerability is not appropriate for armored vehicles. From the point of view of simulation model design, it is desirable to characterize vulnerability in terms of kill probability because, in simulating a shot against a particular target, one must deal with only two possible outcomes: a kill or no damage. If the target is killed, it is simply removed from the battle; if it is not killed, it continues with full capability. Again, this relatively simplistic representation seems counter to real-life experience in armored vehicle combat, but it was probably a necessary simplification given the limited computing power available when simulation models were first developed.

Of course, one obvious question could be raised: why not design simulation models to allow for partial degradation of the performance of a weapon system when it experiences a ballistic attack? Conceptually, this could be done, but the SDAL measure of expected DCU cannot be interpreted in terms of measurable performance characteristics; hence, it does not provide a basis for degrading vehicle performance in a simulated combat action.

It has become common practice to substitute DCU values for probability values in combat simulation models. This practice has continued, apparently, in the belief that, even though it misapplies probability concepts, it would lead to acceptable accuracy in simulation model output. An attempt to rationalize this substitution has been offered by Zeller and Armendt (1987).

6. Contrast Between the Measures of Vulnerability for Armored Vehicles and Aircraft

As mentioned previously, the first formal efforts to study the ballistic vulnerability of military equipment involved tanks and aircraft. Although the subject of this book is ground combat equipment, it is also useful to discuss briefly the measures of vulnerability that were selected for aircraft. This discussion shows that the most appropriate measure of vulnerability for a particular type of equipment depends, in part, on its intended tactical role.

In comparison to armored vehicles, aircraft are "soft" and relatively susceptible to damage. Whereas armored vehicle and aircraft components are, in the main, of similar robustness, it is typically more difficult for threats to get to the components in most armored vehicles. In addition, given the nature of the operating environment of aircraft, maintaining all aspects of flight control is critical to the completion of a mission and the survival of the aircraft and crew. Whether a damaged aircraft aborts its mission or continues to execute is dependent not only on the severity of the damage but also on the need to maintain controlled flight capability to survive; this is a major difference between armored vehicles and aircraft in terms of the effects of combat damage. Given similar mechanical damage, a ground vehicle would be able to continue operating for a time (although in degraded fashion), perhaps long enough to perform some useful combat function.

In addition, aircraft missions usually involve highly specific plans to take off from point A, deliver ordnance or material at point B, and return to point A. If an aircraft is damaged, there is little opportunity to reassign it to a different mission that it could perform in its degraded condition.

Several damage categories have been adopted for aircraft. For each category, the damage required to make the aircraft incapable of performing its mission is defined. These categories are attrition, forced landing, mission abort, and mission available.

The attrition category concerns damage that is so extensive it is neither reasonable nor economical to repair. This category is divided into five damage levels; the first four levels are sequentially inclusive and time dependent. The last level provides a classification for damage that would result in a crash upon landing. The attrition kill levels are as follows:

KK-Kill   Damage that will cause an aircraft to disintegrate immediately upon being hit.
K-Kill   Damage that will cause loss of manned control within 30 s of being hit.
A-Kill   Damage that will cause loss of manned control within 5 min of being hit.
B-Kill   Damage that will cause loss of manned control within 30 min of being hit.
E-Kill   Damage that will cause an aircraft to sustain additional damage upon landing and render it uneconomical to repair.

The forced landing category concerns damage that forces the crew to execute a controlled landing, either powered or unpowered. This category is limited to aircraft that can be repaired either on-site or by removal to another site. The mission abort category concerns damage that prevents an aircraft from completing the designated mission but allows it to return to base. The mission available category concerns aircraft that have landed with combat damage and will require repairs before returning to mission-ready status. The attrition, forced landing, and mission abort kill categories are considered to be mutually exclusive.

The kill categories described previously are the basis of most aircraft vulnerability analyses and are characterized in terms of the probability that each level would occur given a particular combat scenario and weapon threat. In effect, the kill categories define several levels of degradation of the combat utility of an aircraft. By characterizing vulnerability in terms of the probability of occurrence of each level, the entire range of possible degradation is approximated by a few discrete levels. This approach is quite different from the approach adopted for armored vehicles in which the distribution of possible degradation levels is represented only by the expected value.

The reasons for selecting different measures of vulnerability for aircraft and armored vehicles lie in the differences in the operating environments and the tactical roles of the two types of equipment. It is difficult to define tactically significant levels of degradation for armored vehicles. In evaluating the vulnerability of armored vehicles, the context of direct-fire engagements is of most interest. Direct-fire engagements tend to last only a few minutes, and time-to-failure is less critical than in the case of aircraft. Thus, the idea of a set of time-dependent criteria, as used in the attrition category, is not as useful for armored vehicles. Also, in usual tactics, the mission of an armored vehicle is not as narrowly defined as in the case of aircraft. If an armored vehicle experiences damage that is not disabling, it can be assigned to a different mission and continue to operate in degraded fashion. Thus, it is not possible to clearly define the mechanical damage that would make the vehicle incapable of performing its mission.

7. Logistics Vehicles

Transport or logistics vehicles are generally unarmored military vehicles used to transport supplies, military equipment, or personnel and to tow guns, trailers, and other equipment on or off paved highways (U.S. Department of the Army 1977). This type of vehicle received little or no attention before the Vietnam conflict.

As an example, the Soviet truck ZIL-130G represents a family of vehicles that use many common components such as the engine, chassis, cab, and drive trains. Therefore, analytical values derived for the ZIL-130G are applicable to many components of numerous other trucks in the family (Kruse and Brizzolara 1971). Armendt et al. (1972) summarize damage criteria for transport vehicles.

The primary function of transport vehicles is moving materiel and troops. If a truck is incapable of controlled movement, it obviously cannot fulfill its mission. Thus, the defeat criteria for a truck are keyed to loss of mobility or the time required to repair the truck after it is damaged.

A truck is said to have sustained a K-Kill when damage is immediate and massive, and the truck is fit only for salvage. Such damage is caused usually by ignition of the onboard fuel, which develops into an uncontrollable fire, or severe damage caused through violent reaction of the cargo.

To achieve an M-Kill, sufficient damage must be done to a component so that the truck becomes incapable of controlled movement within one of the time categories shown in Table 3.2.

After an A-Kill is achieved, the B-Kill and C-Kill levels are satisfied. Similarly, after a B-Kill is achieved, the C-Kill level is satisfied. However, killing two or more B-Kill components will not (necessarily) result in the truck's sustaining an A-Kill.

Table 3.2 Logistics vehicle kill criteria

Kill criteria	Time to failure
A	0–5 min
B	0–20 min
C	0–40 min

VULNERABILITY ASSESSMENT AND MEASURES OF EFFECTIVENESS

The I-Kill (interdiction kill) criterion measures the success of an attack in terms of both LoF and the time required to make repair. The I-Kill may be used to measure the success of an attack directed against trucks either in or behind the battle area where no immediate ground follow-up is planned. For a component to be capable of sustaining I-Kill damage, the component must be of such a nature that, if damaged, its loss would either cause the truck to stop or materially detract from its operational efficiency.

Dangerous payloads will dominate the vulnerability of logistic vehicles. Those carrying explosives, missile supplies, or fuels are quite susceptible to catastrophic events triggered by blast, fragmenting munitions, and small arms. For logistic vehicles not transporting such dangerous cargo, the predominant threat is small arms and mines. The most susceptible element of such vehicles is the crew. In environments where mines are a large threat, some increased armoring has been employed to protect personnel. Damage to automotive elements often causes time-dependent degradation of mobility function. Examples of damage causing such degradation include holes in tires, holes in engine-coolant systems, and holes in the fuel system. Fires, although not occurring in a majority of cases, are also of concern.

The idea of defining levels of degradation in terms of time to failure, as was done for aircraft, is also appropriate for certain ground equipment, particularly cargo trucks and other logistics vehicles. The combat function of logistics vehicles is delivery of supplies to the front line combat units. These vehicles generally operate beyond the range of direct-fire weapons, but they are subject to attack by aircraft, artillery, and missiles. The purpose of attacking logistics vehicles is to interdict or delay the delivery of supplies to the combat elements. Logistics vehicles, like most combat equipment, sometimes experience damage that is not immediately incapacitating but that will, in time, immobilize the vehicle. As in the case of aircraft, the time between the occurrence of the damage and failure of the system is critical because it determines whether a specified mission can be completed. For analytical purposes, the three kill criteria shown in Table 3.2 were defined based on time to failure, and vulnerability is characterized as the P_K according to each criteria. The numbers in the table are the result of tests on trucks in the 1960s.

8. Self-Propelled Artillery

Although artillery pieces are not intended for face-to-face combat, the physical functions they are expected to perform in combat are similar in many respects to those of tanks. They are required to maneuver and use the terrain for protection, conduct fire missions, and move again quickly to a new position. The vulnerability of self-propelled artillery pieces is commonly evaluated in the following two tactical contexts: 1) while engaged in a fire mission or maneuvering short distances to reach a firing position, and 2) while traveling long distances (generally by roadway) to reach an area where tactical maneuvering and firing can begin.

In the first context, vulnerability is evaluated in terms of fractional DCU in the same manner described for tanks. In the second context, where the intent of an attack is to interdict movement, time-dependent kill criteria are used as described previously for logistics vehicles.

9. Mobile Missile-Launching Systems

Missile systems constitute a substantial portion of combat materiel not already described in the heavily armored or lightly armored vehicle sections. Although such systems are often characterized as unarmored, the extent to which this is true depends on the system's mission. Army missile systems are predominantly ground systems, with the notable exception of rockets deployed by helicopter for air-to-surface and air-to-air confrontations. The ground-based systems are of two types: surface-to-surface and surface-to-air.

Surface-to-surface refers to missiles fired from the ground at targets on the ground. A surface-to-surface system comprises all the equipment involved in carrying out a surface-to-surface attack, including the launching platform, the missiles, and possibly other support materiel. Systems of this type are used in a limited role for perimeter defense and in combat fire support. Because of the battlefield environment, these systems are armored.

Surface-to-air refers to missiles fired from the ground at airborne targets. Surface-to-air systems are used for an air-defense role with a primary mission to deny air space to the enemy. Systems in this category vary substantially depending on the scope of their mission. Forward-area air-defensive systems are part of the front-line combat activity and as such are armored, carry small rockets, and are more self-contained in their overall function. They are likely to be subjected to small arms, fragmenting munitions, large shaped charges, KE penetrators, and bomblets. Intermediate air-defense systems, such as the Phased-Array Tracking, Ranging, and Intercept of Target (PATRIOT) system, protect a wider area, are situated farther away from the front line, have larger missiles, and are softer systems. Indirect fire is one of the likely threats. Theater air-defense systems, such as the Theater High-Altitude Air Defense (THAAD) system, cover an even larger area and consist of numerous unarmored vehicles with specialized functions such as radar detection, communication, power generation, and large-missile launching. Ballistic missiles are the major threat to such a system, but intercepting such missiles is what it does best.

Generally, these systems operate behind the front lines and provide supporting fire against ground or aerial targets as requested by the forward combat units. If a system receives damage to its mobility components, the question of most interest tactically is whether it can continue to operate long enough to reach its designated firing position. Consequently, mobility damage is measured against time-dependent kill criteria. The criteria used for these vehicles are the same as those presented previously for logistics vehicles. Damage to the firepower components tends to be binary (i.e., the damage is immediately incapacitating, or it is insignificant). Thus, time-dependent criteria are not appropriate, and vulnerability is measured simply as probability of F-Kill.

Damage to missiles can arise from many mechanisms induced by fragmenting munitions, blast, or even direct fire. Structural alteration of a missile by blast that buckles or bends the exterior may preclude a launch. Rupture of the external surface of the missile by blast or fragments will also prevent a firing. The effect of fragments and holes in the missile depends on its internal structure. Sufficient loss of propellant to prevent delivery of the missile to its target renders the system ineffective. If the fuels involved are hypergolic (i.e., they can detonate on contact with another material without the presence of a spark), holes may bring about a mixing of chemicals that cause flames or explosions. Leaks of caustic propellant chemicals may cause damage to delicate components.

Various propellant events can also destroy a missile subjected to fragments/direct fire. The propellant may burn, violently burn (deflagration), or explode, and the missile warhead is also subject to potential deflagration or detonation. Missile systems can also be incapacitated without directly damaging the missile. Damage to the launch platform and positioning equipment will delay, if not prevent, launch, as will rupture or bending and perforation, causing internal petaling of launch tubes. In addition, damage to hydraulic or mechanical lifting and positioning equipment will destroy aiming capabilities. This can also occur if the electrical system associated with the positioning motor is broken.

Larger multi-vehicle systems will be degraded by damage to support elements such as radars, communications, and power-generation vehicles. Trucks most often carry these support structures and vulnerabilities of their underlying automotive systems have been addressed previously. Communications equipment is most often housed in panel structures having varying degrees of blast, fragmenting munition, and small arms protection, the level of which is dictated by the threat environment. Shock and penetration are two of the main threats to the electronic equipment contained in such a structure. Radar antennas are also subject to blast and shock damage. Generators are fairly sturdy and commonly unaffected in blast encounters, but they are most susceptible to fragment penetration, leading to either mechanical seizing or time-dependent leak degradation.

Because the degree of protection of missile systems varies depending on their function, a determination of the baseline vulnerability requires an understanding of the specific threat environment. Missile systems are generally most vulnerable when they are in a launch configuration. However, the most common configuration for missile systems is in the stowed position; unless attacks are extremely well timed, the likely encounter will be against a less-vulnerable stowed configuration.

10. Vulnerable Area

A common measure of vulnerability for aircraft and for unarmored ground vehicles is vulnerable area A_V which is defined by

$$A_V = \iint_{A_p} p(x, y) \, dx \, dy \qquad (3.2)$$

where

$p(x, y)$ = the probability that the target suffers a particular type of defeat by a penetrator whose trajectory passes through a point (x, y) in a plane (impact plane) normal to the trajectory of the penetrator

A_p = the area of the target projected into this plane

The probability P that the target suffers a defeat of the type under consideration when subjected to an attack by penetrators with an areal density of r in the impact plane is as follows:

$$P = 1 - e^{-rA_V} \qquad (3.3)$$

(The areal density of penetrators refers to the number of penetrators per unit area over the impact plane – for artillery and other fragmenting munitions, areal density is sometimes expressed in terms of the number of fragments per unit solid angle of the cone containing the penetrators projected onto the impact plane.)

Vulnerable area is particularly useful for area-weapon threats such as fragmenting artillery, which is of considerable concern for unarmored targets.

C. Fuel Fire Damage Assessment

There are two distinct areas of interest in the prediction of vehicle fuel fires: 1) the condition of the vehicle in terms of what is damaged and how badly, and 2) the injury to personnel.

During the Vietnam conflict, the U.S. Army sent battle damage assessment and repair (BDAR) teams into the field to document battle damage to combat vehicles (Dehn 1975). Unfortunately, the teams found no way to correlate degree of damage to components of vehicles with the duration and severity of fires.

On the other hand, WRAIR has provided criteria to assess burn injuries to personnel in vehicles (Ripple and Mundie 1989). One approach is to use free air temperature at each crew member location. Second-degree burns on exposed skin are predicted using the equation

$$T_i = \alpha[T_M - 37] \, dt \tag{3.4}$$

where T_M is the temperature of a thermocouple in degrees centigrade and t is time in seconds. A value of T_i, from an integration over 10 s, which exceeds 1,315°C-s, is predicted to cause second-degree burns on bare skin. A value of 3,300°C-s is required for second-degree burns on skin protected by clothing.

A second approach uses heat-flux calorimetry. The prediction is that thermal exposure to bare skin in excess of 3.9 cal/cm^2 over 10 s will cause second-degree burns. A heat-flux value of 8.8 cal/cm^2 over 10 s is predicted to cause second-degree burns to protected skin.

The original predictive tool for second-degree burns to bare skin was a fire duration of 250 ms. If an explosive fire in the crew compartment can be extinguished within 250 ms, however, second-degree burns are not expected. There is a reasonably good correlation between the 250 ms fire-out times and temperature and flux data.

D. Measures of the Sensitivity of Energetic Materials

Vulnerability analysts need to predict the conditions that lead to ignition of energetic materials and the violence of the resulting reactions. For this purpose, it is useful to have some knowledge of the measures of sensitivity used in the energetic materials community. Some commonly used measures follow. More detailed descriptions of these tests can be found in the Lawrence Livermore National Laboratory Explosives Handbook (Dobratz and Crawford 1985) and the three-volume Navy Band of Explosives Data (Drimmer 1983). Compilations of data are also available in those references.

VULNERABILITY ASSESSMENT AND MEASURES OF EFFECTIVENESS   81

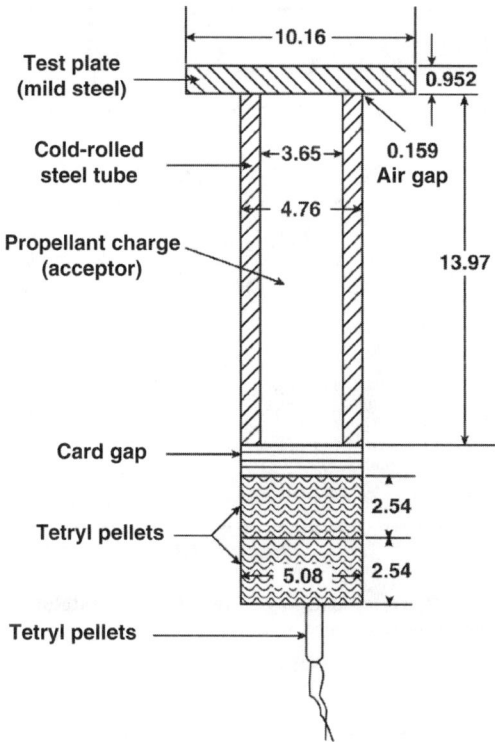

Fig. 3.1   The NOL large-scale gap test (units in millimeters).

Gap test Figure 3.1 illustrates a gap test. A standardized booster explosive is separated from the test charge by a gap filled with an inert material. The thickness of the gap, which just permits detonation of the test acceptor charge, is a measure of the sensitivity of an explosive to shock. The Naval Ordnance Laboratory (NOL) large scale gap test is commonly used and has been required for many years as part of standard hazard-classification procedures (U.S. Department of the Army, Navy, Air Force, and Defense Logistics Agency 1998).

Pop plot A pop plot is a plot of the run distance required for a shock to build to detonation as a function of the initial shock pressure. A pop plot provides more detailed sensitivity information than a gap test. At the same input pressure, shorter run distances imply greater sensitivity. The pop plot is determined from a series of 1-D planar impact experiments and is usually presented in the form of the log of the run distance vs the log of the input pressure. Pop plots for some common explosives are shown in Fig. 3.2 (Dobratz and Crawford 1985). Usually, the pop plots of different explosives are approximately parallel, but occasionally crossovers occur, so that one explosive may be more sensitive at low shock pressure and another may be more sensitive at high shock pressure.

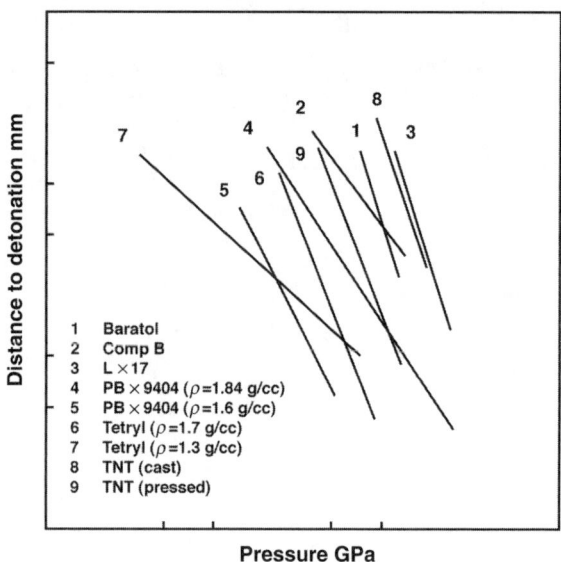

Fig. 3.2 A pop plot, where logarithm of the distance to detonation is plotted as a function of the shock pressure.

Critical energy The minimum shock pressure required to produce detonation depends on the duration of the pressure. Often, but not always, this relation can be represented by the equation

$$P^2 t/(\rho U) = \text{constant} \tag{3.5}$$

where P is the shock pressure, t is the duration of the shock, ρ is the explosive density, and U is the shock velocity in the explosive. The constant (often called the critical energy) has units of energy per unit area and is determined by performing flyer plate impact experiments and/or adjusting the plate velocity or thickness to determine the critical initiation condition. Like the gap test and the pop plot, it is a measure of sensitivity to shock.

Differential-scanning calorimetry or differential thermal analysis These techniques are used to assess the temperature at which an explosive first begins to react (and can also be used to determine rate constants for explosive reactions at relatively low temperatures and heating rates). This temperature, called the first exotherm, is somewhat dependent on the rate of heating or the time for which the sample is held at temperature. Despite this complication, the first exotherm is a good measure of the ability of a material to withstand heat.

Drop-weight impact test In this test, a weight is dropped on a small (typically 20 mg) sample of the energetic material, and the critical drop height, which is just adequate to initiate the sample, is determined. The test results are highly dependent on design details, but the test has been standardized, and the results can be used for roughly ranking the sensitivity of energetic materials.

Friability test   Also called the shotgun test, this test measures how easily an energetic material is damaged in a ballistic event. A sample of the energetic material is fired at a steel wall. The debris is collected and burned in a closed bomb. The rate of combustion in the bomb provides a measure of the amount of surface area generated by the impact. Brittle materials, which produce a large surface area on impact, have a greater tendency to undergo XDT type (delayed detonation) reactions. See United Nations (1990) for a detailed description of this test.

Burn rates   These rates are measured for all gun and rocket propellants, and they have been determined for many explosives. Burn rates increase with pressure. Many materials show what is called a slope break as pressure increases (i.e., at some pressure, the rate of burning dramatically increases). These breaks are often related to deconsolidation of the sample or flame penetration into the material. Materials that show slope breaks are more likely to exhibit violent reactions when impacted.

E. Casualty Assessment

While soldiers are subjected to a variety of damage mechanisms on the battlefield, in general penetrating injuries are the major cause of weapon-induced casualties. Because of this fact, most human vulnerability modeling has been directed toward the development of injury and incapacitation criteria for fragments, bullets, and flechettes. The direct effects of blast overpressure, whole body accelerations, thermal pulses, toxic gases, and chemical exposures have generally been assumed to be secondary effects and have, therefore, received less attention, at least in the U.S. Army. As a result, there are many more tools and data sources available to the analyst for assessing penetrating injury effects.

There are two key ways in which human vulnerability data are employed in the Army analysis: 1) through high-resolution modeling of discrete events (single shotline analyses in the case of penetrating injuries) at the individual soldier level; and 2) through correlation curves that are based on generalizations of discrete data.

The choice of which form to use is normally dictated by the type of analysis being conducted (Neades and Prather 1991).

1. Casualty Assessment Process

A series of casualty workshops was sponsored by the Office of the Secretary of Defense (OSD) to review the data and capabilities for casualty assessment (Office of the Director for Live Fire, Undersecretary of Defense for Research and Engineering 1988). These workshops were not limited to conventional ballistic injuries and addressed the Services' abilities to assess casualties in all areas. As a result of these comprehensive reviews, the need for a standard, comprehensive casualty methodology was identified. This need led eventually to the formation of the Crew Casualty Working Group (CCWG), which, in 1992, became part of the JTCG/ME. Shortly thereafter, the essential common interest of the JTCG/AS in the same crew casualty issues was recognized. The CCWG at that point became a joint JTCG/ME and JTCG/AS working group [now part of the Joint Aircraft Survivability Program Office (JASPO)].

Between 1992 and 1997, the CCWG made excellent progress toward developing and producing new a personnel vulnerability assessment methodology, the Operational Requirement-based Casualty Assessment (ORCA) model (which is discussed in Sec. 4.VII). The foundation of this approach was a more explicit taxonomy for the casualty assessment process (Starks 1991). The taxonomy proved to be most useful for assuring completeness and coordination of the technical tasks. As shown in Fig. 3.3, the taxonomy divides logically into three parts: 1) the determination of the injury resulting from an insult, 2) the resulting impairment of certain human elemental capabilities, and 3) the effect of such impairment on the performance of military jobs. The ultimate objective of the process was to determine if an individual would be an operational casualty as a result of exposure to the insult.

The basic steps involved in this crew casualty assessment process were as follows:

1) Assemble the parameters that quantify an insult.

2) Evaluate the anatomical/physiological injury.

3) Relate that injury to the attendant impairment, expressed as a degradation of elemental capabilities.

4) Independently establish the requirements for satisfactory performance of a military job, also in terms of elemental capabilities.

5) Compare the available (degraded) elemental capabilities to the required elemental capabilities to determine if the specified injury constitutes an operational casualty for an individual in the specified military job.

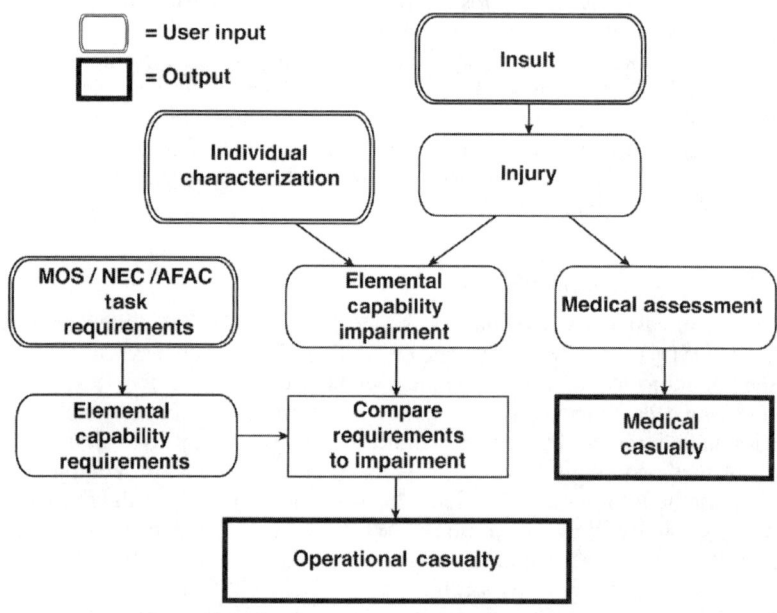

Fig. 3.3 Casualty assessment taxonomy.

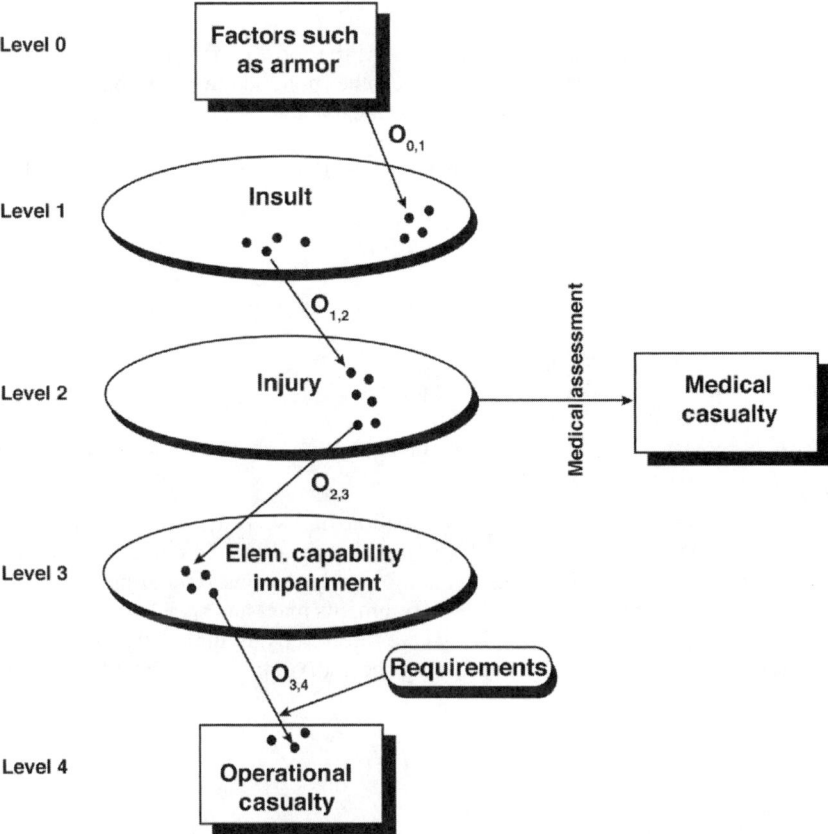

Fig. 3.4 The crew casualty analysis process mapped into the V/L Taxonomy.

It is interesting to note that, although the development of the crew casualty analysis process (as illustrated in Fig. 3.4) was not necessarily based on the V/L Taxonomy (now part of the MMF), the two are practically isomorphic. (See Fig. 3.1 for a comparison.) This fact is yet another demonstration of the broad applicability of the MMF concept.

As is now clear, an operational casualty is defined with respect to a particular task or job. Thus, in addition to defining the insult, the task or job of an individual must be input using MOS, Navy Enlisted Code (NEC), or Air Force Specialty Code (AFSC) job description information. These military jobs have been defined by a list of basic physical, cognitive, and sensory tasks. The task library of the ORCA computer code is described using the elemental capability vector (ECV). For instance, a pilot's job contains several tasks, such as operating controls, reaching above, visual mental processing, communicating, and hearing. Each task is described by the ECV and the job consists of the summation of all the tasks needed by the pilot.

2. Casualty Correlations

Generalized casualty correlation curves are the most common form of human vulnerability data in use. A good example can be found in the curves published by Allen and Sperrazza (1956). In this report, incapacitation potential from random impacts by chunky steel fragments and flechettes were estimated using an MV^α correlation, which relates mass and striking speed of the projectile to the P_{IIH}. The functional form of the P_{IIH} relationship is:

$$P_{IIH} = 1 - e^{-a[MV^{3/2} - b]^n} \qquad (3.6)$$

where

M = projectile mass in grains
V = projectile velocity in feet per second
e = base of natural logarithms
a, b, n = constants that depend on tactical role, time after wounding, and impacted body part

The P_{IIH}'s analyzed to produce this relationship were those observed during evaluation of the projectiles in the four-fragment database. To compute P_{IIH} for a single, random hit on one of the six major body regions (including the entire body) simply requires an estimate of the fragments mass and speed. As mentioned in Chapter 2, the effects of multiple hits are inferred by assuming statistical independence and mathematically combining the individual P_{IIH}'s according to the previously mentioned survivor rule:

$$P_{Total} = 1 - \prod_{i=1}^{n}(1 - P_i)$$

In 1960, Dziemian showed that the conditional probability that a random hit by a sphere, disc, cube, bullet, or flechette would incapacitate an infantry soldier could be related to the amount of energy $\Delta E_{(1-15)}$ lost by the penetrator during its passage between 1 and 15 cm of penetration into a 20 percent gelatin block tissue model at 10°C (Dziemian 1960). The mathematical function used for this relationship was as follows:

$$P_{IIH} = 1/\{1 + e^{a + b\log_{10}(\Delta E_{(1-15)})}\} \qquad (3.7)$$

Dziemian also developed empirical rules for calculating $\Delta E_{(1-15)}$ for spheres, cubes, and stable flechettes, but could find no simple relationship to estimate $\Delta E_{(1-15)}$ for unstable flechettes in gelatin or for bullets. To assess these projectiles, high-speed motion pictures taken as the projectile penetrated a 38-cm-long block of gelatin were analyzed frame by frame to obtain remaining velocities of the projectile at any distance of penetration. Remaining kinetic energies were then calculated from the projectile's known mass. The Dziemian $\Delta E_{(1-15)}$ gelatin criteria were used by the Army and other services to estimate bullet and flechette effectiveness through 1968.

Later, in an attempt to reduce test costs and eliminate some of the technical difficulties associated with inferring precise projectile position in a gelatin target

VULNERABILITY ASSESSMENT AND MEASURES OF EFFECTIVENESS

using a light-photographic method, a ballistic-pendulum method was adopted (Bruchey and Sturdivan 1968). This method required a new relationship, which was obtained by correlating the 15-cm energies and corresponding P_{IIH} for a modified projectile set that included both fragments and flechettes. The function chosen to relate P_{IIH} and energy deposited in a 15-cm cube of gelatin as measured by the BRL pendulum was as follows:

$$P_{IIH} = 1 - e^{-a(\Delta KE)^n} \tag{3.8}$$

The ballistic pendulum and the BRL ΔKE casualty criteria were the methodologies used by the small-arms community to estimate incapacitation for bullets between 1969 and 1975. In 1975, an alternate methodology for estimating bullet or flechette effectiveness was proposed (Sturdivan 1981). In 1977, an expert panel endorsed this new model, referred to as EKE (for Expected Kinetic Energy), as the U.S.-recommended method for the North Atlantic Treaty Organization (NATO) small arms trials and also established it as the official Army model. The new EKE model correlated P_{IIH} with the experimentally determined, incremental EKE deposited in a 20-percent (at 10°C) gelatin target. Experimental projectile paths in gelatin were obtained out to 38 cm and extrapolated, if necessary, to 45 cm (the theoretical maximum horizontal trajectory through the human anatomy in a standing position). The weighted expected EKE deposit is then calculated from the following:

$$\text{EKE} = (m/2) \sum_{i=1}^{45} P_i \left[(v_{i-1})^2 - v_i^2 \right] \tag{3.9}$$

where

EKE = the expected energy deposit (in joules)
P_i = the probability of the projectile being in body tissue at depth i, given a random impact on the body
v_i = the projectile velocity at depth i
m = the mass of the projectile

EKE can be determined experimentally or analytically for stable projectiles (Sturdivan 1981). Probability of incapacitation can then be estimated from the function

$$P_{IIH} = \lambda / \{ 1 - e^{\alpha - \beta \, \ell n (EKE - \gamma)} \} \tag{3.10}$$

where α, β, γ, and λ are constants based upon stress situation and time.

The criteria presently available for assessing the effectiveness of helmets and body armor were derived from the incapacitation criteria described in the previous section. The latter were found to be incapable of evaluating small, but significant changes in protectionslevels from these items and so new criteria were established in 1969 (Waldon et al. 1969).

These new criteria predict the probability of sustaining a serious or lethal fragment wound given a hit and were established by considering only the subset of the wound ballistics database that involved hits that produced the following types of injuries:

Serious Wounds that would "probably result in hospitalization (e.g., wounds to the cranial, pleural, peritoneal cavities, larger muscles; vascular, nerve, and skeletal damage)."

Lethal Wounds that would "probably result in death (e.g., injuries to the head, larger vessels, lungs, central nervous system, abdominal organs)."

While these criteria provide a convenient way to evaluate casualty reduction potential of protective clothing in a relative sense, they are not well-founded and fail to provide the means to assess in a meaningful way, the medical consequences of fragment wounds.

Another useful correlation is ballistic dose. The concept of ballistic dose was introduced by Saucier and Gilman (1996). In simple terms, ballistic dose is the measure of a projectile's ability to produce incapacitating or lethal injuries (i.e., its wounding power). In specific terms, ballistic dose is a measure whose quantity is a function of projectile mass, speed, and the number of projectile impacts. The appeal of this concept is that ballistic dose can be correlated to P_{IIH}, much like MV^α was used to produce the popular criteria described previously. What makes ballistic dose different from MV^α is the addition of n, the total number of impacts involved.

From runs of ComputerMan (which is discussed in Sec. 4.III.E.2), a model that simulates munition fragment wounds and biomechanical degradation, it was determined that ballistic dose takes the form

$$\phi(m, v, n) = e^a m^b v^c n^d \tag{3.11}$$

where a, b, c, and d are constants to be determined. The ballistic dose function is assumed to be related to incapacitation and survivability in the following way:

$$P_I(m, v, n) = 1 - e^{-\phi_I(m,v,n)}$$

and

$$P_S(m, v, n) = e^{-\phi_S(m,v,n)} \tag{3.12}$$

where P_I and P_S refer to the probability of incapacitation and probability of survival, respectively.

Correlation curves of P_{IIH} or P_S as a function of ballistic-dose parameters provide a convenient alternative way of estimating anti-personnel effects from fragment impacts.

Note that a number of attempts have been made to improve on the provisional criteria for the serious and lethal wounds described previously. These attempts have focused on the use of contemporary civilian-injury data that exist in the form of substantial databases, notably, the data associated with the Major Trauma Outcome Study (MTOS), sponsored by the American College of Surgeons Committee on Trauma.

In addition, since the late 1980s, the ComputerMan model, modified to include Abbreviated Injury Scale (AIS) 85 civilian-injury description information, has been used in conjunction with these civilian trauma data in an attempt to develop an improved model applicable to military injuries (American Association for Automotive Medicine 1985). These attempts, however, have met with limited success (Tri-Analytics and the SURVICE Engineering Company 1995). Some of the features of the data in the MTOS database and the conditions under which the data were collected are responsible for this difficulty (Champion et al. 1990). In particular, the noninclusion of data on patients who died at the accident scene and the lack of severe penetrating injuries, comparable to those produced in combat by high-velocity projectiles, make the task a particularly challenging one. Nevertheless, the thoroughness with which the civilian trauma data are collected and the systematic manner in which injury data are collected at major trauma centers make it a valuable information resource that should be further investigated. Thus, a concerted effort to use these data to bring the assessment of military medical casualty classification in line with the civilian world is planned for the future.

II. Fault Trees and Degraded States

A. Fault Trees

In going from level 2 to level 3 in the MMF, it is necessary to relate damage states to capability states, which may be degraded. One could, on a case-by-case basis, generate detailed descriptions of the remaining capability of a damaged target based upon fundamental engineering models. However, such models require extensive development, data generation and input, and output interpretation. Commonly, these requirements may be met only when the underlying models have been developed in support of other activities. For example, development of a tank-suspension model of sufficient detail to allow analysis of the effects of a damaged suspension component generally depends upon the prior development of a suspension model during tank design/development. Clearly, therefore, the application of engineering models to relate damage to capability degradation is limited.

This mapping from damage state to capability state is currently achieved by "reverse engineering" the target to gain a detailed understanding of how components of the target function and interact. The purpose is to identify components critical (i.e., necessary) to the continued operation of the target. As mentioned previously, this reverse engineering of the target is termed a criticality analysis. It is important to note that a criticality analysis does not attempt to define how a component is rendered nonfunctional. Rather, it defines the result to the target of rendering a component nonfunctional.

The results of the criticality analysis are presented as a series of deactivation diagrams (also called "fault trees," "failure diagrams," or "functional failure diagrams"). A deactivation diagram is simply a logic diagram that presents the operating relationship of critical components of the target. As long as an unbroken path can be traced through the diagram from beginning to end, no target capability has been lost. However, if an unbroken path cannot be traced through the diagram, some loss of capability has occurred. In addition to providing the basis for vulnerability analyses, the deactivation diagrams are used by damage assessors in the field to ensure consistency between the results of numerical analyses and analysis of observed damage (e.g., test results or battle damage).

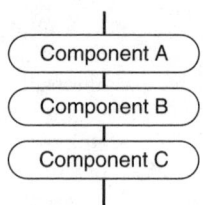

Fig. 3.5 A "series" deactivation diagram.

There are two general configurations of deactivation diagrams. First, there is the "series" diagram, shown in Fig. 3.5. Disabling component "A" or component "B" or component "C" can break the path through the diagram in the figure. Note too that there is no added severity of loss if more than one component is disabled. Stated differently, the function represented by Fig. 3.5 will be lost if "A" or "B" or "C" becomes nonfunctional.

Second, there is the "parallel" path diagram, as shown in Fig. 3.6. To break the path in this diagram, both component "A" and component "B" must be disabled before any loss of capability is sustained. Disabling only one component will not result in the loss of the function being analyzed, although in some instances, one path allows the function to be performed more effectively than the other. For example, a tank may have both an electrical and a manual method for traversing the turret; loss of the electrical method alone does not result in loss of the capability to traverse the turret, but the manual traverse method is less efficient.

Series and parallel constructs can be combined to form diagrams of great complexity, as is shown in Secs. 3.III and 3.IV.

In developing the fault trees for a target, the analyst must consider the capabilities required by the target to accomplish its mission. The required capabilities of a target also loosely define the kill criteria for the target. In the past, kill metrics have been defined for general categories of capabilities (e.g., M-Kills and F-Kills for ground vehicles). There may also be multiple kill definitions in each category. As discussed previously, K-Kills can be thought of as total destruction of the target or LoF of all components in the target. For example, one of the main purposes (and hence capabilities) of tanks and motorized howitzers is to shoot munitions. Thus, one of the principal kill criteria for these targets would be an F-Kill. However, different levels of kill (or alternately, levels of functionality or capability) can exist for firepower. For instance, if a tank loses the use of its laser rangefinder,

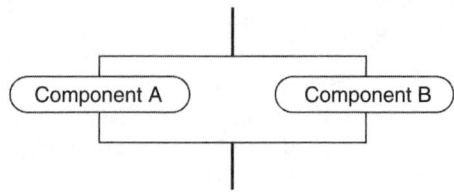

Fig. 3.6 A "parallel" deactivation diagram.

it may still be able to shoot accurately enough at short ranges for particular situations. Similarly, if a howitzer loses the use of an autoloader, it will not be able to shoot as rapidly but may still provide adequate fire rates for certain situations or periods of time. Thus, the analyst must consider the operational modes or rates for a target in defining the fault tree.

As discussed previously, other capabilities that have been translated into kill criteria or metrics include mobility, communications, and specialized "operations" that the target must perform. Considerations for M-Kills may include recognition of situations in which a vehicle must operate at top speed, limp home, move in reverse, make sharp turns, or swim. Depending on the target, the radio or other communications may or may not be essential to successful operation of the target. Specialized operations of a target might include range and accuracy functions for a radar van, communications, computing, or other functionality for a command post or intercept aircraft vectoring control in a command and control facility.

Common functions included in the fault trees for ground-fixed targets (such as buried hardened bunkers) are power, heating ventilation and air conditioning (HVAC), communications, and operations. The reasoning here is that without power, all the other equipment in the facility will not operate, and without cooling (HVAC), the rest of the equipment will eventually overheat and be inoperative. It is also assumed that without communications, the information being processed, or command and control functions being performed in the bunker, cannot be transmitted. Finally, the computers and other equipment that perform the actual function of the facility are commonly designated as operations, which obviously need to be working.

Time considerations generally involve the amount of time during which functionality of a particular target must be disabled to result in a kill. For example, a few minutes of lost capability could be extremely dangerous for a tank, as total time of the engagement may be only 10 min. Likewise, a 15-min window when an air defense system is out of operation could be sufficient to allow an air strike. However, a few minutes of delay would probably not be significant for a supply convoy, but hours and certainly days could have a large impact.

Other special considerations may involve cargo or contents of certain targets. In these cases, the analyst must decide if the objective is to deny the enemy access to the contents, disrupt the activities associated with the contents, or destroy the cargo or contents altogether. In the case of a munitions storage facility, collapsing entrance or exit structures may be sufficient to deny access. In addition, if the munitions are chemical or biological weapons, denial may actually be preferable to destruction of the contents because of the potential for release of the contents into the atmosphere. For an armored personnel carrier, on the other hand, it may be sufficient to prevent the cargo (in this case, soldiers) from arriving at the battlefield rather than causing injuries to each soldier.

As can be seen, the criticality analysis must be approached with the end in mind to some extent. The analyst must factor in the purpose for the vulnerability analysis, the mission that the target performs or fulfills, the time frame over which it must perform that mission, the various aspects of the mission, and any other considerations particular to the target or its configuration.

Because the logic used to develop deactivation diagrams is basically the same for any target, Sec. 3.III develops a deactivation diagram for complete loss of tank mobility within a short time of damage. Section 3.IV then illustrates the development of deactivation diagrams for a lightly armored vehicle.

B. Extension to Nonbinary Forms

Historically, the status of fault-tree elements is expressed in binary form, either available or not available. Because the status of each component must be determined from the damage states discussed in the preceding sections, this representation places a significant burden upon component-damage models. Such models must determine whether a component is functional or dysfunctional, based on the fault trees that involve all other components that support the function of the original component.

The Degraded-States Vulnerability Methodology (DSVM) is a straightforward extension of the traditional fault-tree formalism. Rather than traversing a single fault tree to determine the existence of a particular capability, a set of fault trees is used, with each member of the set related to a given level of the particular capability. For example, the single fault tree in Fig. 3.5 might be replaced by a set of degraded states fault trees. Each fault tree would be associated with the level of system integrity necessary to allow the system to perform at a given level. For example, there may be a tree for the ability of a vehicle to move at least 10 mph and one for it to be able to move at least 20 mph.

Subsequent to determination of a damage state, an implementation of DSVM proceeds as follows. The most demanding (full capability) fault tree is traversed. If successful, the corresponding capability state element is scored as fully capable and the analysis proceeds to another capability. If, however, the full capability fault tree cannot be traversed, analysis proceeds to the next most demanding fault tree for the same capability. Should this tree be traversed, the capability is given the degraded value corresponding to the traversed tree, and then the analysis proceeds to another capability; otherwise, the next most demanding tree is analyzed. This procedure continues until a tree is traversed; if all fail, the capability is scored as totally absent.

C. Degraded States

For a long time, it was generally recognized that there was a fundamental difference between the fractional functional loss as determined in full-scale tests such as the CARDE trials and the probability of total functional loss as calculated in computer models such as the Vulnerability Analysis Methodology Program (VAMP), the Vulnerability Analysis for Surface Targets (VAST), and early versions of the Stochastic Quantitative Analysis of System Hierarchies (SQuASH) model. This difference is demonstrated by comparing the following two statements:

1) The result of a particular encounter with the threat causes a 50 percent loss of speed.

2) There is a 50 percent probability that the result of the encounter is a total loss of mobility.

The need for a new approach to vulnerability metrics led to the development of the concept of degraded states, as described in the following passage (Roach 1990):

> For this new approach, a fuller and more specific set of metrics is developed. The major functions of the combat system are divided into kill categories;

VULNERABILITY ASSESSMENT AND MEASURES OF EFFECTIVENESS

for example, six kill categories were developed for the M1A1, specifically, mobility, firepower, acquisition, crew, communication, and ammunition. Each kill category is further divided by using a number of kill definitions that describe damaged states of the combat system. These damaged states encompass various functional levels (e.g., slight or significant) and include a no-damage state and a killed state. Within all the kill categories, except crew, each possible combination of definitions is also defined separately.

Therefore, for a given damage vector, one and only one kill definition from each kill category will be satisfied. This, in turn, generates a combat-system degraded state. This combat system degraded state, which reflects the damage states for each of the combat system's critical functional capabilities, presents a full picture of the system's specific capability following an encounter with a damage mechanism.

Once combat system kill definitions are described, they are expanded into mathematical fault trees to allow calculation of their probability of occurrence. These fault trees consist of combinations of the system's critical components that, if killed, would result in the satisfaction of that particular kill definition. A kill definition is satisfied if no uninterrupted path from top to bottom exists in the fault tree.

III. An Example of Fault-Tree Development for Traditional Metrics

This section provides an example of fault trees for traditional measures of combat utility of a tank. For the purposes of this example, the following assumptions define the initial condition of the tank and its crew before any damage:

1) The tank has a complete basic load of fuel and ammunition.
2) The tank has a complete crew.
3) The crew is fully trained, capable, and well-motivated.
4) If any crew member is wounded, duties will be redistributed and all individual crew members will continue to operate the tank to the best of their ability.
5) The possible psychological effects of the situation on the crew are ignored (i.e., crew members will not perform superhuman feats, nor will they become immobilized by fear).
6) The tank is in good operating condition.
7) All appropriate systems of the tank are operating (i.e., the crew is aware that combat is likely; therefore, they have the vehicle prepared to engage the enemy).
8) The crew will choose to continue to fight with a damaged vehicle if at all possible, rather than abandoning it on the battlefield.
9) The crew will choose to fight with the tank with all hatches closed as long as possible.
10) Any system degradation will be the result of combat damage, rather than fair wear and tear or manufacturing defect.
11) The tank to be analyzed in this example is powered by a V-12 water-cooled diesel engine.
12) Power is transferred to the tracks through a typical clutch and manual transmission arrangement.
13) Steering is accomplished by differential braking of the track drives.

Stated succinctly, the assumptions define a tank in good operating condition, ready and able to fight. It is crewed by skilled individuals who will react to combat just as other components of the tank react (i.e., the crew will function as long as they are able). The tank itself is a well-proven and battle-tested design using technology common in the 1960s.

Only the M-Kill, as used in the JMEM, is considered in this example, as the concept of mobility is intuitively obvious. Also, as it is intuitively obvious that a catastrophic reaction of stowed main gun ammunition would result in an M-Kill (as well as a K-Kill), and that incapacitation of the crew would result in an M-Kill, neither of these is discussed.

Damage that results in an M-Kill of the tank does not necessarily result in immediate immobilization of the tank. Loss of control of the tank or damage that results in immobilization after a "short period of time" is sufficient to result in an M-Kill. The logic behind selecting a time period criterion for damage to take effect is based on the duration of a tank-to-tank engagement in the set of mission contexts being considered. Therefore, if the tank can continue to function for this "short period of time," it will probably survive to be withdrawn or repaired before being sent into combat again.

A "zerothlevel" analysis of the mobility of a tank reveals that, to have controlled movement, a tank must have a source of power (a functioning engine), a method of transmitting engine power to the tracks, some method of controlling the power and movement of the tank, and support for the hull of the tank. The deactivation diagram describing this initial level of analysis is shown in Fig. 3.7. Loss of any one of the four functions described will result in loss of controlled mobility, provided the crew cannot repair the damage on the battlefield. Each of these functions is now examined in detail.

A. Engine Power

The next level of the analysis typically develops a listing of systems that must operate for the engine of the tank to develop power. The engine must have a source of fuel, it must have cooling, it must be lubricated, the basic structure of the engine

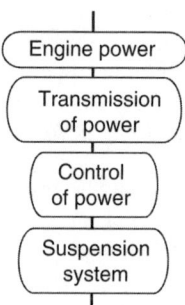

Fig. 3.7   Initial analysis of mobility.

VULNERABILITY ASSESSMENT AND MEASURES OF EFFECTIVENESS

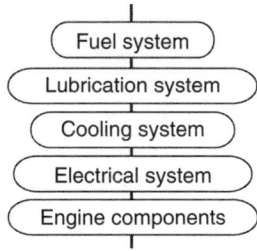

Fig. 3.8   Systems required for engine power.

must be sound, and the moving parts within the engine must retain their structural integrity. Figure 3.8 shows that each of the five systems listed must be functional in order for the tank to retain engine power.

1.   Fuel System

LoF in the fuel system is accomplished by preventing sufficient fuel pressure or volume for the task at hand from reaching the engine. This can be accomplished by the perforation or severing of components containing or transporting fuel or by the deformation of fuel-carrying components to reduce fuel throughput. Ignition of a fuel fire beyond the capability of the onboard fire extinguishers in either the crew compartment or the engine compartment will result in an M-Kill and will likely grow to catastrophic proportions.

Examination of the fuel system, shown schematically in Fig. 3.9, reveals that fuel is supplied to the engine from any one of three fuel cells. External fuel cells are designed to drain into the three fuel cells within the hull of the tank as the engine consumes the fuel. None feeds the engine directly; therefore, these cells will not appear on the deactivation diagram as critical components for the fuel system. Each of the three internal fuel cells is connected to a three-way fuel valve, which is operated by the driver. Fuel flowing from the three-way valve feeds the fuel priming pump. The priming pump is manually operated by the driver before starting the vehicle to ensure that the fuel lines are full of fuel and contain no trapped air. After this priming is accomplished, the fuel priming pump is not operated further while the engine is operating.

A bleed line from the side of the pump provides fuel to the engine preheater. Because the engine preheater operates only when warming the engine before starting in extremely cold conditions, the fuel control valve for the engine preheater is normally closed; therefore, the preheater is not considered a critical component. However, the fuel control valve for the preheater is considered critical. Perforating the preheater will not result in a leak in the fuel line, but perforating or destroying the control valve will. (The preheater is not operated after the tank is started; therefore, the fuel control valve is normally closed.) The concern with a fuel leak in this portion of the fuel line, or the coarse filter, is (in addition to the loss of fuel) the loss of ability of the fuel transfer pump to move fuel through the filters and on to the injector pump under sufficient pressure (an estimated 50 psi) to prevent

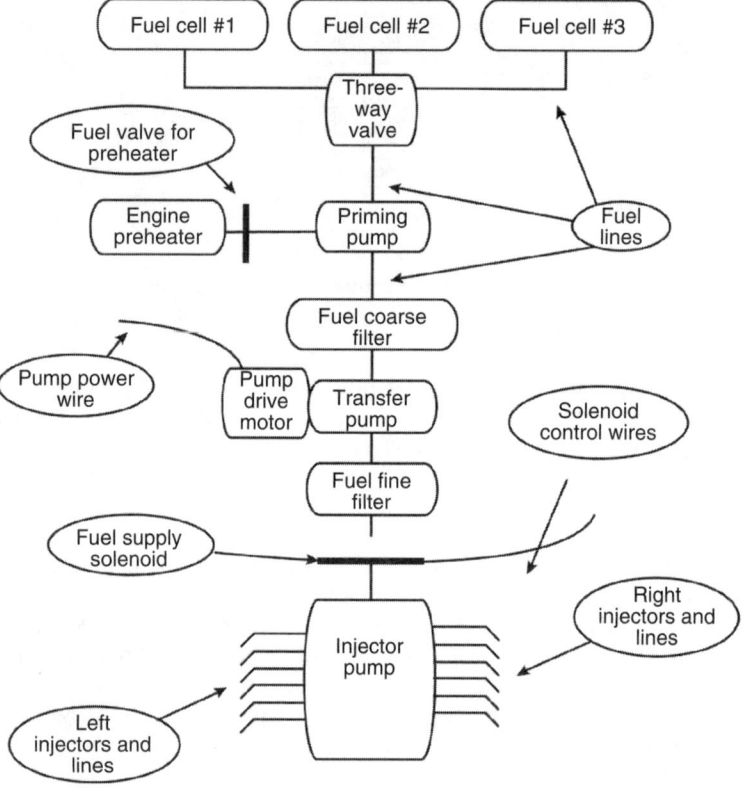

Fig. 3.9 Schematic of fuel system.

vapor lock or foaming of the fuel. The fuel system schematic in Fig. 3.9 shows that the engine preheater is on the suction side of the fuel transfer pump, rather than the pressure side. The integrity of the fuel lines on the suction side of the fuel pump becomes particularly important because the coarse fuel filter and fuel transfer pump are mounted well above the floor of the engine compartment; therefore, gravity will not mitigate any loss of suction. Any perforation of the fuel line in this area will result in air being drawn into the line by the fuel transfer pump, with resultant foaming of the fuel, accompanied by loss of both fuel pressure and volume to the engine.

The main output line from the priming pump feeds the coarse fuel filter and the rest of the fuel system. The fuel transfer pump is driven by an electric motor, as shown in Fig. 3.9. LoF of the motor, as a result of damage to the motor itself, or as a result of loss of electrical power, will result in loss of pump function. Therefore, both the pump motor and its electrical power supply are as critical to the mobility of the tank as the transfer pump itself.

The fine fuel filter is considered critical, not so much for its filtering function (an engine can operate for some time on unfiltered fuel) but because the housing must retain its integrity for a continued, adequate, pressurized flow of fuel to the injector pump.

The fuel supply solenoid is closed when there is no flow of electricity. (That is why turning the ignition switch off results in the engine stopping.) As a result of this design, the electrical control lines to the solenoid are critical, because interrupting the electrical supply will result in the solenoid closing and the engine stopping for lack of fuel.

LoF of the injector pump will result in loss of metered, timed, fuel supply to the engine, with resultant loss of engine operation. The injector pump is powered by a mechanical linkage from the engine, which is driven by a gear train powered by the engine timing gears.

It is intuitively obvious that loss of a single injector or injector line will result in LoF for the associated cylinder; therefore, the injectors and their lines are critical. The decision to be made is the number of cylinders that must remain functional for the engine to develop sufficient power for the tank to retain usable mobility. Because the situation of the tank at the time of damage is unknown, it is assumed that the loss of output from one bank of six cylinders will result in loss of usable mobility for the "typical" case.

The deactivation diagram based on this analysis of the fuel system of the tank is presented in Fig. 3.10. The deactivation diagram shows that each of the three internal fuel cells is connected in series with its fuel line. Disabling either the fuel cell, or its associated line, will disable that leg of the diagram. However, the diagram also shows that there are three of these series diagrams for the fuel cells connected in parallel. To disable the fuel system in this portion of the diagram, each of the three legs will have to be disabled. The fuel system is singly vulnerable from the three-way valve to the injector pump. Stated differently, disabling any single component in the diagram between the three-way valve and the injector pump will disable the fuel system and result in a loss of all mobility. Following the injector pump are two large parallel portions of the diagram describing the criticality of the injectors and their lines. Each injector is in series with its associated feed line, but all six injectors on one side of the engine must be disabled to result in a loss of mobility. Disabling an injector can be accomplished by disabling its feed line from the injector pump or by damaging the injector itself.

2. Lubrication System

LoF of the lubrication system is achieved by preventing sufficient lubricant pressure and volume from reaching bearings, gear trains, and other wear points in the engine. This can be accomplished by perforation of lubricant-filled components (loss of lubricant), breaking drive linkages to the pumps (loss of pressure and volume of lubricant at wear points), or deformation of components, thus preventing development of appropriate lubricant pressure or volume at wear points. A constant supply of pressurized lubricant is necessary at the wear points of the engine because the centrifugal force of the rotating parts tends to throw the lubricant away from the surfaces that require constant lubrication.

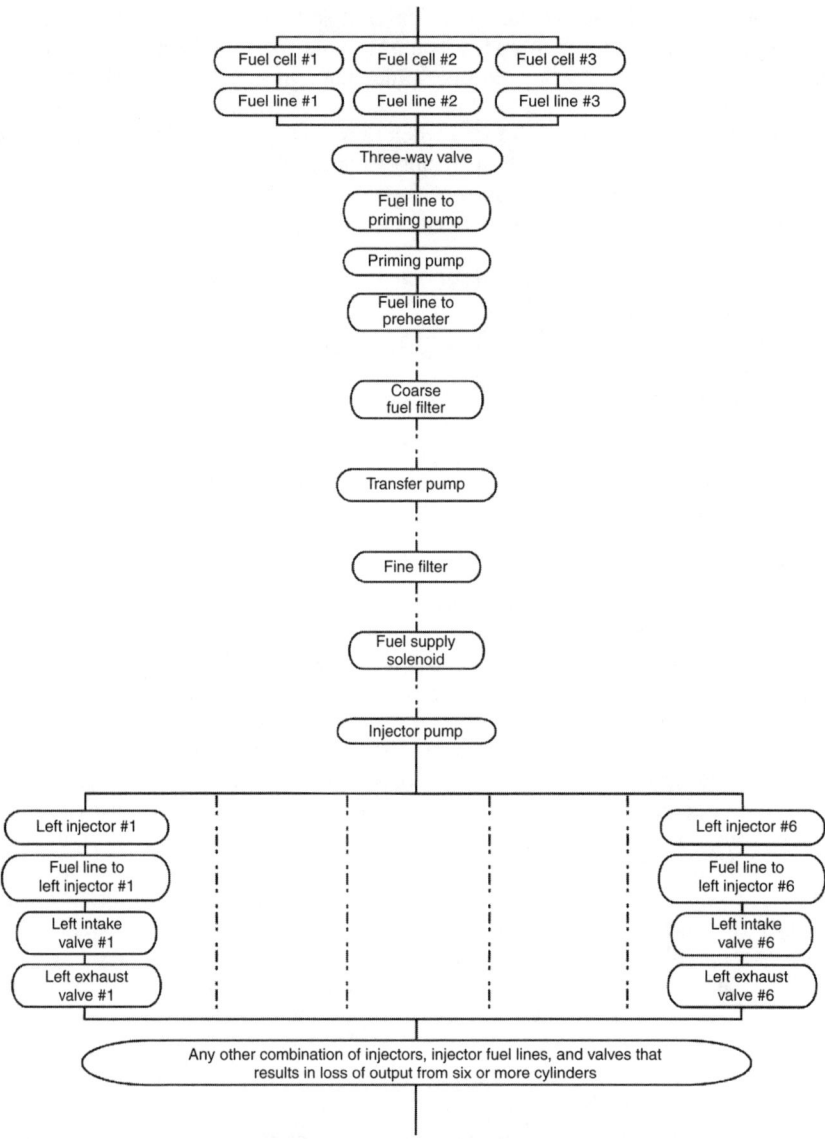

Fig. 3.10 Deactivation diagram for fuel system.

The lubrication system of the tank is a dry-sump design. In this tank, the oil pan of the engine does not serve as a storage reservoir for the oil. Rather, the water-jacketed oil pan collects and cools the oil as it drains from the points of lubrication (e.g., bearings, gears, and other points of wear) before the oil transfer pump returns it to the oil tank for reuse. Figure 3.11 presents a schematic of the lubrication

VULNERABILITY ASSESSMENT AND MEASURES OF EFFECTIVENESS

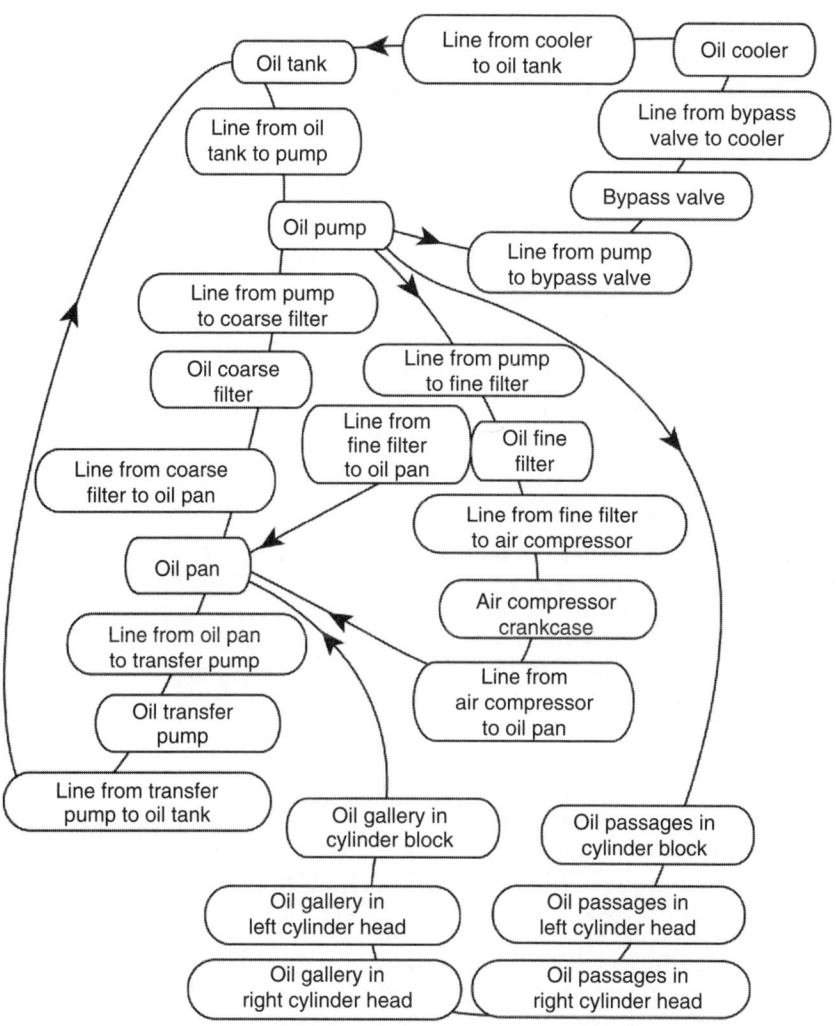

Fig. 3.11 Schematic of the lubrication system.

system of the tank. The oil pump moves oil through four rather distinct "circuits" in the engine of this tank. It pressurizes the oil passages in the cylinder block and the cylinder heads to feed lubricant to the bearings of the crankshaft and camshafts and other points of wear in the engine proper. The return path for the oil is through oil galleys (cast in the cylinder block and cylinder heads), back to the oil pan, where the oil-transfer pump returns it to the oil tank.

From the oil tank, the oil returns to the oil pump where it may be fed to the bypass valve, which serves the function of a thermostat. For this particular engine, it was necessary to design the engine with a water-cooled oil pan (which appears again in the discussion of the cooling system) as well as a separate oil-filled radiator to assist in controlling the operating temperatures of the engine. This oil-filled radiator is referred to as the "oil cooler" in this discussion. If the oil is hot enough, the bypass valve opens and the oil is routed through the oil cooler and back to the oil pan, and again transferred back to the oil tank. The bypass valve is assumed to be open for this analysis.

The deactivation diagram for the lubrication system is shown in Fig. 3.12. This particular deactivation diagram is a simple series diagram, showing that disabling any single component will result in loss of engine function. Examination of the deactivation diagram shows that neither the air compressor nor its associated oil lines are present. Analysis of lubricant flow in the oil feed circuit for the air compressor reveals that lubrication is provided through a flow restrictor to keep the fine filter under sufficient pressure to force the oil through the filter. This flow restrictor reduces the possible volume of oil to the air compressor, so that even completely severing the oil feed line will not result in loss of sufficient volume of lubricant to cause the engine to seize in a meaningful period of time.

3. Cooling System

Draining the coolant can disable the cooling system. Simply preventing circulation of the coolant will not result in loss of engine function within the generally accepted 10-min definition of the kill.

The cooling system of the tank is a closed system (Fig. 3.13). Coolant is circulated through a radiator to cool it, pumped through the cylinder block and heads to cool them, and then circulated back through the radiator. In this engine, each cylinder head has its own water manifold to distribute cool water along the length of the head, rather than just dumping the cool water in at the front of the head. Unless coolant is distributed along the length of the cylinder head by the coolant manifold, the rear cylinders and valves receive insufficient cooling. In early engines without water manifolds, the rear of the cylinder heads were subject to warping and valve problems associated with excess heat.

The oil pan of the engine has been designed with a water jacket, which is a part of the tank's cooling system. The air compressor is also cooled by the cooling system of the engine. In addition, as discussed in the description of the fuel system, the engine of this tank is fitted with a fuel-fired engine preheater. In extremely cold temperatures, the preheater is ignited and heated coolant is pumped through the water jackets of the cylinder heads, the block, and the oil pan before starting the engine. Warming the engine and lubricant in this manner facilitates starting the engine under extremely cold conditions.

Analysis of the engine cooling system is not as straightforward as analysis of the fuel system or the electrical system. It is obvious that interruption of fuel flow to the engine will result in the engine stopping, regardless of whether it is just idling under no load or operating at maximum output. The engineering is simple: no fuel means no operation at any power level. Conversely, the cooling system is situationally dependent. The engine can operate for relatively long periods of time at idle with no coolant in the system. It can operate for even longer periods of time if coolant is trapped in the water galleries of the heads and the block, but not circulated. On the other hand, an engine operating at full power may overheat with a fully functional cooling system. Understandably, this variability presents a dilemma to

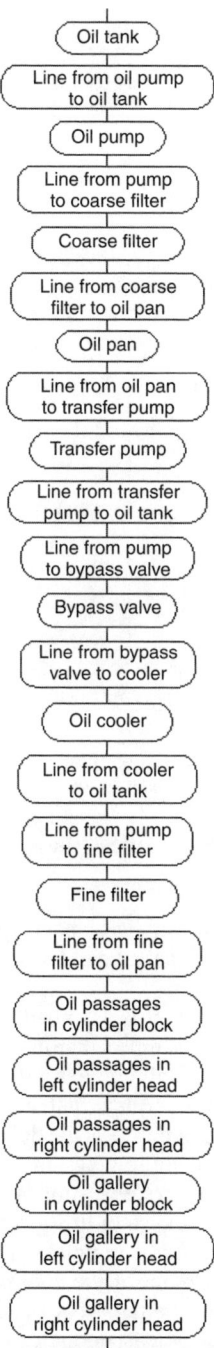

Fig. 3.12 Deactivation diagram for lubrication system.

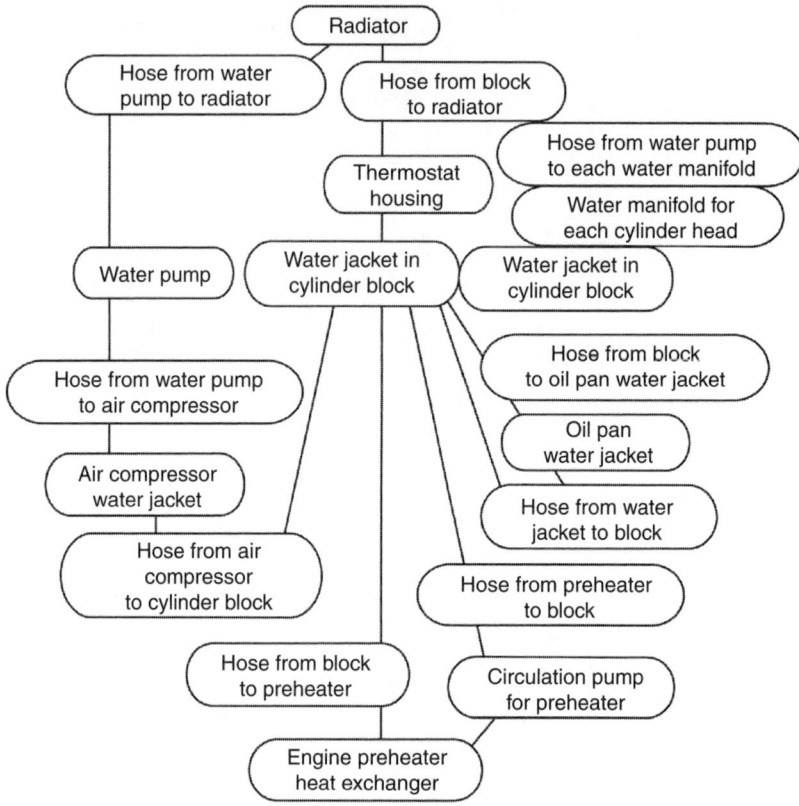

Fig. 3.13 Schematic of the cooling system.

the analyst. There are few data to serve as a basis for decisions that must be made, and there is virtually no way the analyst can know the operating conditions of the tank at the moment of attack.

For the purpose of this analysis, it is assumed that the engine is operating at a moderate power level. It is also assumed that in order for sufficient heat to accumulate in a timely manner, almost all coolant will have to be lost in an extremely short period. These assumptions require that the elevation of each component above the floor of the tank be considered as the deactivation diagram is developed. The deactivation diagram for the cooling system is presented in Fig. 3.14.

The system is a simple series system. Any of the components listed has the capability of draining the cooling system in short order, provided the leak is sufficiently large. Note that the radiator is not classed as a critical component. In this tank, the radiator is the highest point of the cooling system; therefore, the entire radiator could be removed from the tank, but the cooling jackets of the cylinder heads and the block would retain at least enough coolant for the engine to continue operation for about 10 min. The cooling circuit for the air compressor is not considered critical because the diameter of the coolant hoses involved are small in

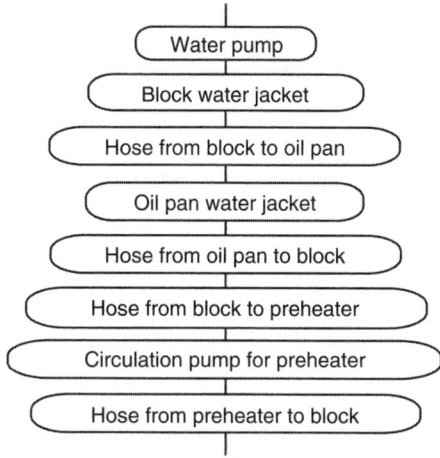

Fig. 3.14   Deactivation diagram for the cooling system.

proportion to the volume of coolant that must leak. Again, the engine would likely continue to operate beyond the time criterion. The water manifolds for the heads and the water jackets for the cylinder heads are not considered critical because the block water jacket would still contain coolant, plus the valves and pistons would continue to receive some cooling benefit from continued lubrication.

Components considered critical are all low in the hull. The water pump and the block water jacket are at the bottom of the hot zone around the swept area of the cylinders, where the heat resulting from combustion of the fuel is appreciable. The water jacket around the oil pan, the preheater, and associated coolant lines all are lower than the block water jacket and are considered critical.

4.   Electrical System

When first considered, the electrical system may not appear to be a critical system for the mobility of a tank with a diesel engine. However, as the fuel system of the tank is considered, it becomes apparent that electricity is necessary for the continued operation of the engine. Loss of electrical power causes the fuel-supply solenoid to close and severs fuel flow to the engine. Loss of electrical power to the fuel-transfer pump results in loss of fuel pressure and fuel volume delivered to the injector pump, with resultant loss of engine operation. Note that only a small portion of the electrical system has any effect on the mobility of the tank, and only that portion of the system is discussed here.

The portion of the electrical system affecting mobility can be disabled by preventing electrical power from reaching the fuel-transfer pump or the fuel-supply solenoid. This can be accomplished by preventing generation of electrical power or by severing or shorting the transmission path.

A schematic of the portion of the electrical system affecting mobility is presented as Fig. 3.15. When the engine is running, there are two sources of electrical

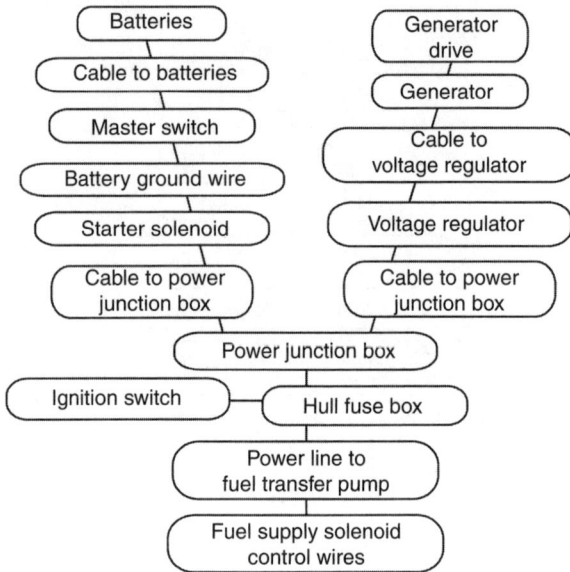

Fig. 3.15 Schematic of the electrical system.

power: the batteries and the engine-driven generator. The batteries are connected to the main power junction box of the tank through the master switch. The starter solenoid is used as a convenient tie point for the heavy current-carrying wires of the engine compartment.

The engine-driven generator provides power to the main power junction box, as controlled by the voltage regulator, based on the state of charge of the batteries and the electrical demand of circuits in the hull and the tank turret. Current is fed from the power junction box to the hull fuse box. Both the fuel-transfer pump and the fuel-supply solenoid are powered from the hull fuse box, as controlled by the ignition switch.

The deactivation diagram for the electrical system is presented as Fig. 3.16. The batteries, associated cables, master switch, and starter solenoid are in parallel with the engine-driven generator, its associated mechanical drive, electrical cables, and voltage regulator. At the power junction box, the diagram becomes a series diagram, with the ignition switch, hull fuse box, power line to the fuel-transfer pump, and the control wires to the fuel-supply solenoid each acting as a single point of failure.

5. *Engine Mechanical Components*

Engine mechanical components can lose function as a result of perforation, deformation, or fracture, any one of which may be the result of impact by a penetrating damage mechanism.

This category is a catch-all category for components and functions in and around the engine that do not fit neatly in any of the other categories discussed thus far. It includes the engine timing gears, the injector-pump drive, left and right

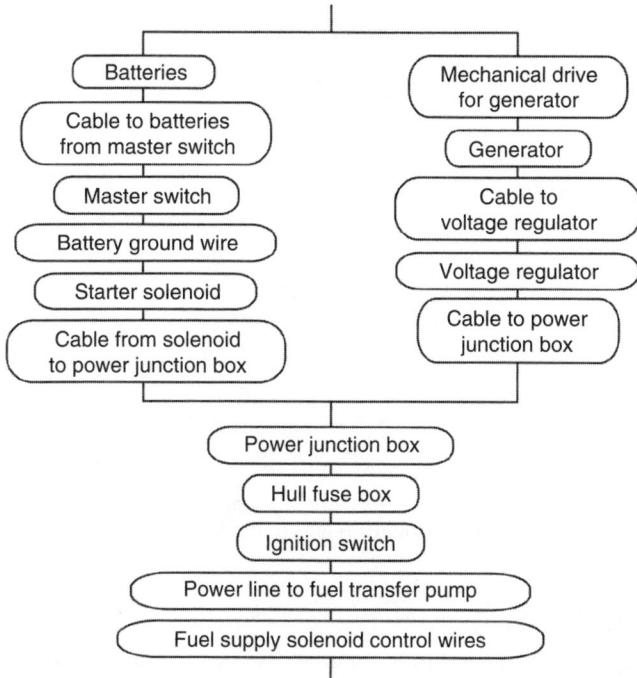

Fig. 3.16 Deactivation diagram for the electrical system.

camshaft drive towers, left and right camshafts, crankshaft bearings, pistons and connecting rods, oil-transfer-pump drive gear and shaft, and the oil-pump drive gear and shaft. The timing gears of this engine provide synchronized power to the injector-pump drive and the camshaft drives. Loss of this synchronization will result in severe degradation of engine power at best and complete loss of engine power at worst. The aforementioned assumption that loss of power from one bank of cylinders prevents the engine from developing useful power dictates that each camshaft is a critical component. Severe damage to any piston or connecting rod will result in the KE of the operating engine causing a cascade of damage with resultant destruction of the engine. Damage to the crankshaft of the engine will also likely result in cascading damage and destruction of the engine.

The deactivation diagram (shown in Fig. 3.17) is a simple series diagram, indicating that a LoF of any component in the diagram will result in a loss of mobility for the tank. It should be noted that neither the injector-pump drive nor the generator drive appears in the deactivation diagram. These components have been listed in earlier diagrams; hence, they are not listed again. It should also be noted that the oil-transfer-pump drive and the oil-pump drive are shown in the figure. Neither of these components has been listed previously because they are internal to the engine itself, just as the crankshaft and pistons are. Unfortunately, there is no hard-and-fast rule defining how one should group components. In this case, it has been found helpful to list the components with others from the same general location, rather than with the lubrication system.

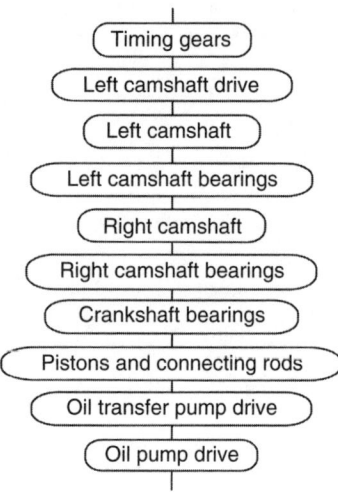

Fig. 3.17   Deactivation diagram for engine mechanical components.

B.   Transmission of Power

This category is intended to include all the heavy components designed to transmit and control engine power from the output of the engine to the final drives of the tank. To this point, only those components and systems necessary to provide engine power have been considered. This analysis begins where the crankshaft exits the engine. Components residing within the hull necessary to transfer engine power to the final drives are the clutch, transfer case, transmission, and left and right drive shafts that connect the transmission output to the left and right final drives. Components providing control, in addition to the transmission, are the left and right steering brakes and the left and right service brakes. The deactivation diagram for these components is a series diagram, as shown in Fig. 3.18.

Disabling any one of the components is considered an M-Kill. The service brakes appear as singly critical because disabling either brake would result in a violent steering to one side either at the time of damage, or later when the brakes are applied to stop or slow the tank.

C.   Control of Power

To control the tank, the driver must be able to see the terrain around the tank, steer, shift gears, accelerate, slow, and stop the tank. Thus, all driver's controls and their linkages must operate as designed. All pivot points must remain fixed, and all yokes that transmit movement of control linkages around corners must not flex when force is applied. Linkages must not stretch, flex, or collapse, or the necessary range of movement will be lost (e.g., if a gearshift linkage flexes, a gear may not be fully engaged when power is applied, resulting in internal transmission damage). As the control systems are being analyzed, the analyst must be sensitive to

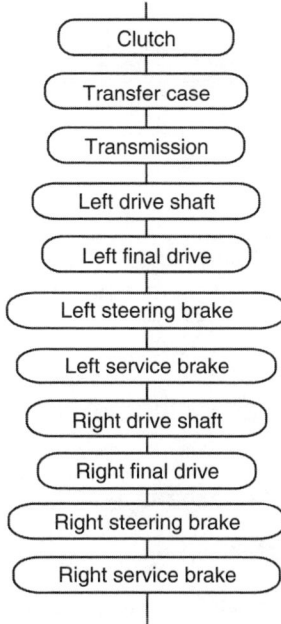

Fig. 3.18   Deactivation diagram for transmission of power.

redundant control linkages. In the tank illustration, the accelerator pedal and the hand throttle are tied to a common linkage from the driver's compartment to the engine compartment. Because it is assumed a skilled driver could operate the tank effectively with the hand throttle if the accelerator pedal were disabled, Fig. 3.19 shows the hand throttle and the accelerator pedal in parallel.

There are also other factors to consider. For example, in order for the driver to be able to operate the controls effectively, he must have rigid, fixed support. Expressed differently, if the driver's seat does not function as designed, the driver will not be able to control the tank effectively in combat. Therefore, the seat and its mounts could be considered critical to the operation of the tank. As another example, the driver must be able to see the surrounding terrain to control the tank. Because the driver has the option of opening his hatch and operating the tank with his head exposed, his vision devices are not necessarily critical.

D.   Suspension System

The suspension system of the tank is the "bottom line" of the mobility system. If the track, drive sprockets, idler wheels, and track tension adjusters do not work, the tank does not move. Moreover, in addition to transferring power developed by the engine to the ground to move the tank, the suspension is designed to support the tank, providing sufficient ground clearance to facilitate tank movement.

The suspension is a torsion bar suspension, designed with five roadwheels (in this case). If a torsion bar breaks or a roadwheel is lost or a roadwheel arm breaks,

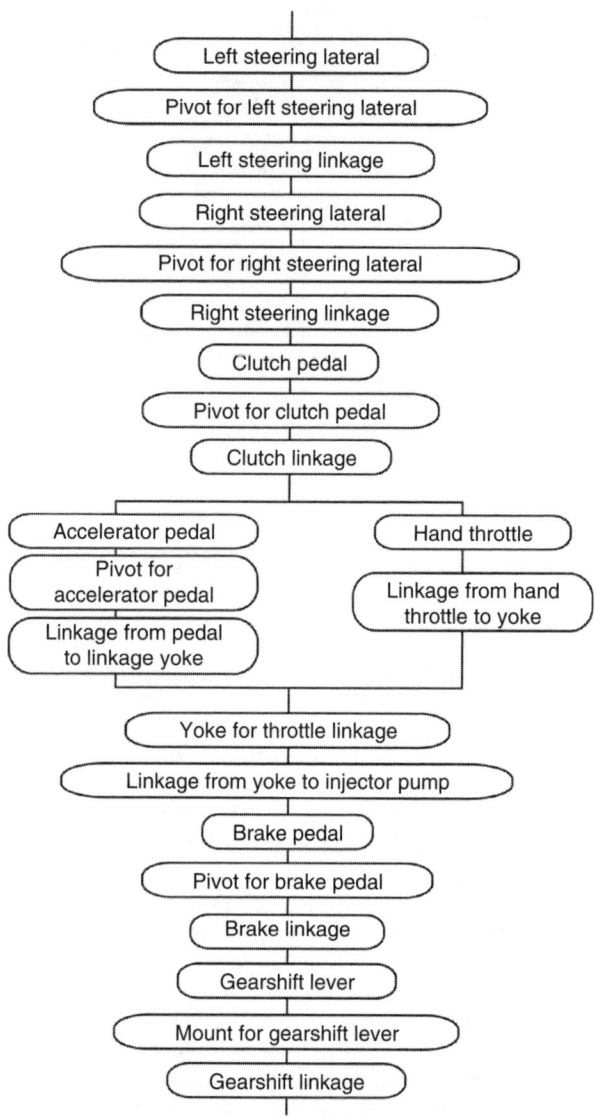

Fig. 3.19 Deactivation diagram for driver's controls.

the tank settles a bit on the side with the damaged component. At some point, the suspension no longer provides appropriate support to the tank, and the tank becomes immobile as a result of a thrown track or the hull resting on the ground and preventing movement. It seems reasonable that this point of immobility will be reached at different points, depending on terrain, mission, and other unknown variables. Therefore, a conservative approach has been adopted for this example. For our

VULNERABILITY ASSESSMENT AND MEASURES OF EFFECTIVENESS 109

purposes, the tank will be declared immobile if the function of the four roadwheels on one side of the tank is lost. This is shown in the deactivation diagram (Fig. 3.20) in the two parallel groupings at the bottom of the diagram. Note that if the front roadwheel on either side is lost, the tank also is declared immobile.

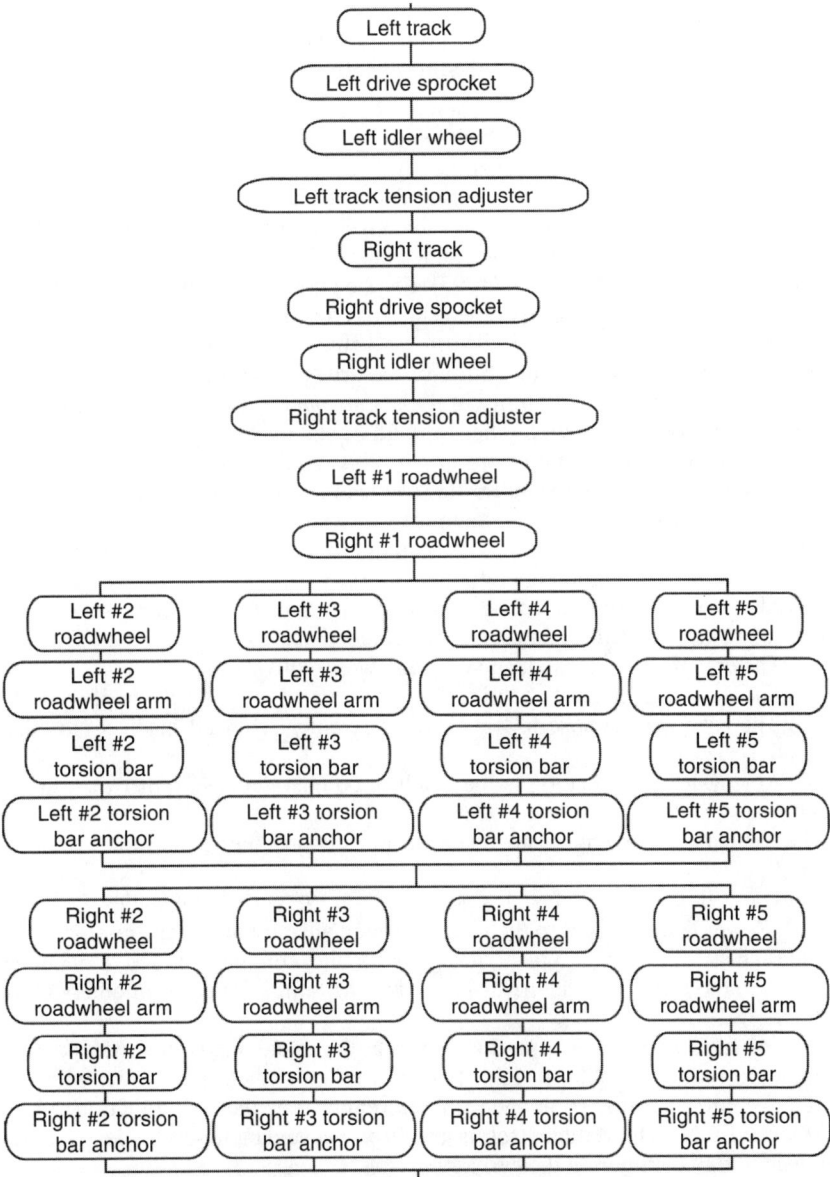

Fig. 3.20 Deactivation diagram for the suspension system.

Each of the deactivation diagrams developed thus far describes one facet of tank mobility. To grasp the real complexity implied by loss of mobility, however, the analyst would expand the deactivation diagram given in Fig. 3.7 with those in Figs. 3.10, 3.12, 3.14, and 3.16–3.20. This would give the entire diagram for the loss of mobility.

IV. An Example of Fault-Tree Development for Degraded States Methodology

This section discusses the development of deactivation diagrams for a lightly armored vehicle, based on level 3 of the vulnerability taxonomy. The fundamental premise of a level 3 analysis is that the residual capabilities of the vehicle (or loss of vehicle capabilities) can be measured physically in engineering units (e.g., horsepower, speed, etc.), rather than subjectively in terms of combat utility. For a more complete view of a degraded states analysis of an M2A1 Bradley Infantry Fighting Vehicle, see Price et al. (1993). For the purposes of this example, only the mobility (in its broadest sense) of the vehicle is considered. Mobility, as considered here, includes not only the ability of the vehicle to move across terrain in a controlled manner but also the ability of the driver to see the terrain around him and react appropriately, the ability of the commander of the vehicle to coordinate movement of the vehicle with others around it, and the ability of the unit to redirect subordinate vehicles in response to the ebb and flow of the situation.

As with the previous example, the following assumptions define the initial condition of the vehicle for this example.

1) The vehicle is fully stowed with ammunition, fuel, crew, and passengers.

2) The vehicle is in good operating condition.

3) All appropriate systems of the vehicle are operating (i.e., the crew is aware that combat is likely; therefore, they have the vehicle prepared to engage the enemy).

4) The crew is fully trained, capable, and well-motivated.

5) The crew will choose to continue to fight with a damaged vehicle, if possible.

6) The crew will choose to fight with the vehicle with all hatches closed, if possible.

7) Any system degradation is the result of combat damage, rather than manufacturing defect or fair wear and tear.

8) Neither the passengers nor their weapons contribute to the mobility of the vehicle.

9) The vehicle to be analyzed is powered by a V-8 water-cooled diesel engine.

10) Power is transferred to the tracks through an automatic transmission.

11) The vehicle is a typical design with the commander and the gunner located in the turret and the driver located in the hull, isolated from others in the vehicle.

Stated succinctly, the assumptions describe a vehicle that is in good operating condition, ready and able to fight. It is crewed by skilled individuals who will react to combat as other components of the vehicle (i.e., they will function as long as they are able). The vehicle itself is a well-proven and battle-tested design based on technology common in the early 1980s.

The results of this example analysis are presented as a series of deactivation diagrams defining the capability state of the vehicle for the stated damage.

A. Momentary Loss of Control

Loss of any one (or all) of the driver's periscopes will result in a momentary (1 min or less) loss of controlled vehicle mobility while the driver takes some corrective action to restore his ability to view the battlefield. The driver has three options: 1) replace the damaged periscope with a spare; 2) reposition the undamaged (or less damaged) periscope to give the optimum field of view; or 3) open his hatch, reposition his seat, and view his surroundings directly. Exercising any one of these options will result in a momentary (1 min or less) diversion of the driver's attention from control of the vehicle.

Loss of any (or all) of the commander's vision blocks affects the mobility of the vehicle by limiting the ability of the commander to observe the surrounding battlefield. This results in a corresponding loss of ability to direct the driver and coordinate vehicle movement with others, avoid obstacles, take evasive action, or take full advantage of available cover and concealment. Naturally, the commander has the same options as the driver for repositioning or replacing damaged vision blocks or opening his hatch to observe his surroundings directly. However, any of these actions will momentarily (1 min or less) divert the commander's attention from vehicle control.

Loss of the commander's intercom will also result in a momentary loss of control of mobility because it prevents direct communication with the driver. If the commander's intercom becomes nonfunctional, the commander must realize that the driver is not responding appropriately to instructions, must diagnose the problem to determine that the driver is still functional, and must determine that the intercom is nonfunctional. The commander must then shout instructions for the driver to the gunner so that they can be relayed over the gunner's intercom. It is estimated that this corrective action can be completed in about 1 min.

Loss of the radio is also classed as a momentary loss of control because the commander must realize that he is not communicating with the others of his unit, must make at least one other vehicle realize that he cannot communicate by radio, must receive instructions, must communicate the new instructions to the driver, and must execute them. If the commander cannot communicate his situation to another vehicle around him, he is reduced to following the unit and reacting to their actions until he can make his situation known. A vehicle commander will likely react to loss of his radio within 1 min.

The deactivation diagram for momentary loss of control is shown in Fig. 3.21. All of the components are in series because damage to any of the listed components will result in a momentary loss of control.

B. Loss of Communication with Driver

Loss of communication between the commander and the driver will also severely hamper command and control of the mobility of the vehicle. The driver sits low in the hull and does not have the commander's perspective of the battlefield. Therefore, loss of communication between the driver and the commander may result in inappropriate movement of the vehicle, particularly because the driver may not realize that communication has been lost. In this case, the electrical connections to the turret and the driver's intercom box are critical components because loss of any one

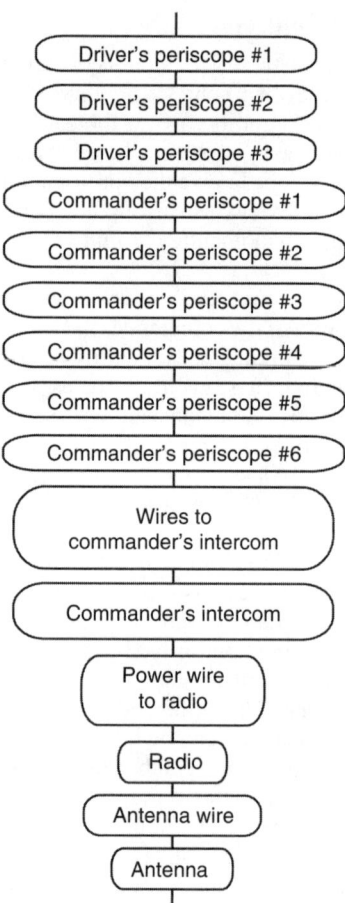

Fig. 3.21 Deactivation diagram for momentary loss of control.

of them will result in loss of communication with the driver. The diagram defining components (which, if damaged, lead to this state) is presented in Fig. 3.22.

C. Loss of Ability to Change Transmission Gear Range

It is assumed the transmission is in the gear range appropriate to the vehicle's mission at the instant of damage. It is also assumed, based on the transmission and shifting linkage design, the transmission will remain in the selected gear range after damage (i.e., it will not automatically shift to neutral). The transmission can be made to downshift by simply "flooring" the accelerator. Thus, the vehicle can generally move and maneuver as planned, although the driver must now avoid situations where it might become necessary to use another gear range (e.g., reverse). Damage to either the transmission shift lever and mount or to the shift linkage and pivots will result in this degraded state, as indicated in Fig. 3.23.

VULNERABILITY ASSESSMENT AND MEASURES OF EFFECTIVENESS 113

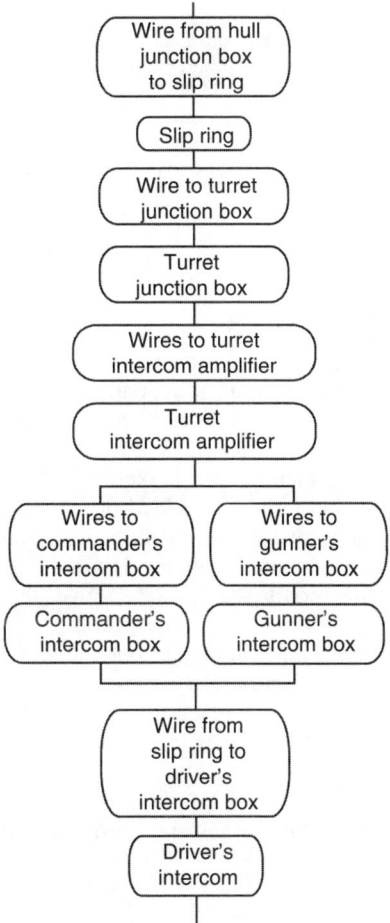

Fig. 3.22 Deactivation diagram for loss of communication with driver.

D. Loss of All Mobility

Loss of all mobility is described as a series of deactivation diagrams, one for each of the subsystems. Presenting the information in this manner enhances the visualization of the subsystems, which include the suspension, control, fuel, engine lubrication, transmission fluid, mechanical, and electrical systems. The track and drive sprockets serve as the final link between the engine output and vehicle's movement. Obviously, the loss of either track or either drive sprocket results in loss of all mobility. Further, the loss of either track adjuster or either track idler will result in such a slack track that movement of the vehicle will result in loss of the track. Figure 3.24 shows the diagram for the suspension system.

The ability to make the vehicle move without the ability to steer, control speed, or stop is useless on the battlefield, thus the concept of controlled mobility. The

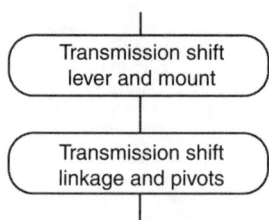

Fig. 3.23 Deactivation diagram for loss of ability to change transmission gear range.

necessity to steer the vehicle is intuitive. Hand- and foot-operated throttle controls and throttle linkages are included because normal design practice is to spring load the linkage to return the engine to an idle condition when pressure is removed from either the foot or hand control. An engine that operates only at idle also provides no useful motive power. Loss of brakes in this example indicates loss of ability to apply the brakes, rather than the brakes being locked in the fully applied condition. In some cases and terrain, a skilled driver may be able to reduce his speed and stop by using appropriate steering maneuvers and transmission shifting. Stopping the vehicle in this manner, however, is viewed as an emergency procedure rather than controlling mobility. Figure 3.25 gives the overall control system deactivation diagram.

Obviously, the engine will not continue to operate without an appropriate fuel supply. Interrupting the flow of fuel to the engine or reducing fuel flow to only the volume required for engine idle will result in loss of mobility. Loss of any seven injectors, their associated fuel feed lines from the injector pump, or their associated intake and exhaust valves will result in loss of mobility. It is believed that the resulting cylinder will produce ~2.5 hp/ton, which is below the power necessary to move the vehicle over native terrain, even at creep speed. Loss of seven

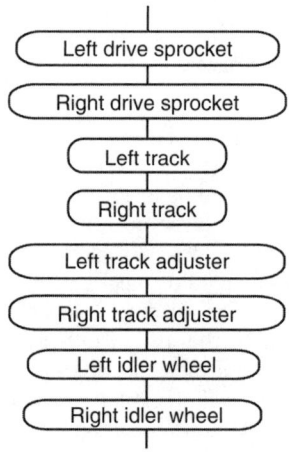

Fig. 3.24 Deactivation diagram for the suspension system.

VULNERABILITY ASSESSMENT AND MEASURES OF EFFECTIVENESS 115

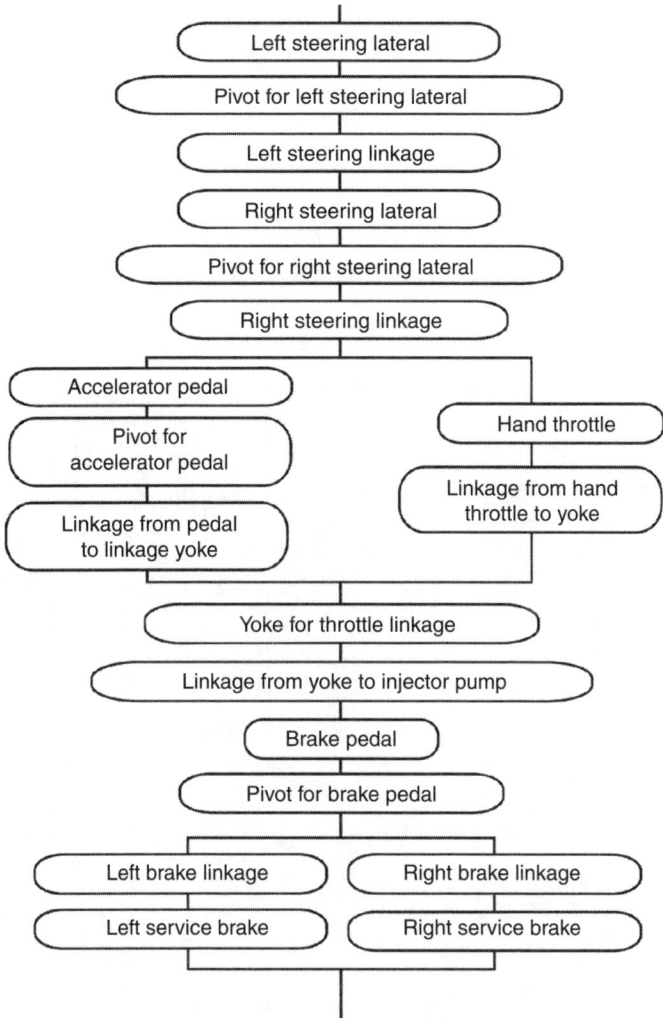

Fig. 3.25 Deactivation diagram for the control system.

cylinders can be achieved by loss of one camshaft or its drive and any three cylinders on the engine's opposite side. The fuel system is diagrammed in Fig. 3.26.

Loss of engine lubricant will result in loss of engine operation within minutes. Some extrapolated test data seem to suggest that the engine might continue to run for 15 min or so after the loss of lubricant; however, incidental observation of engines destroyed from a lack of lubrication strongly suggests that the time to failure is appreciably less than the extrapolated time. This system is diagrammed in Fig. 3.27.

The transmission in the vehicle is an automatic transmission; therefore, the loss of the transmission fluid will result in a loss of motive power. The deactivation

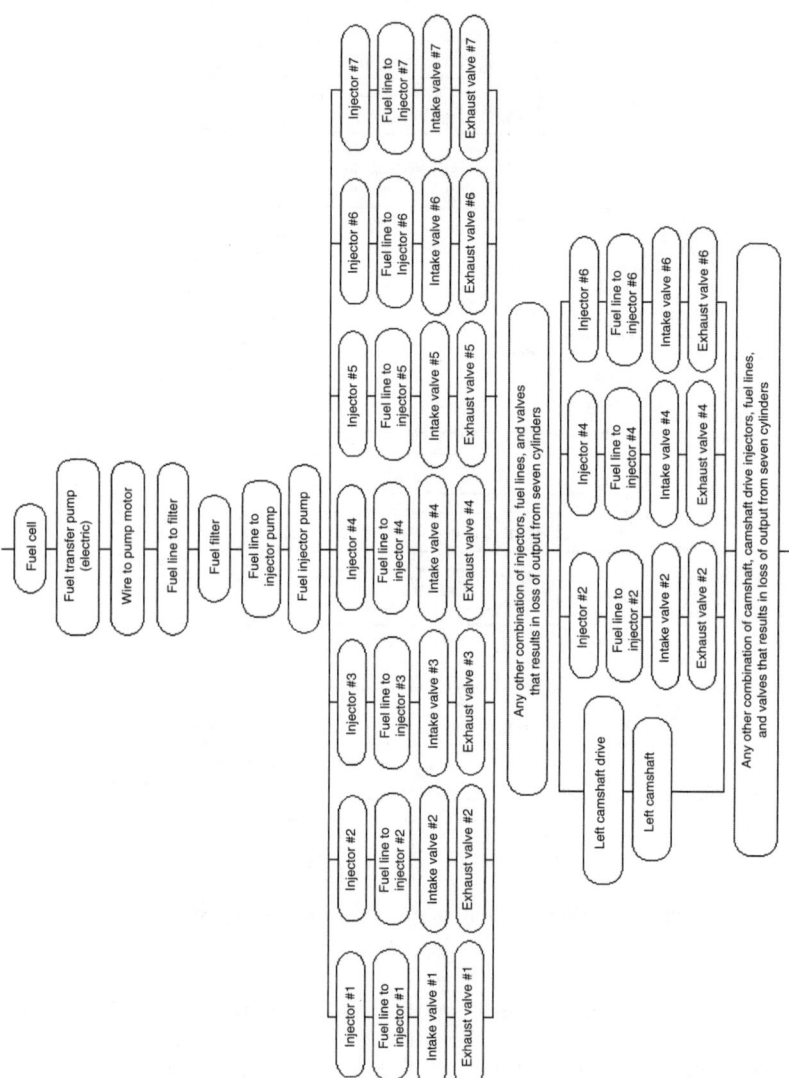

Fig. 3.26 Deactivation diagram for the fuel system.

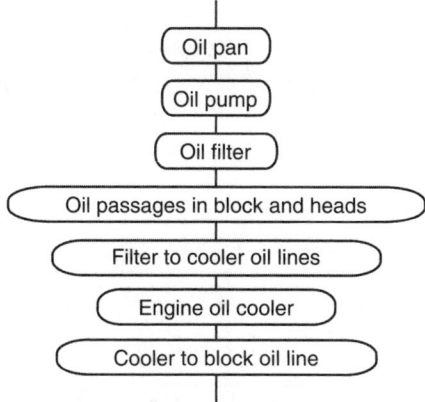

Fig. 3.27   Deactivation diagram for the engine lubrication system.

diagram in Fig. 3.28 identifies those portions of the transmission that can cause a loss of mobility simply by leaking transmission fluid.

The components shown in Fig. 3.29 are those robust, mechanical components within the engine compartment requiring substantial damage before failing, rather than the portions of the respective systems that will leak necessary liquids if perforated.

Loss of electrical power to the hull will result in loss of all mobility because the fuel control solenoid will interrupt the flow of fuel to the engine. Figure 3.30 diagrams the electrical system.

V.   Networks and Networked Combat Systems

Recent advances in wireless communication, including high-rate transfer of data and video, have enabled closer cooperation between combat systems than ever

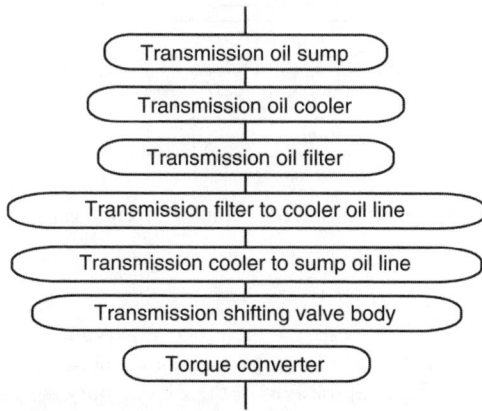

Fig. 3.28   Deactivation diagram for the transmission fluid system.

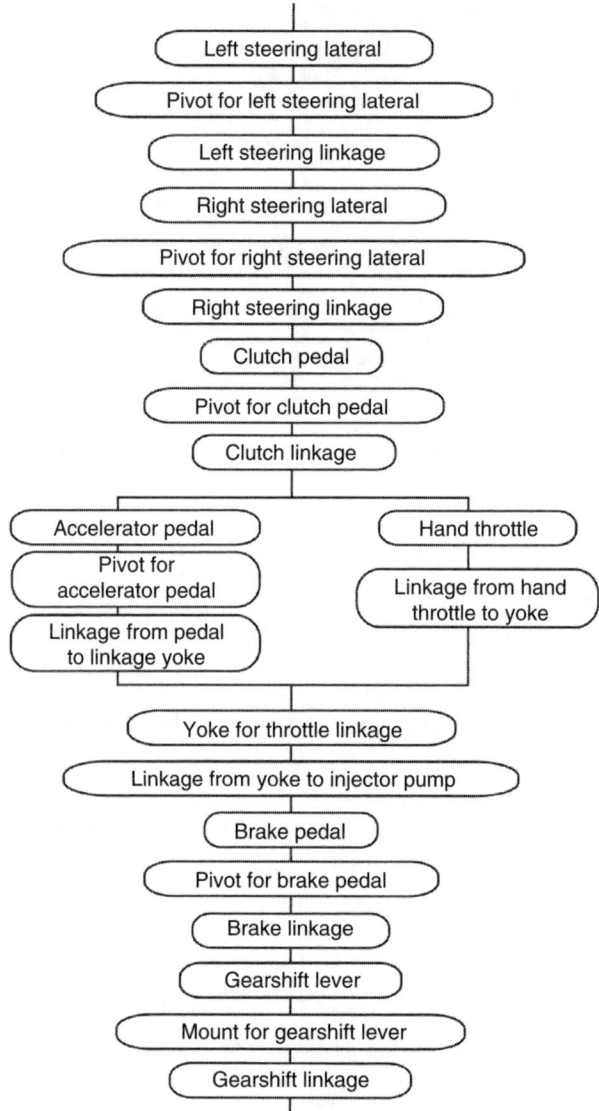

Fig. 3.29 Deactivation diagram for mechanical systems.

before. Remote sensing and the continued progress in development of "smart" munitions (those with the ability to find their own targets and/or adjust course based on communication with other platforms) have enabled networked, long-range fires. Combat system development at the start of the 21st century has been dominated by electronics and the network of communications, which ties nodes (systems and soldiers) together over potentially large distances, enabling these nodes to act as a

VULNERABILITY ASSESSMENT AND MEASURES OF EFFECTIVENESS

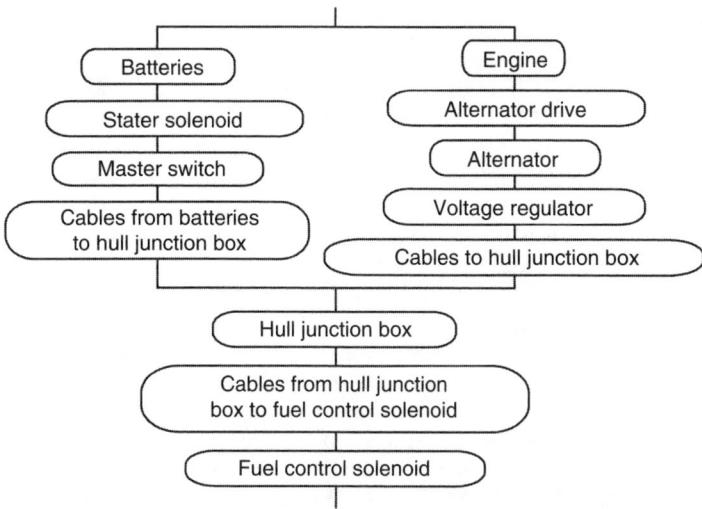

Fig. 3.30   Deactivation diagram for the electrical system.

cohesive combat entity. The trend is away from an individual platform having all requisite capabilities onboard to a set of capabilities being distributed over a number of platforms or systems, some or many of which could be unmanned. In addition, the widespread use of unmanned aerial vehicles (UAVs) has blurred the distinction between ground system and aircraft system vulnerability analysis, because many ground platform functions may be accomplished by these UAVs.

Thus, the challenge for the vulnerability analyst is this: because an ensemble of systems that, taken as a whole, constitutes a set of capabilities previously residing on a single platform, how does one account for the contribution each piece of the ensemble makes to survivability (or lethality or sustainment)? The solution to the problem is twofold: 1) because the network enables distributed combat functions, its vulnerabilities must be analyzed as a subsystem of the ensemble just the same as all the other pieces; and 2) because an ensemble of systems (including its network) is considered a combat unit, the survivability (or vulnerability, lethality, or sustainability) of the unit as a whole must be determined in some way from the contributions of the network and the individual nodes. In this section, methods for performing such analyses are discussed.

Networked systems also present a new paradigm for the vulnerability analyst. Because the first line of defense on the battlefield is avoidance of threat systems or engagement of those threat systems at ranges beyond their ability to return fire, vulnerability of networked combat systems must be conducted in a mission or scenario context. Without such a context, the analyst cannot determine which threats were engaged by which nodes of the network, and to what extent the network contributed to survivability by providing situational awareness and/or networked fires. The analytical toolkit must now include force-level simulations that support modeling network communications. Accordingly, the information in the following two sections is taken from a method developed by Walbert (2004b).

A. Network Vulnerability

The method of fault trees described in Sec. 3.II provides the foundation for analyzing network vulnerability. It is important at the outset to note that the term *network vulnerability* as used here does not mean the same thing as "information assurance." The ensuing discussion is not about electronic vulnerability or the vagaries of communication because of terrain or weather; rather, it is about viewing the network from the point of view of conventional vulnerability analysis. In this sense, the analysis is threat-independent: one analyzes the result of a network functional disruption, whatever its cause.

The mission context also introduces the element of time in a critical way. Because of terrain, weather, antenna orientation, or other factors, nodes may drop out of the network and drop back in or not as the combat scenario proceeds. Networks will have pre-determined rules governing formation and self-healing in an attempt to retain vital links. This means that a fault tree describing the Boolean connectivity among the nodes is subject to change over the course of time. The time-dependent fault tree is the key to analyzing network vulnerability.

A simple example serves to illustrate. Figure 3.31 shows a simple five-node network and the accompanying fault tree describing the connectivity at a particular point in time. Note that two nodes are "connected" if each is within the intersection of the two circles representing their ranges of communication. At some time later, assume something causes node 5 to drop out of the network (it may have gone "behind" a terrain feature that obstructed communications, in which case the loss may be temporary; or it may have suffered a component or total system failure, in which case the loss is permanent), as indicated in Fig. 3.32.

As Fig. 3.32 illustrates, loss of node 5 has broken the fault tree, leaving node 4 out of the network. Either through network rules for self-healing or through human intervention, node 3 is repositioned to be able to communicate with node 4, while retaining its communication with node 2. Figure 3.33 illustrates the new fault tree configuration of the network after communication with node 4 has been re-established.

This example illustrates the dynamic nature of the fault trees describing networks. If the loss of node 5 was temporary, the network may, at some later time, return to the configuration shown in Fig. 3.31. Clearly, as the mission progresses, there will be many such changes in network configuration. The essence of analyzing the vulnerability of the network is to analyze its fault-tree configuration over mission time.

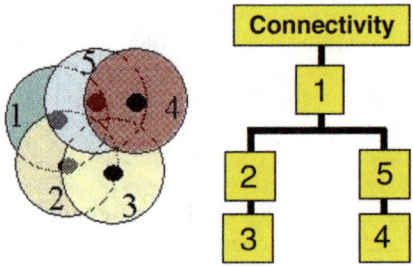

Fig. 3.31 A five-node network.

VULNERABILITY ASSESSMENT AND MEASURES OF EFFECTIVENESS 121

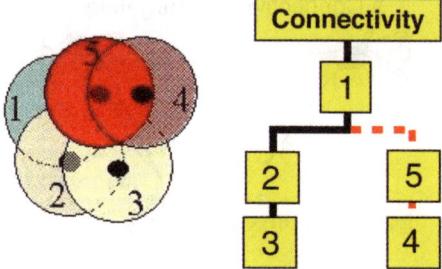

Fig. 3.32 A five-node network with a broken line of communication.

Suppose, for example, that node 4 provides the most critical function for a particular mission, and for 75 percent of the mission time going through node 5 was the only means of communicating with node 4. The vulnerability analyst might suggest adding another communications node (for redundancy) or perhaps suggest changing the characteristics of node 5 to avoid having it drop out of the network. Thus, vulnerability reduction for an ensemble of systems may be accomplished in a number of ways, only some of which have to do with reducing the vulnerability of an individual platform.

The configuration of the network fault tree over time is the "criticality analysis" that provides the ability to determine the vulnerabilities of the network in the same sense and manner as one determines vulnerabilities of the platforms. This is the foundation for determining ensemble vulnerabilities and the relative contribution of the network to ensemble and platform survivability.

B. Analyzing the Survivability of Networked Combat Systems

A key element of the transformation of military forces in the latter part of the 20th century and the start of the 21st century is the network. The relevant question for the vulnerability analyst is how to determine the following: the relative

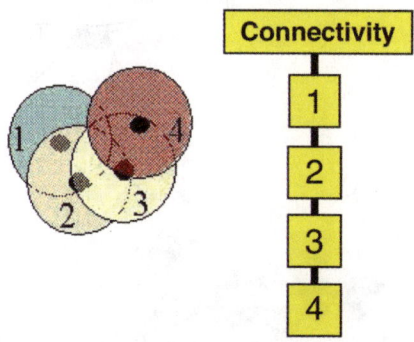

Fig. 3.33 A reconfigured four-node network.

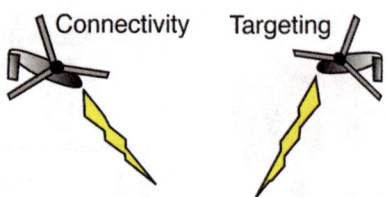

Fig. 3.34 A three-platform SoS (type 1).

contribution of the network to platform survivability; the relative contribution of conventional platform survivability measures (passive, active, and reactive) to platform survivability in the presence of the network; and how to combine platform and network survivability/vulnerability into an ensemble survivability/vulnerability.

The first step in understanding the process is to recognize that simply combining, in some fashion, individual node (or platform) vulnerabilities will not necessarily lead to correct ensemble vulnerability. To illustrate, consider the simple three-platform SoS ensemble shown in Fig. 3.34. Here, a ground platform is supplemented by two UAVs, one serving as a communications and data relay and the other providing targeting information for the platform's weapon system. Suppose the communications UAV is taken out of operation (by enemy action, terrain, weather, or other means). In this case, we assume that the targeting UAV can also serve as a communications relay, and the network reconfigures to the state shown in Fig. 3.35.

If we assume further that, before loss of the communications UAV, both UAVs were operating at full bandwidth, then given the configuration of Fig. 3.35, the ensemble (and the single remaining UAV) is now operating at a degraded state for both targeting and communications, because both are accomplished by the same UAV. This degraded state in one UAV is because of the loss of the other UAV,

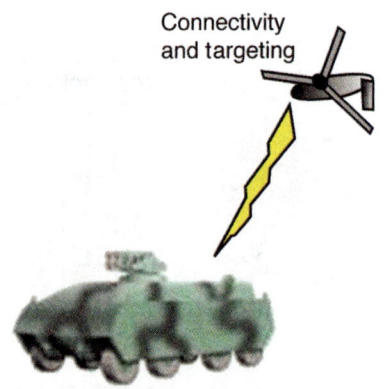

Fig. 3.35 The reconfigured SoS (type 1).

VULNERABILITY ASSESSMENT AND MEASURES OF EFFECTIVENESS   123

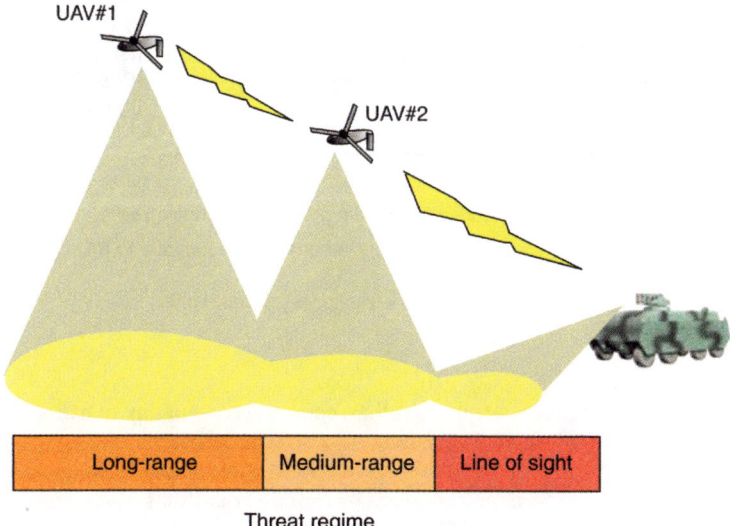

Fig. 3.36   A three-platform SoS (type 2).

considering only the vulnerability of the original communications UAV will not provide the correct analysis of the outcome.

How, then, does one analyze an ensemble, together with the network, to determine vulnerabilities? Suppose one has the ensemble shown in Fig. 3.36. In this instance, a ground platform's sensors are supplemented by sensors aboard two UAVs. The first UAV has responsibility for detecting distant, long-range threats. The second UAV has responsibility for detecting nearer-range threats, but not necessarily those within line-of-sight (LOS) of the platform sensors.

Using a force-level simulation, suppose this ensemble is taken through a particular mission or scenario and certain data (described in the following) are collected. First, one might plot sensor "availability" vs mission time, as in Fig. 3.37. Here "availability" simply means that information is, or is not, received by the ground platform from sensors. If information is not being received, it might be

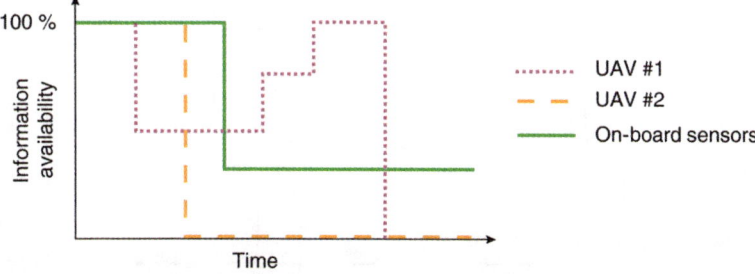

Fig. 3.37   Information availability vs mission time for a three-platform SoS (type 2).

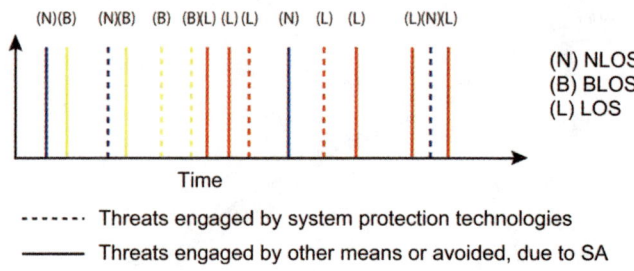

Fig. 3.38 Threat attacks vs mission time.

because equipment failure, terrain, weather, enemy action, or any of a host of other possibilities. In the figure one sees that, at various times during the course of the mission, sensor information from the UAVs and the ground platform is not always fully available, as is to be expected given battlefield conditions.

Figure 3.38 is a plot of threat attempted or possible attacks vs mission time. The vertical axis has no dimension; the vertical lines are simply indicators of the time at which a certain type of threat detected, attempted to engage, or engaged the ground platform. The solid vertical lines indicate those threats that actually engaged the platform [LOS, beyond line-of-sight (BLOS), and non-line-of-sight (NLOS)], from which regime (range) they engaged the platform, and, therefore, which platform survivability technologies attempted to counter the threat.

Superposition of the data in Figs. 3.37 and 3.38 leads to the first set of analytical conclusions. Note that, in this analysis, we assume that if the sensors on the UAVs detect an NLOS threat, the ground platform is able to avoid being engaged by the threat, either through avoidance or by having other networked fires remove that threat.

Figure 3.39 shows information availability from UAV 1 sensors vs NLOS threats. One can conclude that for at least 75 percent information availability, sensors on board UAV 1 detected all threats; at less than 75 percent information availability, platform survivability measures had to engage the long-range threats (50 percent of the total number of long-range threats).

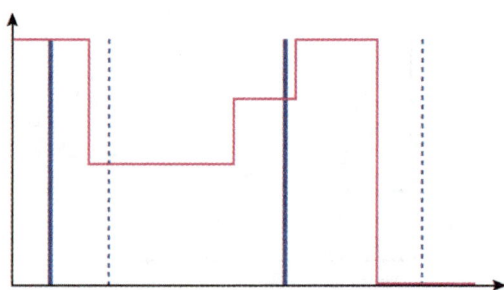

Fig. 3.39 Information availability for UAV 1.

VULNERABILITY ASSESSMENT AND MEASURES OF EFFECTIVENESS 125

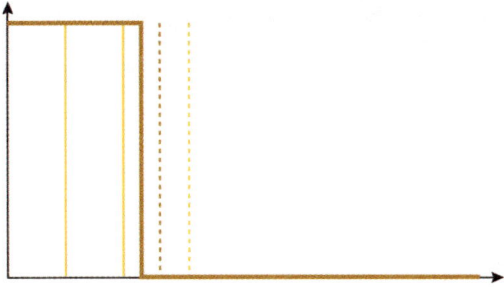

Fig. 3.40 Information availability for UAV 2.

Figure 3.40 shows information availability from UAV 2 sensors vs BLOS threats. From this figure, one can conclude that with full information availability, all threats were detected, while platform survivability measures engaged 50 percent of the threats.

Figure 3.41 shows information availability from ground platform sensors vs LOS threats. From this figure, one can conclude that with 33 percent information availability, five out of seven LOS threats were detected and avoided, while platform survivability measures engaged two out of the seven threats.

This information, when combined as shown in Table 3.3, provides the basis for determining which measures contributed to platform and ensemble survivability. If we assume that different platform survivability measures responded to threats according to the threat type (range from which launched), then Table 3.3 shows the relative contribution of each measure to survivability, and also shows the network contribution, termed situational awareness.

In addition, the introduction of the force-level simulation (mission dependence) has other benefits for the vulnerability analyst. One could re-run the previous

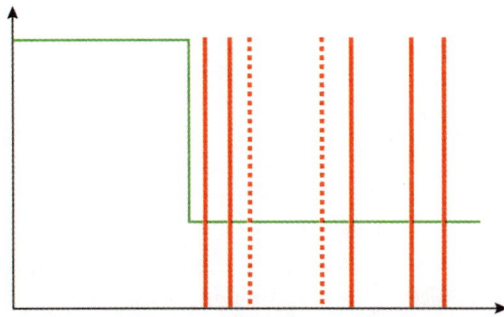

Fig. 3.41 Information availability from the ground platform.

Table 3.3 Protection mechanism vs number of threats for the ground platform

Protection mechanism	Number of threats		
	Line of sight	Medium range	Long range
Technology A		2	
Technology B	2		
Technology C			2
Situational awareness	5	2	2

mission with additional UAVs, with UAVs at different altitudes, or different sensor suites, and see how the survivability changes based on changes in information availability. Trade studies for vulnerability reduction can now include the network configuration, or even the network rules for reorganization, as well as tactics, techniques, and procedures (TTP). The MMF then provides the ability to determine the configuration that leads to optimum combat utility.

Chapter 4

Modeling and Simulation Tools and Methods

I. Background

TO THIS point, we have focused on the building blocks of V/L analysis. Chapter 2 has presented characteristics of ballistic (and other) threats, insights into the damage mechanisms they employ, and principles of mapping from the MMF's level 1 (initial conditions) to level 2 (component damage states). Chapter 3 has focused on defining measures of effectiveness and mappings from level 2 to level 3 (vehicle engineering capability) and from level 3 to level 4 (combat utility) via the SDAL. We now turn our attention to the development and use of models that allow the analyst to assemble these building blocks in a unified way to estimate damage from a ballistic encounter, determine the effect of that damage on system engineering capability, and compute V/L metrics that quantify the system-level vulnerability of a target vehicle or the lethality of an attacking munition.

The approach we take in Chapter 4 is to present a historical perspective on how these models evolved, general principles of how analysts use these models, and ways in which modeling at each level of the taxonomy is integrated in system-level models.

While examples of several historical and extant models are discussed, note that we have not attempted to include all extant models or describe the operation of any given model in great detail. Models at all levels are constantly being improved as our understanding of threat–target interaction physics and engineering grows, as new data become available, and as computational capabilities become more robust. In addition, note that the terms *model* and *computer program* are used somewhat interchangeably. Strictly speaking, a model is a mathematical and logical representation of some physical phenomena or process, and a computer program is merely an implementation of the model. Within the V/L analysis community, however, it becomes cumbersome to maintain these distinctions because a given discussion often inseparably involves both concepts. So, in general, practitioners rely on context to convey the intended meaning, and when a distinction is necessary, it is so noted. This approach is the one we take in this book. Similarly, the term *model* can refer to a specific physical process such as predicting perforation of armor. It can also refer to an integrated collection of submodels within an analysis environment to estimate system-level V/L metrics. Again, we rely on context to distinguish meaning.

Finally, the general class of ground systems (sometimes called ground-mobile systems) includes several types of vehicles. AFVs generally refer to heavily armored vehicles that are intended to engage in close combat operations and fire or that require protection against attack by medium- and large-caliber anti-armor munitions. Lightly armored vehicles refer to vehicles engaged in combat operations anticipated to involve small-caliber weapons or to vehicles that require protection from small-caliber weapons. Unarmored vehicles have little or no armor and are most typically exposed to small-caliber weapons. (Artillery and other indirect-fire weapons are threats to every type of vehicle on the battlefield.) Because the development of system-level V/L analysis programs has been motivated by the need to analyze each type of ground vehicle and threat munition, AFVs have received the most focused attention over the decades. In the discussion that follows, AFVs are most commonly used to illustrate the use of V/L models. Generally, the major principles of analysis and the models implementing them apply to all types of vehicles, although details vary. For example, lightly armored and unarmored vehicles are usually not concerned with BAD, use different kill definitions, and do not rely on SDALs to map component damage or subsystem LoF to combat utility. We do not attempt to address these nuances in this book.

II. Empirical vs Semi-Empirical Models

We begin this section with an overview of models describing various physical effects and phenomena produced directly or indirectly by certain munitions against targets. Detailed discussions of blast; BAD and BAE; and penetration by KE projectiles, SCs, EFPs, and fragments are presented in the appendices and in various references.

Empirical models are actually just simple functions that allow statistically good fits to data with the empirical determination of a few parameters, which are often case-dependent. The world of V/L analysis is replete with such simple rules, many of which are outlined in this book.

Semi-empirical models are similar except that the functions used are based on a simple physical model that, by itself, is inadequate to describe the phenomena without parametric fitting. In these cases, the fitting might be done to comparatively simple analytic models or to closed-form functions.

Classic empirical V/L models are the Thor equations, which describe fragment penetration. The name Thor designates a series of reports that resulted from Project Thor at Johns Hopkins University during the 1950s and 1960s. These reports document fragment firings against plates and the resulting empirical equations used to predict penetration. The Thor equations are power-law equations that relate fragment penetration with fragment mass, presented area, hardness, and so on. Empirical coefficients and exponents are used to represent various types and shapes of fragments and conditions (Johns Hopkins University 1961; Malick 1968, 1969; Holloway et al. 1978; Bely et al. 1992).

Newer and expanded data were included in a more complete semi-empirical fragment-analysis system called Fast Air Target Encounter Penetration (FATEPEN). Figure 4.1 gives an overview of the FATEPEN system (Yatteau et al. 1991).

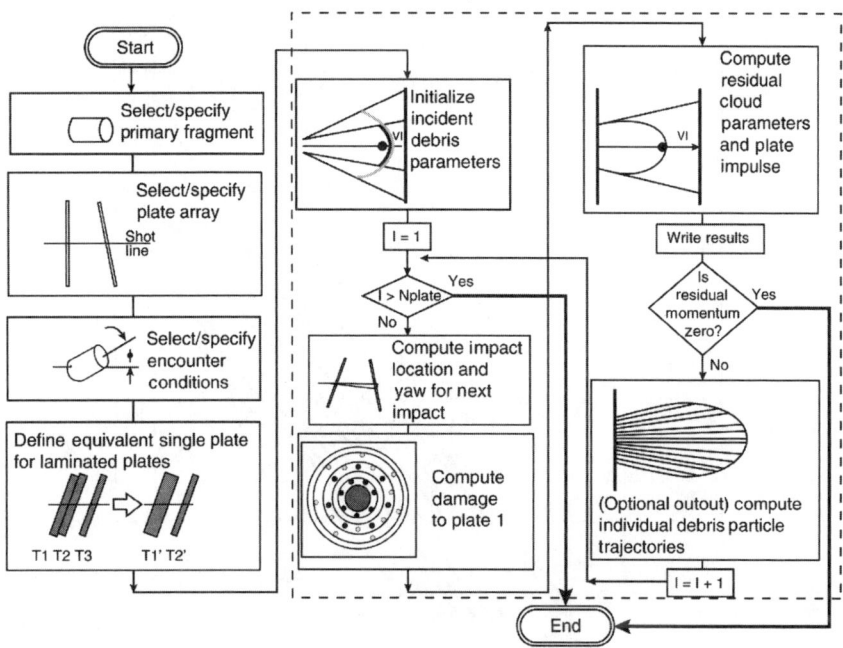

Fig. 4.1 FATEPEN overview.

In another semi-empirical model, correlations among P_K, mean armor thickness, and weight of charge were established for SC warheads in contact with armored vehicles (Fig. 4.2). Such correlations are particularly useful to assess how much blast and shock from the detonation add to effect of the jet alone. To derive these blast-effects models, dimensional analysis is combined with actual test data to relate charge weight and range for given levels of damage (U.S. Army Materiel Command 1971).

The characterization of BAD is usually accomplished with some form of fitting process to an assumed distribution such as the Weibull function. For a discussion of the statistical processes, see Shnidman (1988). However, there is an approach to this problem in which a shape is assumed for the cloud of debris to which kinematics are applied to describe the growth of the cloud. Although the assumed initial cloud shape is based on experimental observations, this approach is semi-empirical because the subsequent cloud growth is based on physical kinematics (Saucier et al. 1995).

III. Phenomenological Models

A. Closed-Form Models vs Engineering Simulations

Phenomenological models use basic physical principles and basic material properties to simulate physical events (e.g., perforation of armor, ignition of

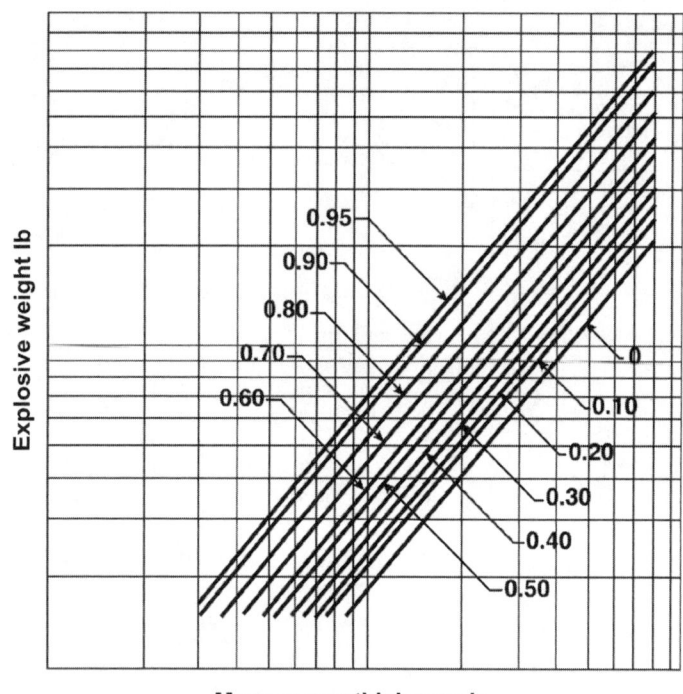

Fig. 4.2 Probability of M-Kill because of blast, case fragments, and shock.

combustibles, and damage to components). There is a variety of such models that can be used to simulate events associated with ballistic threats and related V/L effects. The models range from quite simple closed-form equations to huge computer programs that can run for weeks on a single case. These models might be used directly and routinely in vulnerability calculations, they might be used to develop parameters for simpler empirical models, or they might be used to simulate certain specific cases to answer particular V/L questions or develop insights into the physical processes involved.

The simple closed-form models are typified by the penetration model based on the Bernoulli principle shown schematically in Fig. 4.3 [which is taken from Walters and Zukas (1989)] for which SCJ penetration may be estimated from the following equation:

$$P = L(\rho_j/\rho_t)^{1/2} \tag{4.1}$$

where

 P = penetration into homogeneous target
 L = length of jet
 ρ_I = density of jet (j) and target (t)

MODELING AND SIMULATION TOOLS AND METHODS

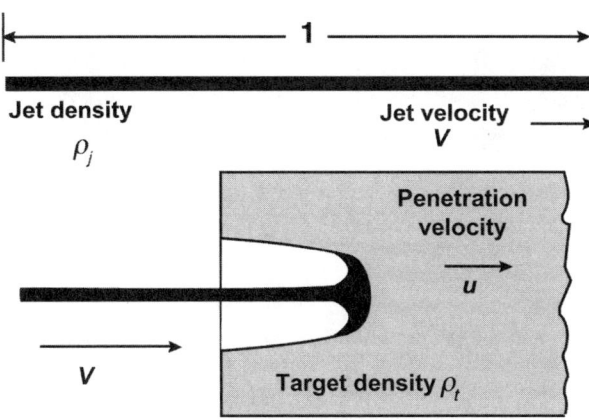

Fig. 4.3   SCJ penetration.

Such models allow good first approximations of extremely high velocity penetration typical of SCJs (Walters and Zukas 1989). At the other extreme of complexity, engineering simulations such as hydrocodes and finite element models are used to model interaction physics.

Blast waves in air against complex targets, for example, can be simulated by hydrocodes, which are numerical approximations to the partial differential equations of fluid mechanics. In some cases, though, simple closed-form models can sometimes be used rather than the hydrocodes (Bingham et al. 1996; Courant and Friedricks 1948). Likewise, hydrocodes can be used to simulate the formation of SCJs because the material properties other than density are often secondary. Hydrocodes are also useful for EFP design and variants (Randers-Pehrson and Juriaco 1978) such as the pellet warhead concept shown in Fig. 4.4 (taken from Reed 1992a).

Finite-element codes in which the penetrator and target are simulated by small interconnected elements can be used to simulate the penetration process, but the simulation can become quite complicated and lengthy if the target and/or projectile have complex geometries and if the impact is oblique (requiring 3-D representation).

Finite-element models can also be used to model entire vehicles (or subassemblies) and their response to impact. Again, these models are quite extravagant in using computer time and are generally not practical for the myriad cases usually required in a V/L analysis. They are useful, however, for developing simplified empirical models, looking at special cases, and generating response data that are used in V/L analyses.

Various hydrocodes have the capability to treat SDT events in energetic materials (Lee and Tarver 1980; McGlaun and Thompson 1990). Although these are too time consuming for the vulnerability analyst to use routinely, they are used to guide the development of simplified models that can be used in everyday V/L analyses.

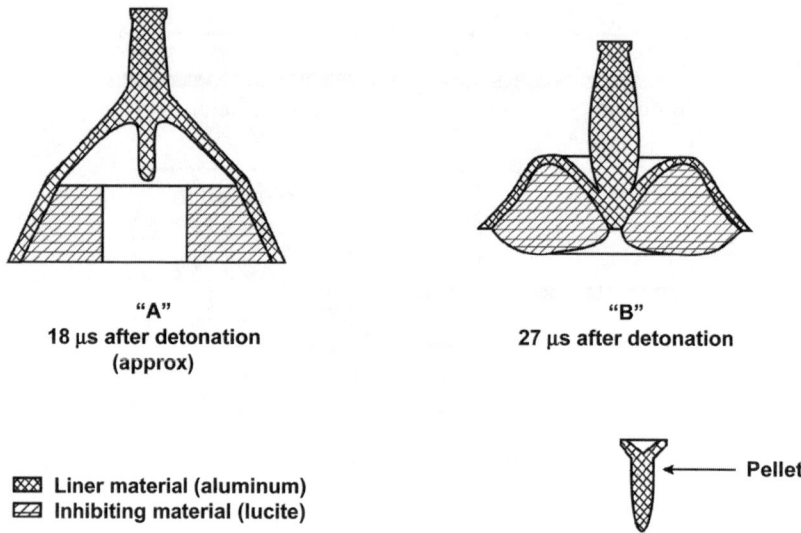

Fig. 4.4 Kronman's hypervelocity pellet warhead.

B. Predicting Fire

Although fire (which has been discussed briefly in Chapters 2 and 3) is a common occurrence in combat vehicles hit by threat munitions, the ability to predict fire initiation, sustainability, and extinguishing based on hit location and threat type is a nontrivial matter. Indeed, repeated experiments where the same threat enters a vehicle in the same place will not produce consistent results. Only the right combination of heat, oxygen, and combustible material will result in fire. Each of these elements is a function of the stochastic nature of threat–target interaction, and predictive models must account for this stochasm.

1. Illustrative Experimental Results

Since the early 1950s, the Army has conducted firings at diesel-fuel-loaded aluminum and steel vehicles, containers, and surrogates, with fuel at various temperatures. Although it is clear that, at higher fuel temperatures, sustained fires are more likely given a hit by an SCJ, it is also possible to have sustained fires with fuel temperatures from just above 32°F to over 165°F (Hanna and Goodwin 1955; Beichler 1956; Skinner 1959).

For example, the results of the 1959 CARDE tests are often cited as evidence to assign a low probability of sustained fire, given a hit on diesel fuel (Canadian Armament Research and Development Establishment 1959). In fact, the CARDE tests involved two distinct series of SCJ challenges to diesel fuel. In one series involving diesel fuel temperatures of 44°F to 67°F, using a simulated engine compartment covered by a grill, there were only 13 sustained fires in 64 tests using SCs from 5 to 8 inches in diameter. In this case, it was pointed out that there might have been a problem with the availability of sufficient air for sustained fires. Examination of the original data shows that there were only two sustained fires in the 42 tests in which the grill remained on the fixture for the entire test, and there were 13 sustained fires in the 22 tests in which the grill was blown off the fixture. Although the larger SCs had a greater tendency to blow off the grill, there is no question that there is a finite probability of sustained combustion, even with cold diesel fuel, if there is an adequate initiating source and air supply.

In the other series of CARDE tests that involved running engines in tanks, 5- to 8-inch-diameter SCs were fired into cells containing cold diesel fuel. There were 11 sustained fires out of 15 tests. All four tests using 8-inch-diameter SCs gave sustained fires, whereas only one out of four tests using 5-inch-diameter SCs gave a sustained fire. This reinforces the concept that an adequate initiating source and air supply are necessary to achieve sustained combustion with cold diesel fuel.

Data have been collected on over 400 SCJ and KE round firings at diesel fuel containers, both inside and outside vehicles. Curves have been constructed to represent the probability of sustained fire, given a hit on the fuel cell of a steel vehicle and on the fuel cell of an aluminum vehicle as a function of fuel temperature (Skinner 1959; Groves 1986). Like the CARDE data, the results of these firings suggest the probability of sustained fire is highly situation-dependent.

2. Fuel Fire Start Algorithms

The question of whether or not a fire is initiated when a weapon strikes a fuel container in a combat vehicle is currently answered using simplistic look-up tables or by relying on the judgment of personnel experienced in the area of combat vehicle fires. The simplistic approach says the probability of a sustained fire when a weapon perforates a fuel cell containing diesel fuel is 0.1. The probability of a sustained fire when gasoline is the fuel is 0.9. For fuels other than diesel or gasoline, the analyst must use best judgment. When experienced personnel estimate the probability of sustained fire when a fuel cell of a vehicle is perforated, factors such as fuel type and temperature, vehicle construction (steel vs aluminum vs composite), the penetrating ability of the weapon, and the presence of an automatic fire-extinguishing system are all considered.

Until recently, there was no algorithm that takes into account all the important chemistry and physics to predict probability of sustained fire in a combat vehicle. Dehn developed a physical model that takes into account the necessary physics and chemistry required to predict sustained fires in combat vehicles under any set of conditions. This model is presented in both descriptive form and mathematical form in Dehn (1979). According to Dehn, the probability of a sustained, destructive fire in a vehicle is the product of several probabilities: the probability of fuel-spray formation leading to the probability of a sustained fire and the probability of failure of an extinguishing system. Unfortunately, the mathematical form of the Dehn model is difficult to use because the correct values of the parameters and variables that go into the model cannot readily be obtained experimentally.

Wright and Slack (1980) presented a simplified form of the Dehn model that allows prediction of the probability of a sustained fire given a particular fuel temperature using an equation requiring only two experimentally determined temperatures and one constant. However, the equation into which these are inserted applies only to a specific type of vehicle and a specific attacking munition. For example, a particular temperature and constant set applies only to a steel type vehicle containing an interior steel splash surface, with an aluminum fuel cell containing diesel fuel attacked by a 90-mm SC. If any of these conditions changes, the equation must be changed to find the correct probability of a sustained fire. Even with the simplified equation, a large number of tests must be carried out at low and at high temperatures to determine the two limiting temperatures required for the equation. In addition, more tests are required at intermediate temperatures to determine the constant of the equation. Neither form of the Dehn model is currently used because of the difficulty in determining values for model parameters.

Detailed combustion algorithms such as those developed at Los Alamos National Laboratory (Repa) include features such as 3-D, time-dependent fluid flow, turbulence, heat transfer, fuel sprays, chemical reactions including heat release, and multi-component species transport, with a flexible geometry. These models, combined with other models, address the combustion phenomena and added complexities of weapon attack, hydrodynamic ram, spall formation, ignition processes, and fire suppression. The fire protection model, a simplified model based on the results from this and other complex engineering simulations, is currently under development for use in V/L analyses. In fact, some features have already been used for analyses of aircraft. In addition, an existing model, FIRESIM, uses Monte Carlo techniques to predict probability of fire given a hit (U.S. Department of the Army 1995). This model does not predict ignition based on the interaction of threat and vehicle. Instead, it predicts fuel fires based on probability inputs.

Rules of thumb for predicting sustained fuel fires have been given in Sec. 2.II.H.

C. Predicting the Results of Ballistic Impacts on Ammunition Components

From the aforementioned information, it is clear that initiation phenomena, like many other vulnerability phenomena, are complex. Nevertheless, the analyst needs simple algorithms for predicting the results of ballistic impacts on ammunition components, and each situation should be critically examined before any algorithm is applied. For analysis purposes, in many cases it may be sufficient to predict the ignition point (where burning begins) and the detonation point. The ballistic limit of the motor or shell case can often be used as a rough approximation of the former, at least for composite (ammonium–perchlorate-based) propellants and propellants based on nitrocellulose and nitroglycerin.

Detonation usually occurs as an SDT transition, and several techniques are available for describing SDT events. Hydrocodes, as noted previously, can be used, but they are much too time consuming for most vulnerability analysis. The Jacobs–Roslund (J–R) equation (sometimes called the Ledsham equation) is a simple procedure for estimating SDT thresholds after projectile impact. As originally formulated, this equation states that the critical velocity required to detonate an explosive (or rocket propellant) by a projectile is

$$V = A(1 + B)(1 + CT/D)\sec(\theta)/(D)^{0.5} \quad (4.2)$$

where

V = critical velocity
T = the thickness of any inert covering over the explosive
D = the projectile diameter
θ = the obliquity

A, B, and C are parameters. In principle, A depends only on the shock sensitivity of the explosive and the projectile material (which, along with impact velocity, determines the strength of the shock that is generated); B depends only on nose shape; and C depends only on the cover-plate material. In actual practice, it is found that these parameters are not really independent (B depends on cover-plate thickness, for instance). Liddiard and Roslund (1993) and Boyle et al. (1996) discuss modifications that have been made in this relation to account for these interdependencies. These references also give values for A, B, and C for many cases. If a value for A cannot be found, it can be estimated from the gap test value, as described in Liddiard and Roslund (1993). In any case, if at all possible, the equation should be calibrated under realistic conditions.

Based on the data in Liddiard and Roslund, it appears that one should be cautious in using the J–R equation for projectiles whose length-to-diameter ratio is greater than about 3. The J–R equation also has some problems when the cover plate is more than about 1 projectile diameter in thickness or when the projectile's relative velocity becomes greater than the speed of sound in the explosive.

James proposed another approach to dealing with the initiation of detonation by projectile impact (Cook et al. 1989). This approach is based on a generalization of the critical energy criterion mentioned previously. The criterion is a little more involved than the J–R equation, but it is still simple enough to be implemented in a vulnerability analysis program if desired. Like the J–R equation, the James approach probably runs into trouble when the cover plate is more than 1 projectile diameter in thickness.

These approaches deal with the initiation of detonation by an SDT process. This mechanism is most likely to lead to detonation after projectile impact, but other mechanisms are also possible. Finnegan has shown that projectile impacts that pass through the hollow bore of a rocket motor can lead to detonation by an XDT process (Finnegan et al. 1995). Although Finnegan has developed a general understanding of this process, there is still no simple algorithm to predict it. In these cases, detonation occurs after the projectile has crossed the bore of the rocket motor and under conditions well below those that lead to prompt SDT.

Korba (1974) has suggested a third approach to computing the detonation threshold based on computing the residual projectile velocity and mass as it enters the explosive and then using bare charge data to determine if the residual projectile can initiate the explosive. However, there is little theoretical justification for this approach because it is normally the shock, not the penetration of the projectile, that produces detonation. In addition, for situations where the energetic material is protected by thick covering materials, this technique may significantly underestimate the detonation threshold. For some other cases involving highly sensitive explosives, the technique may overestimate the detonation threshold because the transmitted shock may cause initiation for cases where the projectile cannot reach the explosive. Nonetheless, the Korba approach is easy to implement and may be used if all else fails.

Warheads are somewhat less likely to undergo violent burns or XDTs at lower impact velocities, especially when they are lightly confined (as they may be in a missile warhead or a demolition charge), because they lack a center-bore-like rocket motor. Also, warhead explosives are not quite as easily ignited as are typical rocket propellants and are not as prone to burn for impact velocities just above the ballistic limit of the case. When impacted at lower velocities, lightly confined explosives often simply break apart without producing sustained reaction. However, heavily confined explosives will produce violent, nondetonative explosions at impact velocities well below the SDT threshold (Frey et al. 1979). This is particularly true of composition B, Octol, and LX14, which have been extensively used by the Army. When composition B and Octol are confined by more than about 6 mm of steel, a reasonable algorithm for the occurrence of explosion (not detonation) is to equate the ballistic limit of the case with the explosion point. However, many of the newer and less brittle plastic bonded explosives are quite resistant to this type of initiation, and this algorithm should not be used for them.

Hot embedded fragments can lead to burns in either rocket motors or explosive charges. However, in solid explosives, these events are likely to be quite mild (the gaseous products vent out entrance holes, and significant pressure buildup does not occur). Pyrophoric fragments seem to significantly promote the occurrence of burning, but they do not seem to promote the occurrence of violent explosions, at least not in solid explosives. The violence of the event is controlled not by projectile temperature but by the shock it generates and the degree to which it fractures the charge.

A direct impact by a typical SCJ on an explosive produces a shock pressure much higher than detonation pressure. Detonation normally occurs almost instantaneously unless the failure diameter of the material is extremely large (larger than the motor/warhead diameter or much larger than the jet diameter). When a thick covering material is present, the impact shock dissipates in the cover and does not reach the explosive (or propellant). However, when the jet reaches the explosive, it penetrates the explosive supersonically and generates a strong bow shock in the explosive. This bow shock can produce an SDT (Chick et al. 1989). Consequently, cover plates do not provide anywhere near as much protection against SCJs as they do against ordnance velocity projectiles. For jets, it has been observed that the quantity V^2d, where V is the jet velocity and d is the jet diameter, is approximately constant at the threshold for detonation. This criterion is often used to predict when jets will initiate explosives. However, there are really two critical V^2d constants, one of which applies to impacts on bare explosives or explosives with extremely thin cover plates (impact initiation), and another, larger value that applies to bow-wave initiation. There is evidence that the impact mode of initiation ceases when the failure diameter is about five times the jet diameter, but the bow-wave mechanism persists for even smaller jets. Normally, it is necessary to condition a jet by passing it through a considerable thickness of armor to reduce its velocity to the point that it will not initiate detonation in warhead explosives. Some critical V^2d values may be found in Chick et al. (1989).

The analyst should be aware that there are many situations where an SC penetration will *not* initiate detonation or even deflagration of an HE charge. For example, during World War II, explosive ordnance disposal (EOD) personnel developed, and widely used, small lined cavity and linear charges to render safe explosive ordnance in need of deactivation. Liner materials that were tested successfully included copper, steel, brass, and lead. One linear device was able to cut a TORPEX-loaded depth charge into two halves, exposing the explosive across a diameter but without detonation or deflagration. Based on this reported EOD work, the U.S. Navy, in the late 1980s, sponsored the investigation of a mine neutralizing system using small SCs to attack mines buried in a beach or shallow water. Initial tests were conducted using munitions similar to the M46 dual-purpose improved conventional munition (DPICM) grenade against Mk 82 bombs. Because the grenade was capable of perforating the bomb's casing and the tar-like lining of the cavity, and of penetrating deeply into the explosive filling, detonations were rarely obtained. A significant development program followed from which a 38-mm SC design emerged that was capable of a high probability of initiating the TNT/RDX explosive of the Mk 82 from long standoff and through sand. The cost of production, however, prevented full-scale development and fielding of the system.

In their use condition, most gun propellants have extremely large failure diameters (50–150 mm for the propellants used in tank ammunition), and their resistance to impact-induced detonations is presumably related to this fact. There are, of course, exceptions to this rule, especially when fine-grained propellants are considered; fine-grained M9 propellant behaves like a sensitive explosive. Although most gun propellants are not easily detonated, double-based (nitroglycerin and nitrocellulose) gun propellants are easily ignited, and by design, flame spreads quickly through the bed once ignition has occurred. Nitramine-based (RDX and HMX) gun propellants are more difficult to ignite. When double-based propellants are in steel or brass cartridge cases, or stored in steel packaging, ignition occurs at velocities just above the velocity required to perforate the casing. These low-impact-velocity ignitions are believed to be the result of heat conduction from hot embedded fragments, and the required residual velocity is somewhat greater when the packaging and cartridge case are plastic (Wise et al. 1980). As projectile velocity or size increases, the violence of the reaction, as measured by blast output, increases. This is true for nitramine-based as well as double-based materials. However, experiments show that even 40-mm steel projectiles fired at 1.8 km/s do not cause detonation in common tank-gun propellants (Baker and Stegall 1997). Reactions equivalent to 1 kg of TNT explosive (in terms of blast output) were observed. The blast output of the reactions increases with both projectile diameter and projectile velocity.

When an SCJ impacts solid gun propellant, some of the propellant, directly in the path of the jet, probably reacts almost instantaneously. More is fractured by the impact and reacts rapidly. When the propellants are cooled below their glass transition points, however, their reactions become much more violent, and in some cases detonations occur. Above the glass transition point, Watson et al. (1991) observed, the violence of the reaction, as measured by blast output, increases as an approximately linear function of the kinetic energy that the jet deposits in the propellant. The blast output is often stated in terms of TNT equivalency (i.e., the amount of TNT that would have to be detonated to give the same blast). If a good description of the jet is available, the deposited KE can be calculated from well-known equations; and with a little propellant response data, the TNT equivalency can be determined from the deposited KE relation (Dispersio et al. 1965). At least one gun propellant shows an unusual tuned response that causes an extremely violent reaction, sometimes a detonation, when the jet is conditioned by certain amounts of armor, however, this behavior is not well understood. Armors that disperse jet particles, driving them off axis, may also increase the violence of propellant reactions, but this effect has not been carefully studied.

D. Predicting the Effects of Ballistic Shock

Predicting the effect of ballistic shock involves three distinct steps. First, it is necessary to determine the loading function. When a penetrator or blast wave hits a vehicle, KE deposited in the vehicle structure initiates shock waves that travel through the structure. Loading functions that define these shock waves can be determined experimentally or estimated from KE deposition calculations.

Second, the propagation of the shock waves through the structure must be modeled to determine the loading (amplitude and frequency) experienced by structural elements and shock-sensitive components. This is relatively straightforward for extremely simple structures, however, for actual vehicles, structure seams and joints, gaps (e.g., between the hull and turret), different materials, and other design complexities make this part of the problem quite involved. In addition, shock mounts and other means of shock mitigation must also be considered. Several techniques have been used to model shock propagation. One of the most promising techniques involves the use of transfer functions commonly used in electrical engineering. Another method currently being developed uses finite element modeling of the structure to define the shock response at many points (nodes) throughout the structure. The shock loading at specific component locations is determined from the response at the closest node.

Third, the damage to the structure of sensitive components must be determined. This area is yet another one that is currently far too complex for theoretical analysis. Ballistic-shock methodologies must rely upon experimental data to define failure thresholds. Unfortunately, most data available from the testing community is from acceptance tests, providing sure-safe information only. This means testing provides a frequency-amplitude loading threshold above which shock-induced component failures *could* occur. Thus, current efforts are aimed at gathering vulnerability data. Indications are that component failures to shock are markedly stochastic. Further, manufacturers use different manufacturing techniques to produce components that are functionally equivalent. Some designs will be more prone to failure than others. As a result, many measurements will be required.

The following are examples of typically observed damage because of ballistic shock:

- Damage to optics.
  - Optical wedges displaced.
  - Brackets holding sight mirrors distorted.
  - Lenses cracked.

- Mounting brackets for electronic/hydraulic/optical units damaged.
  - Mounting bolts sheared.
  - Bolt heads ejected.
  - Fasteners damaged.
  - Hydraulic couplings damaged.
  - Defects in castings fracture.
  - Circuit cards displaced.
  - Components on circuit cards fractured and displaced.
- Basic structure distorted.
  - Welds fractured.
  - Bulkheads displaced.
  - Mechanical linkages distorted.
  - Hatch lifting and locking mechanisms damaged.
  - Brake pedal attachment points fractured.
  - Hydraulic line attachment points fractured.
- Any sensor in contact with the excited surface damaged.
  - Position resolvers damaged.
  - Gyroscopes damaged.
  - Fire suppression sensors damaged.
  - Hatch-open/closed sensors damaged.
- Slip rings and hydraulic reservoirs damaged.
- Cooling mechanisms for night sights damaged.
- Display unit elements damaged.
  - Electronic contacts fractured.
  - Power-supply connectors damaged.

E. Wound Ballistics

Wound ballistics involves the study of projectile impact on tissue. For the purposes of weapon or munition anti-personnel effectiveness or individual protection effectiveness, a desired result of wound ballistic studies has often involved quantitative means to characterize the debilitating effects on a soldier to conduct his operational mission, or his incapacitation.

Early efforts in wound ballistics arrived at rules of thumb such as incapacitation results from the application of 58 ft lbs of work. By the early 1950s, wound ballistic studies often made use of a combination of hand-plotted tracks through detailed human anatomy charts, and personnel operational casualty estimates were derived for several standardized post-wounding times and for military tactical combat roles.

For many years, BRL and the Biophysics Division of the Chemical Warfare Laboratory at Edgewood Arsenal, Maryland, conducted cooperative experiments to establish anatomical and physiological damage and to measure penetration into all impacted tissue types. Ordnance or ballistic gelatin blocks were used to simulate projectile retardation characteristics, and human anatomical cross sections from *A Cross Section Anatomy* (Eycleshymer and Schoemaker 1911) were used to assist with the evaluation of these types of impacts.

Over the past 100 years, various metrics, mathematical connections models, and approaches have been developed and used to derive estimates of soldier incapacitation from projectile impacts, including development of the U.S. Army ComputerMan model in 1980 (Clare et al. 1980). [See Sec. 4.III.E.2, Saucier and Kash (1994), and Clare et al. (1980) for a discussion on ComputerMan.] In addition, since 1992, the ARL's Survivability/Lethality Analysis Directorate has been developing the "next-generation ComputerMan" modeling and simulation method, the ORCA modeling system (see Sec. 4.VII.D).

1. Modeling Conventional Ballistic Injuries

While soldiers are subjected to a variety of damage mechanisms on the battlefield, penetrating injuries are the largest cause of weapon-induced casualties. Because of this, most human vulnerability modeling has been directed toward the development of injury and incapacitation criteria for fragments, bullets, and flechettes. The direct effects of blast overpressure, whole-body accelerations, thermal pulses, toxic gases, and chemical exposures have generally been observed both on the battlefield and in operational testing and evaluated to be secondary effects; therefore, they have, in general, received less attention. As a result of this, there are many more tools and data sources available to the analyst for assessing penetrating injury effects.

In general, there are two key ways in which human vulnerability data are employed in the Army analysis community: 1) through high-resolution modeling of discrete events (single shotline analyses in the case of penetrating injuries) at the individual soldier level; and 2) through correlation curves that are based on generalizations of discrete data. The choice of which form to use is normally dictated by the type of analysis being conducted.

Within the DOD, virtually all penetrating-injury weapon-effects estimates make use of the Army's extensive wound ballistics database. The information in this database was generated over roughly a 40-year period with the assistance of the Army Surgeon General's Office, beginning around the early 1940s. In 1952, the Army's wound ballistics program was formally begun, and a considerable body of information on penetrating wounds was produced. The wounds were assessed by medical personnel and their results projected onto theoretical human targets to estimate the effect on human arm and leg functionality. Simultaneously, military tacticians determined the requirements for limb functionality for soldiers performing in several combat tactical roles (assault, defense, reserve, and supply). The tacticians assigned subjective incapacitation levels for various combinations of arm and leg functionality. By combining the observed wound data with the theoretical limb requirement information, expected incapacitation levels could be computed. Finally, by encoding all of this wound and tactical information onto cross-sectional charts of the human anatomy, analysts could trace thousands of theoretical shotlines (initially by hand and eventually using computers) for each of the projectiles evaluated and compute an average incapacitation value for impacts anywhere on the human body or for impacts on selected anatomical regions (e.g., thorax, arm, leg).

Incapacitation assessments were conducted for (theoretical) infantrymen performing typical infantry tasks, and thus computed incapacitations were closely linked to those military roles. The injuries and incapacitation levels thus established for the series of four fragments that were chosen to be representative (as well as

MODELING AND SIMULATION TOOLS AND METHODS 141

Table 4.1 Four-fragment database

Projectile	Mass (g)	Striking velocity (m/s)
0.85-grain steel sphere	0.055	305, 914, 1524
2.1-grain steel cube	0.136	305, 914, 1524
16.0-grain steel cube	1.04	305, 914, 1524
225-grain steel cube	14.58	152, 305, 762

across the range) of those produced by fragmenting artillery munitions became known as the four-fragment database. The four-fragment database projectiles and the velocities at which they were evaluated are shown in Table 4.1 [taken from Kokinakis and Sperrazza (1965)].

These and other similar data have been used in one form or another as the basis for weapons effects calculations over the last approximately 50 years. They have also been used to develop predictive relationships that allow hypothetical projectiles and/or untested impact conditions to be modeled mathematically. These analytic assessments are, of course, limited by the projectile sizes and speeds that are represented by actual experimental data. Because weapon analyses involving projectile weights as high as 1000 grains or more are common, these assessments require extrapolations beyond the 225-grain cube upper limit of the database. However, projectiles larger than 225 grains that impact at velocities faster than 762 m/s often have similar results with respect to estimating soldier performance incapacitation in Army analyses.

2. *High-Resolution Modeling*

As mentioned previously, ComputerMan is a model used to simulate wounding and the resulting biomechanical degradation caused by impacts from munitions. It can be used to establish the wounding power of fragments in weapons assessment studies and to address survivability issues such as the vulnerability of crew members.

ComputerMan was originally developed at the Army's Chemical Systems Laboratory (Clare et al. 1980). Maintenance and further development of the program eventually became the responsibility of the ARL, where it has undergone extensive change since about 1987 (Saucier and Kash 1994). Major changes in the ARL version include the following:

1) The conversion of the program from FORTRAN to C.
2) The addition of methodology for treating multiple wounds so that all injuries, not just the one causing the highest degree of incapacitation, are accounted for.
3) The addition of a graphical interface to make the model easier to use.
4) The addition of civilian injury classification codes to allow the prediction of human survivability for a given set of injuries.

The ComputerMan model consists of a detailed computer representation of the human anatomy that is based on the work of the aforementioned Eycleshymer and Schoemaker (1911). The anatomical data are stored in the model in the form of 113 horizontal cross sections, each of which further subdivided into parallelepiped cells (the size of which varies with body part), resulting in a tissue database of 100,000 cells. Approximately 250 different tissue codes are identified with the level of detail including nerves and blood vessels. In addition, the anatomical description can be articulated so that the man can be seated in a crew compartment and thus be considered in an integrated combat vehicle system assessment.

The model simulates the wound ballistics of fragment penetration into human tissue, including depth of penetration and size of permanent cavity, by drawing on the empirical data and resulting mathematical models from the Army's wound ballistics database. In addition, the experience of combat physicians is built into the model to relate the resulting wound tract description to injury severity and limb dysfunction codes. Performance degradation is determined based upon tactical role and time after injury. Several modes of operation are possible, including single shot, grid shot, or point burst.

ComputerMan is primarily designed to predict degree of incapacitation. It does this by accounting for all injuries in terms of their effect upon the use of the limbs. Because almost all tasks performed by a soldier require the use of the limbs, to the degree that the limbs are impaired, the soldier is necessarily incapacitated. It is important to note, however, that a penetrating injury need not occur in a limb to cause an effect upon that limb. Injuries to the torso, for example, may cause systemic effects to all four limbs. The most common measure of incapacitation in use today is P_{IIH}. P_{IIH} reflects a combination of tactical requirement (the importance of a limb in performing the assumed combat tasks) and limb state (the effect of the computed injury on arm and leg function). As originally computed, P_{IIH} was actually an expected value as it corresponded to the weighted average of five levels of incapacitation (0, 25, 50, 75, and 100 percent), which a series of random impacts were assumed to produce. Thus, since 1990, these P_{IIH}'s (which are not true statistical probabilities) have often been referred to as E_{IIH}'s, for expected level of incapacitation given a hit (Davis 2005). Later refinements allowed computation of actual statistical probabilities of total (100 percent) or other intermediate levels of incapacitation, and these are often termed P_{IIH}'s.

ComputerMan also computes survival probability by relating AIS [American Association for Automotive Medicine 1985 (AIS-85)] codes to survivability via either the Injury Severity Score (ISS) (Baker et al. 1974; Baker and O'Neill 1976) or Anatomic Profile (AP) methods (Copes et al. 1990), as specified by the user. Notable limitations of the ComputerMan model include the following:

1) It addresses penetrating injuries only.
2) It is unable to address unstable projectiles (bullets).
3) It primarily focuses on infantry-related combat roles.
4) It does not address wound effects on other physical and mental abilities.
5) It has outdated injury type and severity characterization, based on use of AIS-85. (This characterization is planned for update with AIS-2005.)

The need for valid, relevant personnel-vulnerability criteria is as great today as it was 50 years ago at the beginning of the Army's wound ballistics program. Although a lot of information has been generated in the interim, understanding of the mechanisms of injury and the human physiological response to that injury is far from complete. Moreover, because weapons have become more sophisticated, the number of ways in which they can produce casualties has increased greatly. Although thermal injuries, inhalation injuries, whole-body acceleration injuries, and blunt trauma injuries have always been casualty producers, weapons that exploit these injury mechanisms have been perfected.

In addition, advances in individual protection against these threats (e.g., gas masks, flame-resistant clothing, and, most notably, body armor capable of defeating bullets, fragments, and flechettes) complicate the analysis. Advances in protection for one area can change the relative importance of the various injury mechanisms. The situation is further complicated if one introduces unconventional weapons (NBC) into the discussion. Even if one excludes the very real influence of nonphysical factors, such as motivation and fear, and limits attention to the physical, mechanical response of the wounded individual, there are still many unresolved technical issues.

Understandably, data and models that allow these factors to be taken into account are crucial for meaningful assessment of vulnerability and survivability of military materiel. Accounting for the crew's contribution to system vulnerability must be done analytically. The current methodology for this is the previously mentioned ORCA model (see a detailed discussion in Sec. 4.VII.D), which was initially created by groups such as the ARL, the OSD, the Office of Live Fire, and others in the JTCG/ME and JTCG/AS CCWG. Now maintained by the ARL, ORCA represents a significant improvement in the situation by providing (for the first time) standard, accredited methodology for the DOD. This methodology, which combines the best available injury algorithms from Army, Navy, and Air Force research and development (R&D) efforts with a new way of interpreting the effects of injuries, is well under way. Improvements in all areas of the methodology, however, are necessary to keep pace with the threat and emerging human response data.

IV. Engineering Models

Engineering models are necessary to relate physical damage to component dysfunction, subsystem degradation, and system capability. (The meaning of the term *engineering model* also depends on the context in which it is used. Practitioners, in the context of the principles that are applied, may legitimately consider finite-element and other continuum-mechanics codes to be engineering models. Similarly, mobility models that analyze system performance over various types of terrain and compute payload limits, fuel consumption, hill-climbing ability, and overturn potential may also be considered engineering models. In the context of V/L analyses, an engineering model refers to a model that relates physical damage in components to the functional performance of those components and related subsystems and, subsequently, to the capability state of the system.) Engineering models must account for all relevant damage mechanisms and all pertinent failure modes. Different component failure modes can have drastically different effects on subsystem operation. For example, a brake control linkage that is severed may not preclude at least limited operation of the vehicle, but a bent and jammed linkage may prevent any controlled vehicle movement. Similarly, different damage mechanisms can precipitate different component damage and, thus, different failure modes. A penetrator may sever a wire bundle, causing open circuits, whereas a fire may melt the insulation, causing short circuits. The point is that modeling component dysfunction resulting from specific physical damage requires an understanding of the construction, function, and operation of a component, as well as the interaction the component has with other components in the same system.

Finally, engineering models may also be based on engineering judgment and even speculation (it is hoped that this is based on analyst experience). Although this may seem extremely crude, it is a necessary part of the V/L analysis process, particularly when new design features are encountered or when damage mechanisms, previously of little significance, suddenly take on increased importance. V/L analyses are deadline-driven processes. It is often necessary to produce V/L estimates before rigorous underlying methodologies can be developed or even before the underlying physical phenomena are well understood. In such cases, there are but few options to using the experience, expertise, and training of V/L analysts and subject-matter experts. In fact, many of the rules of thumb, guidelines, and even models discussed in other sections of this book can be traced back to engineering judgments that were applied out of necessity and later refined to their present maturity.

It is unfortunate that the development of engineering modeling is far from complete. In fact, for modeling ballistic damage, it is still in its infancy, particularly for emerging systems. System designs are becoming increasingly complex, especially with regard to the use of optics, electronics, and advanced materials. On the lethality side, there is less reliance on munitions that overwhelm the target. Many submunitions are marginal in their ability to produce traditional types of damage and rely on damage mechanisms that were once considered of minor importance. For example, large tank-fired penetrators rely on defeating target armor and causing massive internal damage to internal components from penetrator pieces, armor spall, or secondary effects from impacts on stowed fuel and ammunition. These will continue to be important damage mechanisms; but, in addition, ballistic shock, damage to external components from casing fragments, and other effects are taking on major roles for some munitions. To account for these effects, it is necessary to devote more resources to the development of engineering models.

Two examples of engineering models are illustrated in Figs. 2.22 and 2.23.

V. Geometric Representation of Targets

Geometric representation of targets has always been a necessary part of V/L analysis. During the 1940s and 1950s, when testing provided nearly all vulnerability data, there was a need to use test results to estimate component or subsystem vulnerability for conditions not tested. Accordingly, the analyst relied on sketches or engineering drawings to trace the path (shotline) of the projectile through the vehicle to determine components likely to be hit and to forecast the effect of component damage on the system's operation. As analytic techniques became more sophisticated and models began to emerge, however, higher resolution geometric representations were required. Today, extremely high-resolution geometric target information is needed for tracing shotlines through the vehicle, creating on-demand shotlines for secondary penetrators, displaying analysis results, and performing analysis diagnostics. Additionally, highly detailed geometric target descriptions are used to predict target signatures and support other nonballistic

MODELING AND SIMULATION TOOLS AND METHODS   145

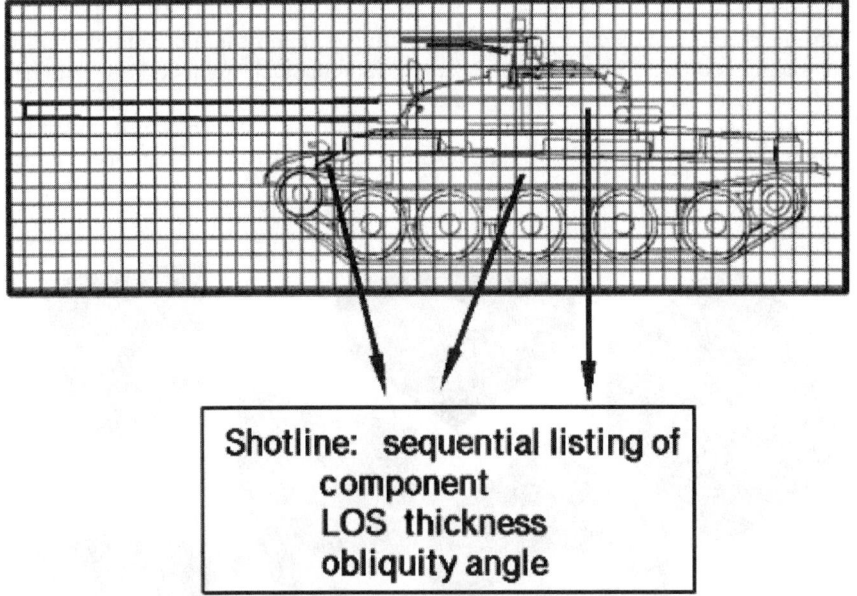

Fig. 4.5   Manual shotlines from engineering drawings.

analyses. This section contains descriptions of the basic processes and models used to create target descriptions, perform raytracing and other geometry processing tasks, and display V/L analysis results.

Typically, V/L analysis computer programs consider large numbers of shot locations on a target from various attack aspects. For each attack azimuth, shot locations are evaluated for a grid completely covering the target. Geometric information is required about each component encountered for each grid cell (or shot location), including data for LOS thickness, entrance and exit obliquity angles, and type of material (as indicated in Fig. 4.5).

Originally these data were generated manually. The manual nature and subjectivity made the process error-prone and time-consuming. In addition, only those attack views for which engineering drawings existed could be evaluated. So, as computer technology emerged, this process was computerized in a twofold manner: a technique was developed to represent/store the target's geometry in the computer; and an algorithm was created to enable the computer to interrogate the stored geometric representation and calculate the shotline data (usually called raytracing).

A.   Computer-Aided Modeling, Analysis, and Design (CAMAD)

One solution to the geometry problem is constructive solid geometry (CSG) (sometimes referred to as combinatorial solid geometry), which uses Boolean combinations of simple solid geometric shapes (or primitives) to model components at any level of detail (Muuss 1991; Deitz and Ozolins 1989). In 1967, the MAGI introduced the CSG technique for representing geometry in a computer and the raytracing geometry interrogation scheme (Mathematical Applications Group Inc. 1967). Figure 4.6 shows renderings of a set of simple primitives available in CSG packages such as BRL-computer-aided design (BRL-CAD), which is discussed in following text. Figure 4.7 shows the results of several basic Boolean operations.

Fig. 4.6 Simple CSG primitives.

MODELING AND SIMULATION TOOLS AND METHODS

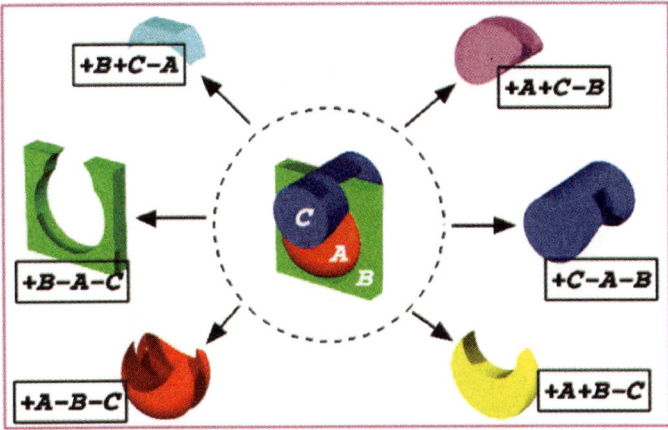

Fig. 4.7 Examples of Boolean operations.

The first geometric modeling system using the CSG technique required three separate files. The first file contained the parameters of the individual primitives, such as the shape, size, location, and orientation of each. The second file defined the regions, which are the Boolean constructs using the primitives from the first file. The third file further identified the regions in terms of the target the region represented. These CSG files constituted what has become known as a target description. The target description was required input to the raytracing program to produce the shotline information.

As its name implies, the raytracing technique mathematically intersects rays, or lines, with the CSG target description. Parallel rays are initiated from a grid plane oriented at the desired attack direction. These rays are intersected with the regions of the description. As a ray encounters a region, the 3-D coordinate locations and surface normals are calculated at the intersections with each of the defining primitives of that region. The primitive-ray intersections are then combined according to the Boolean formula for that region (see Fig. 4.8) to produce the actual intersections for that region. These intersection coordinates are used to calculate thicknesses, which, along with surface normals and other information further identifying the region, constitute the shotline information.

The raytracing program eliminated the shortcomings of the hand-generated shotline process and greatly increased the overall V/L capability. Parallel shotline information could now be quickly generated from any attack direction, including nonzero elevations. Furthermore, divergent rays could be used to simulate the bursting phenomenon. Raytracing continues, to this day, to be a flexible geometry interrogation tool and has been used to simulate many natural phenomena.

The development of the raytracing technique, as well as new capabilities and more detailed analyses, immediately turned the spotlight of the V/L community on to the target description. Soon, the construction and validation of target descriptions became the most crucial and time-consuming element in the V/L process.

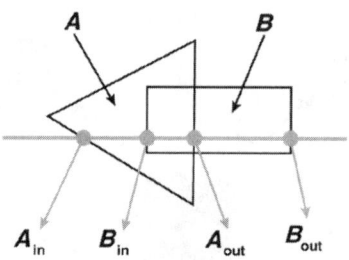
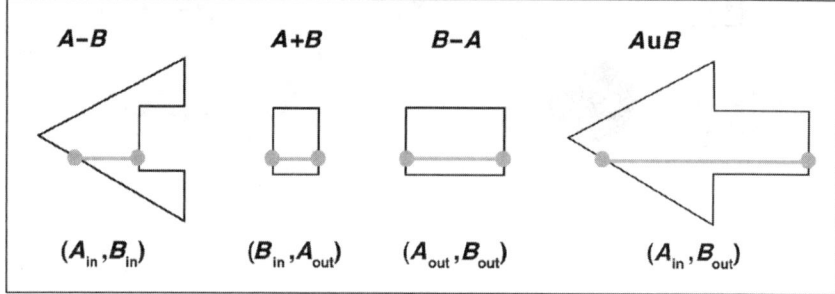

Fig. 4.8 Raytracing a simple region.

The demand for highly detailed, accurate, and timely target descriptions quickly outdistanced the capability to produce them. The main reason was that the construction procedure itself remained a manual process, mired in a mainframe/batch computing environment. With the promise demonstrated by the emerging interactive computer graphics field and the move toward the open environment of the UNIX operating system, a long-term project was initiated in the late 1970s to address this problem. The goal of this project was to create an interactive CSG geometric modeling system within the framework of a more flexible, portable computing environment.

In 1983, the first interactive CSG modeling system was introduced (Deitz and Ozolins 1989). This system greatly reduced the time required to construct and validate CSG target descriptions. The computer hardware necessary to support this interactive modeling system, which included a mini-computer driving a single display device, was rather limited. Soon, however, the graphics workstation entered the market, providing a tremendous boost to the target description preparation process. The workstation provided target modelers with a powerful dedicated computing platform, including excellent interactive graphics capability, and at a relatively low cost.

About the same time, the algorithms of the early raytracing programs were rewritten in the C programming language and organized into a library. This library made the development of new raytracing-based programs much easier. The interactive modeling system and a large volume of associated software were then bundled into what is known as the BRL-CAD package. Available worldwide since 1987, BRL-CAD is the Army's solid geometry, solid modeling system,

Fig. 4.9 Example of a BRL-CAD model rendering of an aggregate combat system.

which includes an interactive geometry editor, raytracing support for rendering and geometric analysis, network distributed framebuffer support, and image and signal-processing tools. The package is now available as an open-source project. For more information on the features and functionality of BRL-CAD, see the four-volume BRL-CAD tutorial series and other pertinent information at http://sourceforge.net/projects/brlcad/ and http://brlcad.org (Butler et al. 2001; Butler and Edwards 2002; Butler et al. 2003; Anderson and Edwards 2004).

Figure 4.9 gives a model view of an aggregate combat system from the front-left aspect angle as rendered by BRL-CAD. For this display, which is from an early target-description file, much of the exterior armor has been made transparent to display the interior components. Figure 4.10 illustrates how raytracing is used to perform penetration modeling and identify affected components along a penetrator's path. In some cases, hundreds or even thousands of replications are run to properly represent penetration and component possibilities.

Such modeling techniques can be extended to using the geometric and the reflective and emissive properties of the target to model its signature properties. This allows a unification of endgame analysis for homing or fuzing systems. It also would be related to modeling the effects of directed-energy weapons upon a target with regard to penetration and interior distribution of the radiation.

Fig. 4.10 Penetration modeling to identify affected components. (Note that component performance is notional and is included for illustrative purposes only.)

B. Target Modeling Practices

There are two starting points for a vulnerability analysis: 1) the description of the target, and 2) the description of the physical processes involved in the damage mechanisms. The first is the subject of this section. In the early days of BRL, targets were often described by using photographs and measuring subareas with a planimeter (or even cutting out pieces of a picture and weighing them) to estimate presented areas. Later, meticulous measurements were taken from drawings and entered into a computerized geometric model. Of course, these approaches were all highly labor intensive. The technology of computer-aided design (CAD) and the related computer-aided manufacturing (CAM) brought two powerful new capabilities: 1) the ability to create systems interactively and rapidly on computer, and 2) the ability to use computer files (the modern blueprint) from the manufacturing-design process as input to the target-description process. Thus, not only has the CAD process made the detailed description of targets much more economical, but it has reduced human error and helped advance the development of highly sophisticated models for other applications (e.g., computing predictive signatures).

1. Analyses and Geometry Requirements

In following sections, we examine the major types of analyses that use BRL-CAD geometry, presenting a brief synopsis of the capabilities, limitations, and requirements of each. Then we discuss how these characteristics translate into demands on the geometric target description required to support that level of analysis. The selection of analysis programs considered is not intended to be all-inclusive, but representative of the major types of analyses most often encountered. These analysis programs are discussed in a somewhat chronological order of development.

Compartment-level analysis. The compartment-level vulnerability methodology was the first approach to V/L assessment of AFVs and, as mentioned earlier, was the driving force behind the development of geometric modeling and raytracing. There were two early compartment-level V/L programs, one for KE warheads and one for SC [sometimes referred to as chemical energy (CE)] warheads. In 1979, these programs were combined into the VAMP program (Nail et al. 1979).

MODELING AND SIMULATION TOOLS AND METHODS 151

Fig. 4.11   Exterior of a compartment-level description.

The compartment-level programs are the least sophisticated of the V/L programs. True to their name, these programs consider the target as a series of compartments contained within an armored shell. Penetration calculations determine the armor defeat criteria. Upon perforation, the residual energy (represented by a hole size) is used to predict damage from a compartment kill and correlation curve. The only other internal components that are considered on an individual basis are those that would contribute to a K-Kill of the vehicle.

There are several important considerations when preparing a compartment-level target description. The armor shell must be described as accurately as possible because the thickness and obliquity angle are crucial in penetration calculations. Other exterior components that would effect penetration (such as the main gun, suspension components, and roadwheels) must be represented. Figure 4.11 is a rendering, taken from Deitz and Ozolins (1989), of the exterior of a compartment-level description.

The complete interior volume within the armor shell must be modeled and divided into compartments. The compartments of interest are the crew, engine, and ammunition. These compartments are defined by modeling air regions and assigning air codes to differentiate the various types. There must not be any space between the inner surface of the armor and the air regions (compartments). There is interior space that must not be included as part of the crew, engine, or ammo compartments. These areas, such as between the hull belly and the floor and between the armor wall and an adjacent fuel tank, are modeled and identified as separate compartments. Figure 4.12, also taken from Deitz and Ozolins (1989), shows these compartment air regions along with the enclosing armor shell.

Any K-Kill component, such as fuel tanks and stowed ammunition, must be represented. In addition, large shielding components, such as engines and transmissions, are usually modeled at an extremely low level of detail. Other interior components need not be modeled. Figure 4.13, also taken from Deitz and Ozolins (1989), depicts the interior detail of a compartment-level target description.

Fig. 4.12 Air regions representing internal compartments.

Component-level analysis – parallel ray. The component-level vulnerability programs represent the next step up from the compartment-level programs in that the target is analyzed at the individual component level. The first such program was known as the VAREA program and was used for such items as lightly armored vehicles and trucks (Joint Technical Coordinating Group for Munitions Effectiveness 1971). The COVART (Computation of Vulnerable Areas and Repair Times) family of programs performs similar tasks for aircraft and some ground vehicles (Joint Technical Coordinating Group for Munitions Effectiveness 1985). These programs compute the vulnerability estimates in terms of vulnerable areas

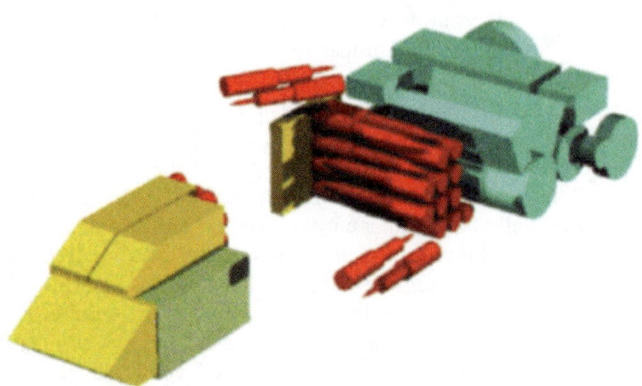

Fig. 4.13 Interior of a compartment-level description.

for specific penetrators, usually fragments and small-caliber direct-fire weapons. This technology requires component-conditional probabilities of dysfunction given a hit for every critical component. The vulnerable area of the target from a specific view is the sum of all the individual cell vulnerable areas. The cell vulnerable areas are determined by calculating a cell P_K and multiplying it by the area of the cell. The view vulnerable area data are presented in a table for the various fragment mass and velocity combinations.

It is no surprise that the geometry requirements for the VAREA programs shifted emphasis to the individual component. However, the exterior shell must also continue to be modeled as accurately as possible. Other exterior components must include any critical components plus any component that could contribute to a K-Kill. Any exterior component providing significant shielding or affecting the penetration capability of the warhead should also be included.

In addition, all critical internal components must be modeled in sufficient detail to support a component kill analysis. The component conditional kill analysis uses presented areas of the component from several aspects, forming ratios of the projected areas of sensitive regions with the total presented area. Thus, the presented area of the model of the component should accurately represent the real component. Any other noncritical internal component that provides effective ballistic shielding should also be modeled. At this level, components such as wiring harnesses and fuel and hydraulic lines are generally also modeled.

Component-level analysis – point burst. The point-burst assessment programs are really just an extension of the VAREA level programs, except the damage resulting from BAD (or spall) is explicitly estimated. Parallel shotlines are used to simulate the main penetrator while divergent shotlines (or spall rays) simulate the spall debris. The spall rays are initiated whenever a burst point is encountered along the path of a main penetrator. A burst point is defined whenever a main penetrator exits an armor component directly into an interior volume. The first point-burst V/L program was the VAST program (Nail 1982), which evaluated KE and SC warheads vs AFVs and light vehicles.

The geometry requirements necessary to support a point-burst V/L analysis are nearly identical to the VAREA requirements as far as the components are concerned. The one major difference is the need to identify burst points. Recall that burst points are located on the inner surface of armor components that are adjacent to interior volume. Hence, to locate the burst points, it is necessary to know when one has entered the interior. As in the compartment analyses, this is accomplished by representing all interior volume as air regions and then identifying those where spall rays should be initiated. Note that there cannot be any undefined volume between the exterior shell and the air regions, or the burst points could be missed. Figures 4.14–4.17, all taken from Deitz and Ozolins (1989), compare the details of several subsystems of compartment-level and component-level target descriptions.

2. Modeling Philosophy

Probably one of the most difficult parts of geometric modeling is determining where and how to begin the process. As with most procedures, the first step should consist of a review, analysis, and planning phase. As discussed in Butler et al.

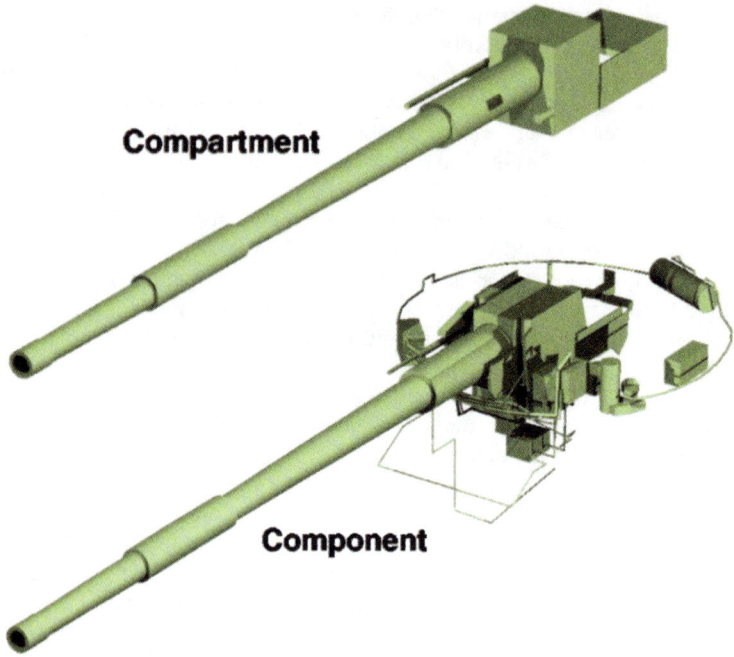

Fig. 4.14 Compartment-level and component-level gun systems.

(2003) the modeler should think about the task that lies ahead, considering such questions as the following: 1) What is the final expected product? 2) To what use will the model be put? 3) How much descriptive information is available and in what form? 4) What prior work may exist? 5) What existing library components can be used?

Next, one should devise a general plan on how to accomplish the actual construction of the model, including at least rough time estimates. This general plan should include all major subsystems of the target that must be modeled. For most military targets, the most basic elements of the general plan would consist of modeling; 1) the exterior shell, 2) the internal air, and 3) all of the remaining components.

Note that the goal is to get a correct exterior shell (with internal air if necessary) *before* any components are added. Thus, if a tank were being modeled, the first two steps might become 1) model the hull shell, and 2) model the hull air; or perhaps 1) model the turret shell, and 2) model the turret air.

Not surprisingly, much of the modeling time and effort typically goes into modeling the components themselves. The BRL-CAD Tutorial Series (especially Volumes II and III) discusses this process in detail (Butler et al. 2001; Butler et al. 2003). As always, initially, time should be spent analyzing and planning: first analyzing the component to be modeled and deciding on the detail required, then

MODELING AND SIMULATION TOOLS AND METHODS

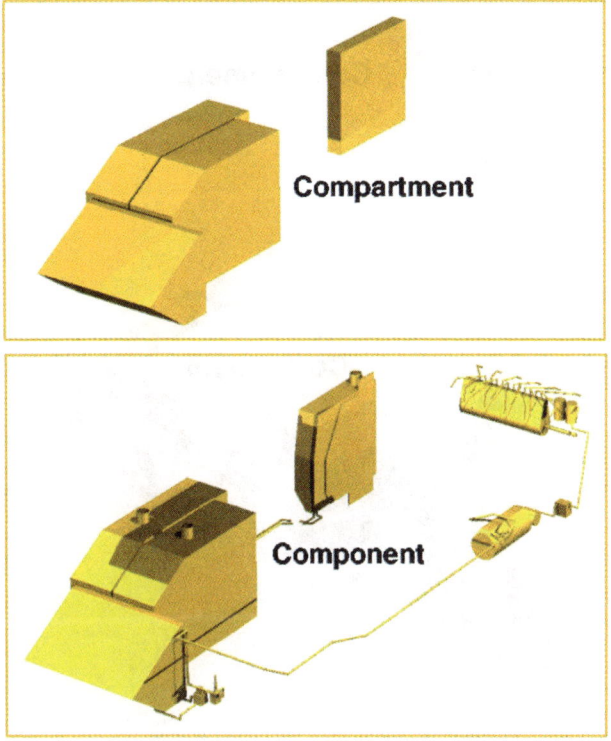

Fig. 4.15 Compartment-level and component-level fuel systems.

formulating how to represent the component, including what solids to use and how to combine them into regions. Then a typical approach is to create the solids with the desired shape, size, and orientation. To take advantage of any symmetry, these solids are often created at the origin and positioned later. Next, the solids are combined into the regions, and the regions are grouped into a combination representing the completed component. At this point, any interferences within the component should be identified and corrected, and fixed pictures should be created for visual verification. Note that both of these actions are performed using raytracing. Finally, after components are located within the target description, interferences between components are identified and corrected. In addition, the simulated description weight can be checked against known actual weights to identify possible modeling errors.

C. Further Uses of Geometry and Graphics Tools

In addition to such classical applications as shotlining and signature prediction, there are many other ways in which geometric-modeling and graphics tools (such

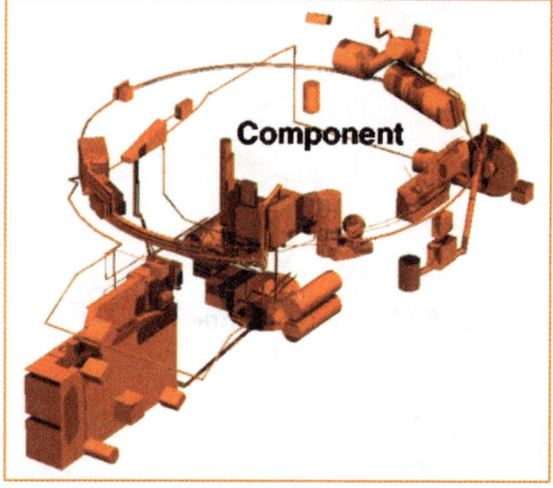

Fig. 4.16   Compartment-level and component-level electrical systems.

as BRL-CAD) are used in vulnerability analysis. To illustrate these tools, and some representative ways in which they are used, we present a variety of vulnerability analysis tasks. Although all the examples presented here concern conventional ballistic threats against armored ground vehicles, the analytical problems they raise and the techniques used to address them are similar to those found in any other area of vulnerability analysis.

The examples are grouped into two classes. First, we discuss various analysis tasks that need to query geometric models. Second, we consider tasks that rely more heavily on useful graphical ways of presenting data to support the analysis process.

1.   Geometry Interrogation

The developer of the M992, an ammunition support vehicle for field artillery, was concerned about the likelihood that various ballistic threats might cause a fire in the vehicle when they encountered either stowed ammunition or combustible

MODELING AND SIMULATION TOOLS AND METHODS 157

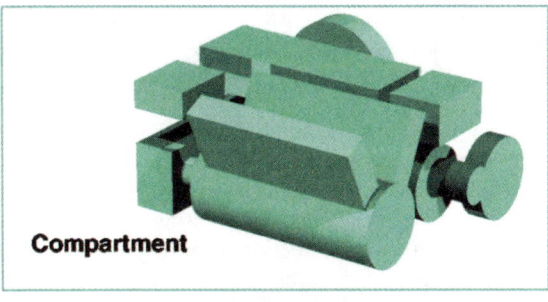

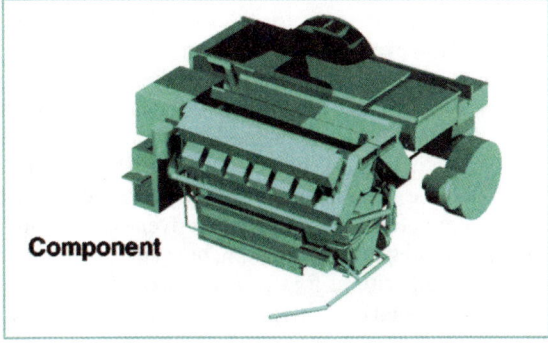

Fig. 4.17 Compartment-level and component-level power trains.

fluids (fuel or hydraulics). Accordingly, to estimate the likelihood of such an encounter, the following tasks were conducted: 1) a target description of the vehicle was shotlined from various directions; 2) the shotlines were categorized in each view according to the nature and sequence of the components encountered; and 3) the categories of shotlines were then analyzed statistically.

The developer needed an estimate of the discrete probability function for the event that the vehicle was hit by a threat. It was agreed the space of possible outcomes of such an event would be modeled as follows:

Wet hit The shotline passes through any component of the fuel or hydraulic systems and through the internal volume of either the crew or engine compartment. In the case of the crew compartment, the shotline does not first hit a massive component of the power pack.

Ammunition The shotline passes through any projectile, propellant, or fuze, and through the internal volume of either the crew or engine compartment. In the case of the crew compartment, the shotline does not first hit a massive component of the power pack.

Blocked The shotline encounters either the engine, transmission, or transfer, and otherwise would have qualified as a wet hit or ammunition.

None The shotline does not qualify as a wet hit, ammunition, or blocked.

The shotline categorization method was used in which a ray structure describes one segment of a ray though some geometry. The structure that was created records the following information:
1) Identification of the object intersected.
2) Shotline serial number identifying the ray containing this segment.
3) Model coordinates of the segment origin.
4) Vector difference between origin and terminus (in model coordinates).
5) Azimuth and elevation of the outward-pointing entry normal.
6) Azimuth and elevation of the direction of principal curvature.
7) Principal and secondary curvatures.

As another example, the developer of a fly-over shoot-down missile needed view-averaged effectiveness estimates for the weapon that took into account missile delivery accuracy. However, the missile's top-attack trajectory caused a complication; its penetrator shotlines were not in the direction of missile flight. Even the position of the missile (i.e., its distance across the target) at warhead detonation was variable, a function of the sensor fuzing logic, among other parameters. The BRL Anti-armor Sensor Simulation (BRASS) model was thus created to incorporate the sensor logic and the complexity of the missile/target geometry.

The weapon developer needed a source of effectiveness estimates that addressed all the complexity of the end game. In addition, he wanted a simulation in which he could test developmental algorithms for the fuze sensor. Existing lethality estimates for the missile had been created by applying the conventional grid-plane methodology to the fly-over shoot-down scenario, with delivery accuracy being accounted for via distributions for the missile's cross-range and altitudinal position. However, using weighted averages of results from several missile altitudes was an inadequate approach, not only because such interpolation is unjustified on methodological grounds, but also because the third component of missile position (i.e., the down-range component) could not be predicted once the target had been replaced by a collection of horizontal planes floating above it (as shown in Fig. 4.18).

BRASS accepts the following as input: the detailed target geometry; a simplified version of the missile geometry, including position and orientation of all sensors and warheads; and the attack geometry, including the azimuth and elevation of the missile's velocity vector and the yaw, pitch, and roll of the missile with respect to its velocity vector.

BRASS also has a plug-in software module to simulate the fire-control logic by reading sensor output and deciding whether and when to fuze the warhead(s). The primary output of BRASS is a warhead-function surface (WFS), the surface that for any cross-range and altitudinal position of the missile passes through the corresponding down-range location of the missile at the time of fuze functioning.

BRASS begins by combining the geometry of the sensor with that of the attack scenario to arrive at the effective direction sigma of sensor search, expressed in (target) world coordinates. The bounding box of the target geometry is then projected parallel to sigma to obtain the sensor prism (Fig. 4.19), the portion of space through which the sensor must pass to see the target. BRASS then produces a set of parallel missile flight paths through the sensor prism. It proceeds along each flight path, periodically firing rays in direction sigma to simulate sensor pulses. The range to target is measured along each pulse ray and then fed into the fire-control logic to determine the location along the missile flight path at which the warheads are fuzed. Among the results that can be extracted from a run of BRASS are those shown in Figs. 4.20–4.23.

Fig. 4.18 Horizontal planes. (All quantitative results are notional and are included for illustrative purposes only.)

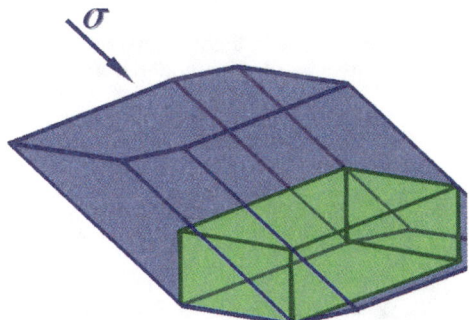

Fig. 4.19 Sensor prism.

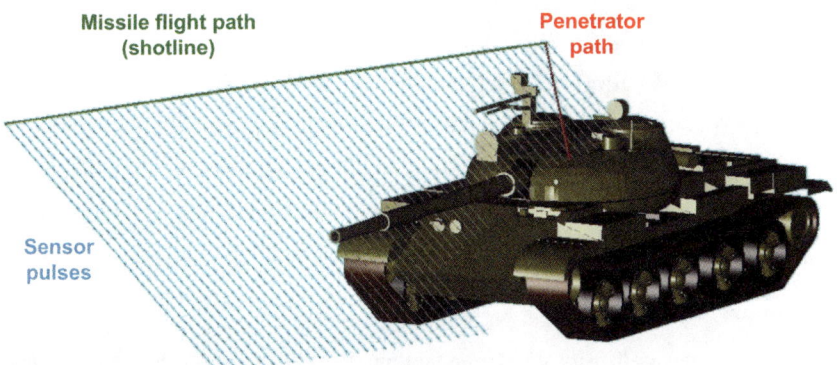

Fig. 4.20 Categorization of missile flight paths according to the behavior of the sensor logic.

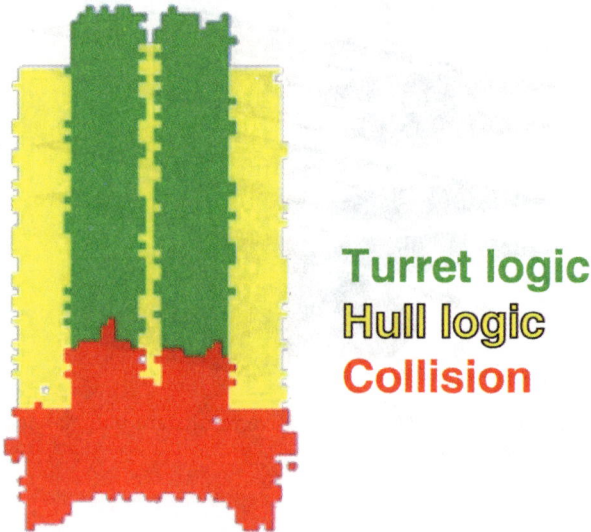

Fig. 4.21 Categorization of missile flight paths according to the behavior of the sensor logic. (All quantitative results are notional and are included for illustrative purposes only.)

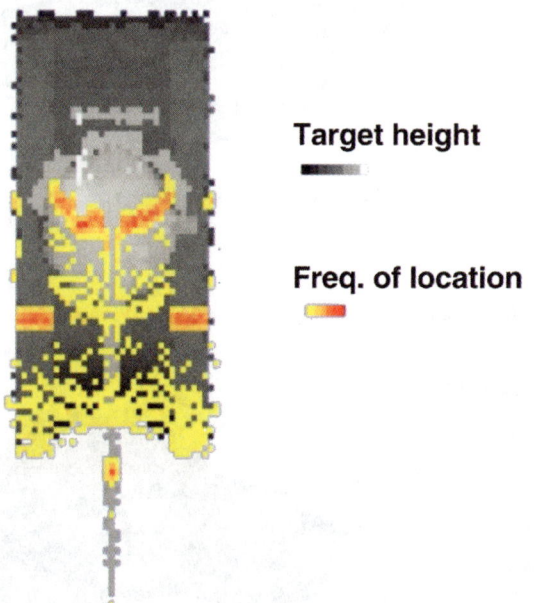

Fig. 4.22 Frequency density of warhead position (in the plane) at the time of detonation. (All quantitative results are notional and are included for illustrative purposes only.)

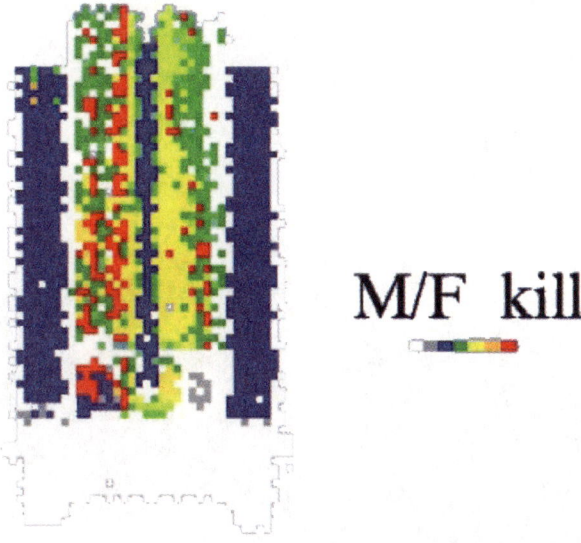

Fig. 4.23 Cell-by-cell weapon lethality data by combining BRASS with a lethality model. (All quantitative results are notional and are included for illustrative purposes only.)

2. *Scientific Visualization and Presentation Graphics*

Probability of kill given center of impact $P_{K|COI}$. Of the techniques for presenting and studying data related to or generated by vulnerability analyses, many of the most effective (and thus widespread) involve, at the core, production of images. They present information corresponding to spatial locations by projecting it onto a viewing plane. One obvious example is the photo-realistic rendered view of a scene (see Fig. 4.24). Others are silhouettes of cell-by-cell data, shown in Figs. 4.25 and 4.26, for example.

Such cell-by-cell data often represent point-estimates of some random quantity, the conditional probability $P_{E|Ci}$ of an event E, given that one is "in the ith cell." When E is a favorable outcome, such as the degrading of a target vehicle to some specified extent, these images can be useful to answer the question, "Where on the target is the best location for me to hit?" But the likelihood P_{Ci} of hitting the ith cell can only be known, in general, statistically.

Because it is impossible to guarantee that any weapon system will hit its target exactly where intended, one often needs to ask a different question: "Given the delivery accuracy of my weapon, where on the target is the best location for me to aim?" A scalar that is often used to summarize a data set is the view average, or sum of the product $P_{E|Ci} \times P_{Ci}$ over all cells, where P_{Ci} is computed as the integral over the ith cell of a probability density function f. The aimpoint question can be rephrased in terms of the sensitivity of view average to changes in the mean of f, and it is this form of the question that we will address.

Fig. 4.24 Computer rendering of a scene.

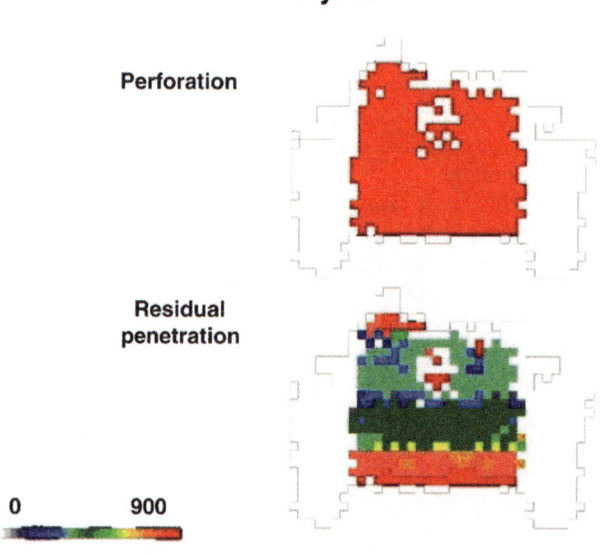

Fig. 4.25 The results of armor penetration models. (All quantitative results are notional and are included for illustrative purposes only.)

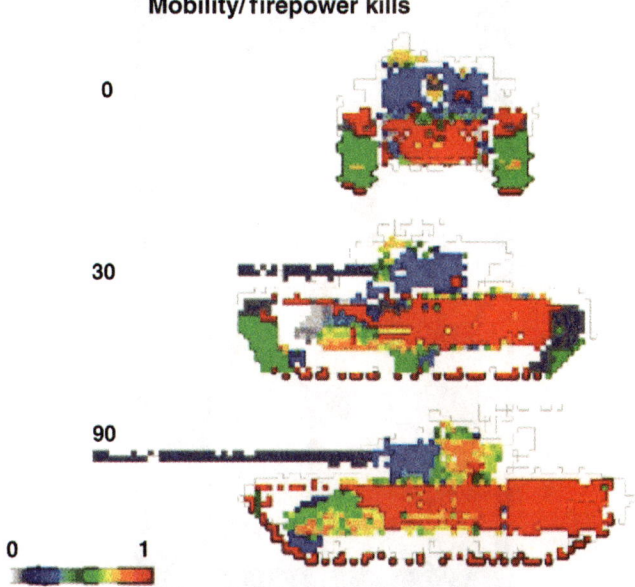

Fig. 4.26 Estimated loss of vehicle function. (All quantitative results are notional and are included for illustrative purposes only.)

We filter cell-by-cell data to convert it from $P_{E|C}$ data shown in Figs. 4.27 and 4.28 to what we call $P_{E|COI}$, or the probability of E given a center of impacts on the ith cell, which is given by

$$P_{E|COI} = \sum P_{E|Ck} \times P_{Ck,i} \qquad (4.3)$$

where $P_{Ck,i}$ is the probability of hitting cell k, when aiming at cell i.

More precisely, given the standard deviation (sx, sy) of a bivariate normal distribution f, we successively step the mean (mx, my) of f through each cell, replacing the cell-by-cell value $P_{E|Ci}$ in the ith cell by the view average obtained by setting (mx, my) to the center of the ith cell.

Pix weight. As seen in the previous section, one must sometimes take weapon-system delivery accuracy into consideration when evaluating raw analog image data. There is another example that can also be illustrated in Fig. 4.29.

Note the large portion of the hull (between the fourth and fifth roadwheels) that is assigned the value 1.0, here represented in red. One might expect that this portion of the vehicle, having a large area and the highest possible value, would pull the entire view average up to a high value. But this may be something of an optical illusion. The human visual system has no trouble picking out that large patch of red as the dominant feature of the image, but it is less effective at appreciating the distance of the patch from any particular aimpoint, like the centroid of the image.

Because the view average depends on both the individual point values and the likelihood of hitting each point, it can be extremely sensitive to the mean and standard deviation of the hit distribution function. Thus, a way to correct images

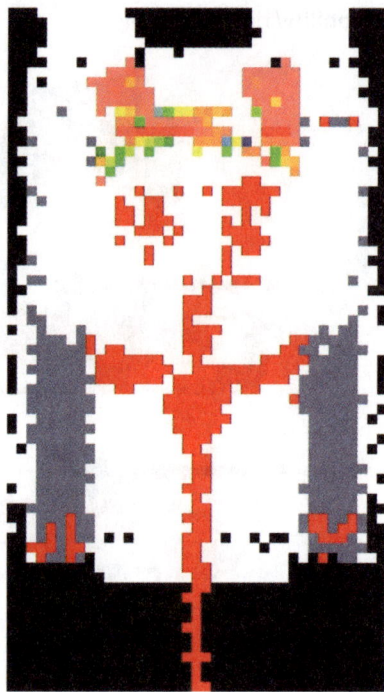

Fig. 4.27 $P_{(E|C)}$. **(All quantitative results are notional and are included for illustrative purposes only.)**

for delivery accuracy was sought, so that an area of an image would draw attention only if its value and the probability of hitting it were both high.

An approach that takes advantage of the visual system's ability to discriminate size and does not rely on precise perception of position was developed. The result was a tool called pixweight, which, given the mean and standard deviation of a bivariate Gaussian distribution f, reads a stream of pix data and geometrically distorts it according to f, so that the probability of hitting any arbitrary region R in the input image is proportional to the area of the region f(R) in the output image. As can be seen, a large area in the output image may correspond to a small portion of the input image near the aimpoint or a larger portion further away. Note in particular in Fig. 4.30 that the red portion appears fairly small in this fish-eye image, which indicates its relatively minor contribution to the view average.

Polar-fb. A final example technique for presenting vulnerability data are a means of comparing phenomena that vary as a function of the angle of view or angle of attack. Such a polar graph is sometimes called a butterfly plot, illustrated in Fig. 4.31.

A typical application for these polar plots is the investigation of the sensitivity of view average to attack azimuth, the angle in the horizontal plane. One often needs to study this sensitivity as a function of several other parameters.

BRL-CAD includes a tool called polar-fb to produce such plots. As an example of its use, suppose one wished to study the view-averaged loss of a vehicle's

MODELING AND SIMULATION TOOLS AND METHODS 165

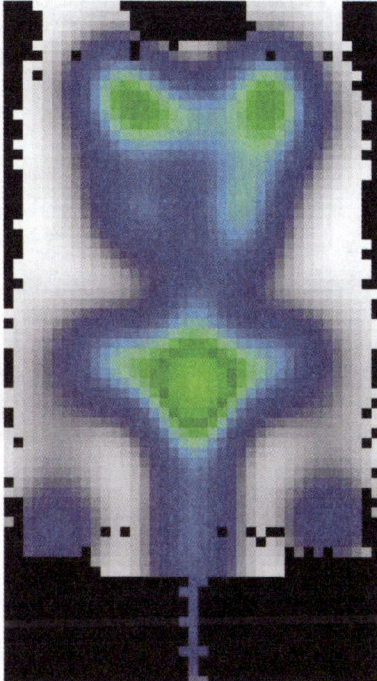

Fig. 4.28 $P_{(E|COI)}$. (All quantitative results are notional and are included for illustrative purposes only.)

mobility/firepower capabilities given the following: two different levels of armor protection on the vehicle; attack by two different munitions; and varying attack azimuth. Making the standard simplifying assumption of left-right symmetry in the view averages, we used polar-fb to display all the data in one combined polar plot, shown in Fig. 4.32.

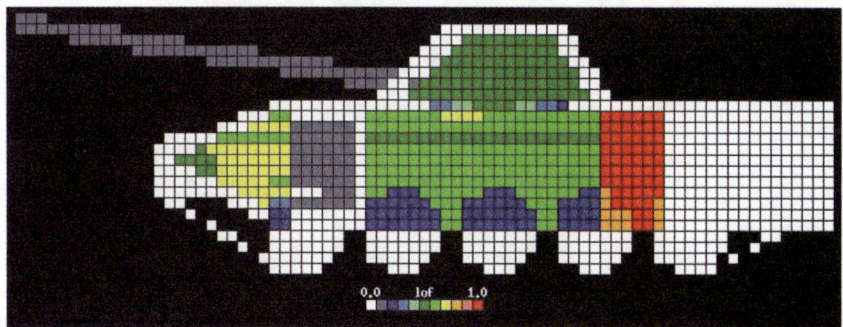

Fig. 4.29 Cell-by-cell view of lethality data. (All quantitative results are notional and are included for illustrative purposes only.)

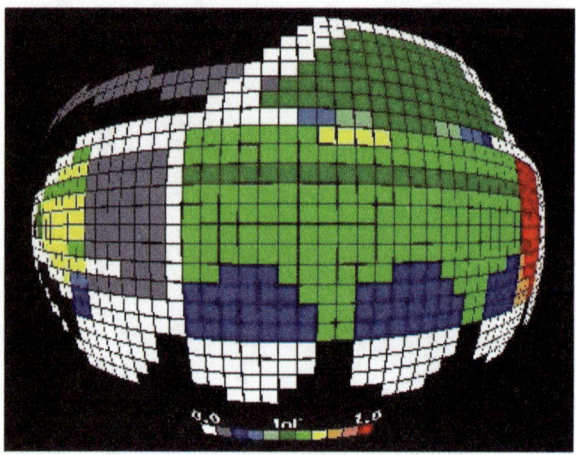

Fig. 4.30 A weighted view of the same data. (All quantitative results are notional and are included for illustrative purposes only.)

View averages were obtained for attack azimuths of 0 deg, 30 degs, ..., 180 degs, and a smooth curve was interpolated through these values with spline. The polar-fb utility can also simply plot the step functions. Then the results for the lower level of armor protection were plotted in shades of green and the higher-level protection in red, with one munition on the left side of the

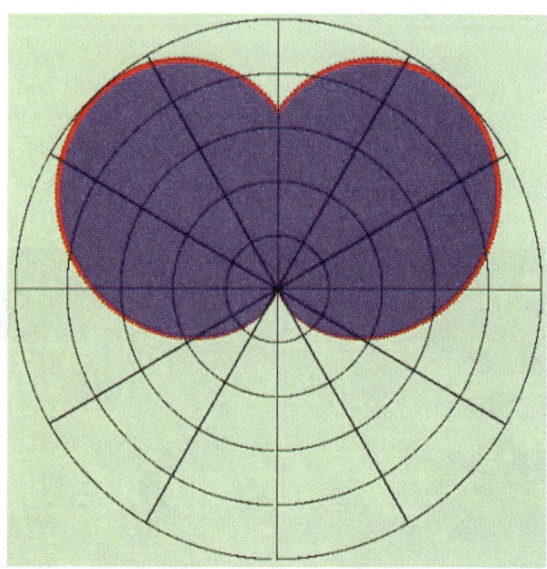

Fig. 4.31 Butterfly plot.

MODELING AND SIMULATION TOOLS AND METHODS 167

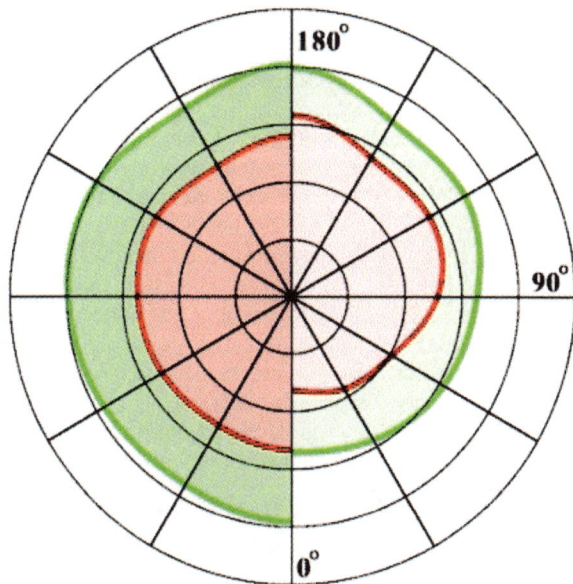

Fig. 4.32 Polar plot.

vehicle and the other on the right. Another application for polar-fb is to create delivery-error ellipses for inclusion in analog target images, as indicated in Fig. 4.33. These curves illustrate the standard deviation (sx, sy) mentioned in the previous discussion.

VI. System Models
A. Introduction to System Performance Models

Ultimately, the goal of combat system vulnerability analysis is to help determine the performance of systems on the battlefield. In the absence of mission

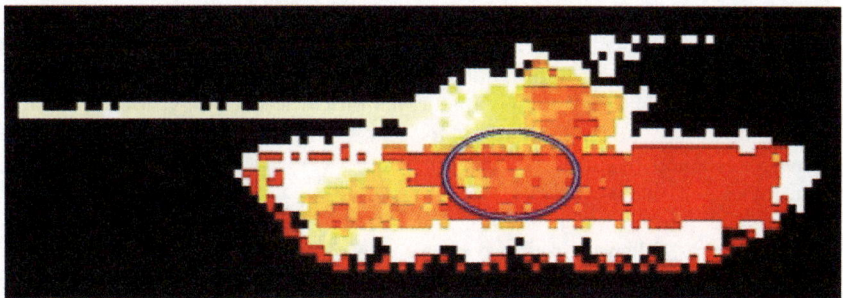

Fig. 4.33 Delivery error ellipsis. (All quantitative results are notional and are included for illustrative purposes only.)

considerations, however, to extend vulnerability analysis past level 3 of the MMF is, at best, difficult and, at worst, misleading. As discussed, a "night sight" on a tank may be critical for nighttime operations but totally useless in daytime operations. Thus, assigning a single number describing its combat utility over *all* types of missions is an interesting exercise, but the resulting number may have no practical application. After all, knowing that on any toss of a coin, the probability of achieving "heads" is 0.5 in no way helps someone trying to predict the outcome of a single coin toss. The outcome is always heads or tails, never half of one and half of the other. That is, the probability never accurately reflects the outcome of a particular event. There are also many unknown factors contributing to combat performance in a specific instance. Different engines drained of lubrication will run for different lengths of time before seizing; some soldiers with debilitating injuries will continue to fight much longer than their compatriots with similar injuries; and Murphy's Law will always be in effect.

System performance models use physics and engineering to simulate threat encounters, predict physical damage, determine resulting loss of engineering capability, and, in some cases, predict a probability of remaining combat utility. Statistics based on results of simulations from level 3 of the MMF are often used as input to force-level simulations designed to predict outcomes of scenarios on the battlefield. Where physics and engineering do not provide fully developed descriptions of phenomena, random draws are often used to provide a specific instantiation of an event taken from a population of possibilities constrained by what *is* known. The process of determining the extent to which a model performs as expected and produces credible results is quite difficult and must frequently be done in the absence of sufficient real data. The next several sections provide principles of performance modeling. However, as noted in the introduction to Chapter 4, new models and simulations continue to be developed, and, consequently, modeling processes continue to evolve. Thus, by the time this text is published, newer models and processes will likely exist.

B. Legacy Methodologies for Armored Vehicles

The earliest attempts at ground systems V/L assessments were concerned with tanks being attacked by direct-fire weapons and relied heavily on subjective judgment. For heavily armored vehicles, the major concern was perforation of the armor; hence, the only geometric information needed was armor thickness and obliquity angle. The penetration capability of the attacking munition was matched against the armor. If perforation occurred, then estimates were made concerning damage and residual system combat capability. Methodology soon began to emerge, however, and by the late 1950s, computer programs existed to estimate damage sustained by armored and unarmored vehicles attacked by direct-fire munitions.

From 1948, when a formal program to study the vulnerability of armored vehicles was established, until the emergence of models in the 1950s, the only method available for evaluating the vulnerability of armored vehicles, or the lethality of anti-armor munitions, was ballistic testing. If one wished to know the vulnerability of a certain vehicle to certain weapons, it was necessary to fire the weapons against that particular vehicle. At best, the analyst could extend test results to conditions not tested, but few predictive techniques were available to apply test results to different threat–target combinations.

At best, this method permitted only approximate comparisons among weapons or vehicles under a narrow set of attack conditions; it was generally impractical to obtain enough data to represent the inherent variability of ballistic phenomena.

The principal reason for evaluating the vulnerability of armored vehicles was to assist the development of new vehicles and new anti-armor munitions by establishing reasonable performance goals and evaluating competing prototypical designs before final selection. On many occasions, in developing new systems, it would have been helpful to have been able to evaluate postulated systems without actually building them and to be able to evaluate foreign systems that were not available for testing.

The experimental approach was too limited and slow to adequately support the development of new vehicles and new munitions. It was thus clear that an analytical evaluation method was needed that would allow more timely response to the developers' needs and permit evaluation of vehicles and munitions that were not available for testing.

C. Early Efforts in Vulnerability Modeling

The first rudimentary attempt at an analytical approach to estimating the vulnerability of an armored combat vehicle or the lethality of a munition provided only estimates of the probability of perforation of the armor given a random hit on a target vehicle under some specified attack conditions. The armor of the vehicle was characterized by a statistical distribution showing the probability that a random hit would encounter a given equivalent thickness or less. This equivalent thickness was defined as the thickness of armor that would offer the same resistance to penetration at normal impact that the actual thickness would offer at the actual impact obliquity. If the penetrating capability of a particular projectile was known or could be estimated, the probability of perforating the armor of a vehicle of interest could be determined simply by referring to the distribution curve. Because armor perforation is the threshold condition for inflicting interior damage, this relatively crude model provided useful guidance to munition designers by establishing minimum penetration performance goals.

This approach provided no explicit information on the extent of damage that a projectile might inflict on a vehicle's components, but it seemed reasonable to assume that any perforation of the armor would cause significant damage (i.e., enough damage to have some effect on the fighting capability of the vehicle). Thus, an armor distribution curve prepared for a particular vehicle might be considered an elementary vulnerability model for that vehicle. This approach was used, on some occasions, to derive quantitative comparisons among projectiles and among armored vehicles. However, the usefulness of the approach was limited because it could not discriminate among projectiles that had similar penetrating capability but apparently were different in terms of the ability to cause damage if they reached the interior of a vehicle.

It was thought that if the approach could be extended to account for differences among projectiles in the potential for inflicting damage or differences among vehicles in resistance to damage, an adequate vulnerability model would result.

With this perspective, analysts approached the task of deriving a better vulnerability model for armored vehicles.

D. The Compartment Vulnerability Model

As discussed in Sec. 4.V.B.2, a compartment-level model is a low-detail model usually used for AFVs in the design stage or for vehicles where little detailed information on internal components is available. The modeled air space inside the target (internal volume) is partitioned to represent major compartments of the AFV (i.e., the crew, engine, ammunition, and passenger compartments). In addition, a few specific components (including armor, fuel, ammunition, gun tube, and suspension) are modeled as distinct items.

Empirical data from actual test firings are used to relate quantifiable damage parameters for the components to remaining utility values for the target. For "compartment" components, these utility values take into account damage to internal physical components in the compartments that were not explicitly modeled.

During the 1950s, the Army conducted a number of programs to develop new anti-armor munitions, and various proposed designs were evaluated experimentally. The prototype projectiles were fired into actual tanks, and the damage was evaluated in terms of fractional loss of mobility and firepower function. As this cumbersome, but necessary, process of evaluation continued, a considerable amount of data accumulated, and analysts began to notice trends that offered a basis for predicting the effects of a ballistic attack on tanks. These trends eventually formed the basis for the analytical model known as the compartment model.

The compartment method dates back to around 1960 when Howard Ege of BRL developed four separate computer programs, using the FORAST programming language, for determining the vulnerability of tanks and APCs to KE projectiles and SC warheads. This model made it possible for the first time to estimate the vulnerability of armored vehicles or the lethality of projectiles entirely by analytical methods. In 1979, VAMP, a computer simulation that combined the four separate vulnerability computer programs into one, was developed by the Computer Sciences Corporation (Nail et al. 1979).

With the compartment model, the path of the penetrator is traced through the target to determine if external components (such as running gear) are damaged. If the penetrator enters a compartment, the model estimates the probability of a functional kill (M-Kill, F-Kill, or K-Kill) based on the parameters of the penetration (e.g., the size of the hole created in the vehicle structure by the penetrator). This correlation function accounts for damage by spall from the penetration as well as from the residual penetrator and from secondary effects such as blast and shock. Finally, the penetrator is followed through the compartment to see if ammunition or fuel is struck to determine the probability of a K-Kill from a fire or explosion.

1. The Experimental Database

The compartment model is entirely empirical; it is based on data obtained from several experimental programs, each fairly limited in scope and intended to answer some specific question. All of these test results, taken together, constitute an orderly data set in which the strength of the attack, attack direction, hit location, and the armor protection of the target are all varied in a reasonably systematic fashion.

The Third Tripartite Conference on Tank Armament held in 1958 in the U.K. (see Chapter 1) is worthy of mention again here, not only as a significant source of data, but also because it typifies the dependence upon the experimental evaluation method that existed at that time and the need for analytical methods. In the mid-1950s, when the United States, United Kingdom, and Canada were independently developing ATGMs, there was little basis for determining the warhead weight needed to achieve acceptable terminal effectiveness against a tank. The CARDE trials were thus conducted jointly by the three nations to answer this question. SC warheads of 5-, 6-, 7-, and 8-inch diameters were fired against tanks and full-scale simulated parts of tanks. This program involved about 400 shots against targets, including the U.S. M-47 and M-48 tanks.

Soon after completion of the CARDE trials, an experimental program of similar magnitude was conducted at BRL to investigate the terminal effectiveness of SC bomblets against tanks and armored personnel carriers. Warheads of 1-, 2-, 3-, and 4-inch diameters were fired in vertical orientations against the target vehicles. The results of these tests are unpublished but constitute a significant part of the compartment model database. [Benjamin (1960) and Benjamin and Gholston (1960) have summarized the results of various smaller test programs conducted before 1960.] The damage to the target resulting from each shot was recorded in great detail, including listings of the damaged vehicle components, an assessment of the extent of the damage to each component, the coordinate location of each damaged component, and an indication of which damage mechanism caused the damage.

2. Development of the Model

The fundamental goal of the model development was to derive a method to postulate an arbitrary impact point on a target and estimate the damage that would be inflicted by a specified projectile. By repeating the process for a large number of impact points and attack directions, it would be possible to arrive at an average measure of vulnerability. Although this idea seems obvious from our modern perspective, it should be noted that it was a departure from earlier thinking where the probability of perforation was calculated from a frequency distribution and specific impact points were not explicitly represented.

When an armored vehicle experiences a ballistic attack, several mechanisms inflict damage. The experimental firings conducted in the 1950s suggested that, in the case of tanks and typical anti-tank weapons, the principal damage mechanisms are the penetrator and the debris produced by armor perforation. Some blast and shock damage was also observed, but these mechanisms seemed relatively weak compared with the penetrator and fragments. Thus, it seemed that penetrators and fragments were the principal damage mechanisms to account for in developing an analytical vulnerability method for armored vehicles.

At the time the model was being developed, vulnerability analysis concentrated on relating combat system damage directly to combat utility. Although later work has shown that there is no mathematically viable direct relationship between damage and utility, the earliest efforts in formal vulnerability analysis nevertheless attempted to make this connection directly. The SDAL was formulated to allow analysts to infer combat utility from damage by assigning a number between 0 and 1 to represent a loss of combat utility because of loss of the component or subsystem (see Chapter 3).

The first step in the development of the model was to verify that the SDAL values for each of the components damaged on a particular shot could be combined mathematically to give a valid measure of the overall DCU for the vehicle. To this end, an analytical procedure that used the component LoF data for each test shot and the SDAL to calculate values of DCU for each shot was devised. The calculated values were then compared with the judgmental values estimated by the assessors for each shot. (The assessors' estimates were considered the standard against which calculated values should be compared.) With this procedure, damage data recorded for each shot was grouped so that singly and multiply vulnerable components were properly treated and the combined DCU for the shot was calculated. Table 4.2 shows a sample of the analysts' calculation sheet.

In Table 4.2, the component damage recorded by the assessors is grouped according to related component function. Adjacent to each damaged component,

Table 4.2 Sample of data in an analyst's $P_{K|H}$ calculation sheet

For interior damage to crew compartment – Perforation by a HEAT round: 1.92-in. profile hole diameter; 23.2-in. residual penetration – Estimated kill: mobility – 67%; firepower – 73%

Component	Jet, frags, blast	% Component loss	% Loss to tank			Location		Repair time, h
			M	F	K	r	Q	
90-mm ammo	F	5				62	12	2 1/2
90-mm ammo ready round	F	100				71	72	
90-mm ammo ready round	F	100				74	26	
90-mm ammo ready round	F	100				70	20	
90-mm ammo ready round	F	100	0	6		71	25	
90-mm ammo ready round	F	0				70	35	
90-mm ammo ready round	F	0				70	32	
90-mm ammo ready round	F	0				70	40	
90-mm ammo ready round	F	0				70	41	
Gunner's periscope	B	100	0	0		20	71	1/2
Loader's intercom	F	100	0	0		79	6	1
Loader's intercom cable	F	100				88	8	1
Radio	F	100	5	5		21	18	2
Radio power unit	F	5	0	0		41	10	
Loader-upper	F					68	13	
Loader-middle	F	100				60	18	
Loader-lower	F					53	31	
Commander-upper	F		65	70		11	52	
Commander-middle	F	100				15	45	
Commander-lower	F					23	40	
Driver-upper	F	0				64		
Turret left wall	J,F	0	0	0		88	0	

the source of the damage and the percent loss of component function suffered are indicated. Next, the percent of loss to the vehicle (percent M or F utility) as a result of the component damage is given based on the SDAL.

The analyst must account for the percentage loss of component function as a result of component interdependence. For example, if the rotary base junction (slip ring) is destroyed, the loss of electrical power to the turret renders other components useless. After grouping damaged components in this manner, the percent LoF for each group is obtained. If more than one of the interdependent components in a group is damaged, the percent loss of component function used for the group is appropriately calculated. They are then multiplied by the percent tank kill value from the SDAL, and the product is entered in the % Loss to Tank columns.

Because of the manner in which damaged components are grouped, the values contained within a given column are independent and may be combined to give a total value for the shot (see Chapter 3). The results of this computation are given in the heading of Table 4.2. For this shot, the calculated DCU values are 67 percent mobility and 73 percent firepower. These values agree well with the assessors' estimates of 70 percent mobility and 75 percent firepower. When all of the calculated values were compared with the assessors' estimates, good statistical agreement was found to establish the validity of the analytical use of the SDAL and the component LoF data. Thus, with the analytical procedure validated and the target damage data reduced to a manageable number of numerical values, the analysts were able to turn their attention to the derivation of an analytical model.

At the time the compartment model was developed, it was not feasible to account for the effects of up to several thousand spall fragments discretely in an analytical model, nor was it practical to predict damage to each of the hundreds of components that make up a vehicle. However, analysis of the available data showed that if a vehicle is divided into a small number of regions (or compartments), the aggregate damage observed within each region could be related to certain characteristics of munitions and the armor. This observation is the basis of the compartment model.

For the purpose of analyzing the data, tank targets were divided into the regions shown in Table 4.3, which is taken from Zeller (1961). Damage correlations were then generated for hits into each of these regions. A damage correlation shows the expected DCU of the vehicle resulting from the total damage occurring within the particular region.

Figures 4.34 and 4.35, which are taken from the unpublished notebooks of R. L. Kirby of ARL, illustrate the nature of the damage correlations. Figure 4.34 applies to perforations into the crew compartment by a HEAT warheads. The profile hole diameter (PHD) referred to in this figure is the diameter of the hole (excluding consideration of spallation) in the armor at the surface where the jet exits the armor and enters the crew compartment. Figure 4.35 shows the currently used damage correlations for high-velocity, long-rod KE projectiles against track and suspension components.

A damage correlation indicates the DCU caused by the penetrator and the fragments within the particular region of the vehicle; the DCU values are averaged over all hit locations within the region that occurred in the tests. In the cases of the crew compartment and engine compartment, the damage correlations

Table 4.3 Divisions of tank presented area for compartment-kill methodology

Ammunition, subdivided into:	Ammunition in direct path of projectile
	Ammunition not in direct path of projectile but struck by fragments
Crew compartment, excluding ammunition	
Engine compartment	
Fuel	
Gun tube	
Track and suspension, subdivided into:	Track face
	Track edge
	Idler wheel hub
	Idler wheel (not including the hub)
	First roadwheel hub
	First roadwheel (not including the hub)
	Last roadwheel hub
	Last roadwheel (not including the hub)
	Sprocket hub
	Sprocket (not including the hub)
	Other (support rollers, shock absorbers, etc.)

contain no information on which particular components or crewmen were damaged. Thus, given a target description, warhead characteristics (e.g., size and penetration capability), and the firepower damage correlations, the expected DCU could be calculated for a single hit by the projectile on the target.

3. Implementation of the Model

The way in which the compartment-kill methodology is implemented is first to overlay a grid (usually of 100-mm cell size) on the target, as shown in Figure 4.5.

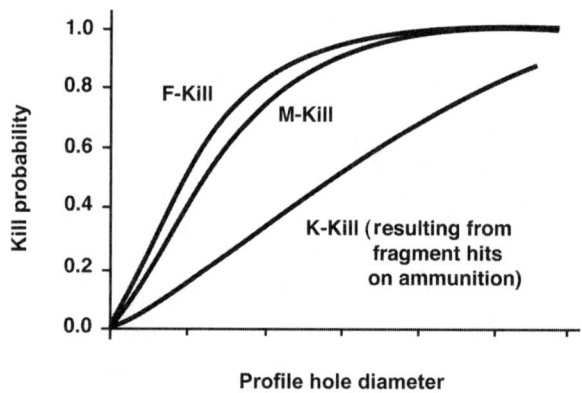

Fig. 4.34 P_K vs profile hole diameter for perforation by a HEAT projectile in the crew compartment.

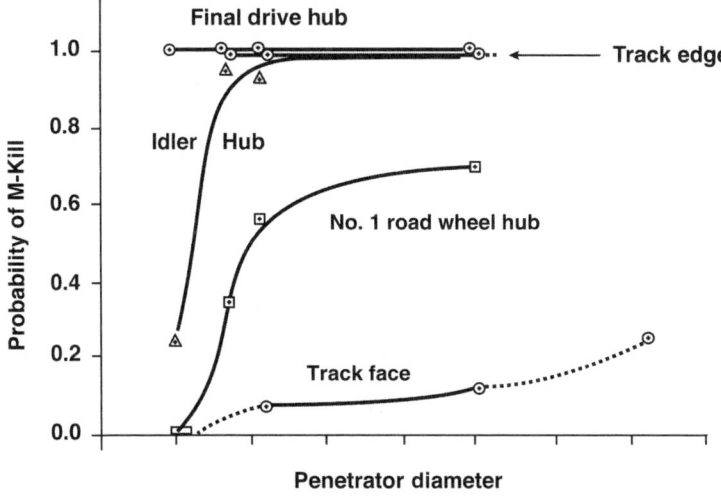

Fig. 4.35 Probability of M-Kill for KE penetrators against track and suspension components.

Within each grid cell, a point is chosen at random. A shotline is passed through each selected point in a direction parallel to the direction of attack and traced completely through the vehicle. As each line is traced, the name of each component encountered is recorded and the component composition, thickness, and surface obliquity are noted. This information, along with warhead penetration data, enables the computer code to determine how far into the vehicle the warhead penetrates. Also recorded along the shotlines is any compartment or component encountered for which damage correlations were prepared (as listed in Table 4.3). As the warhead penetrates into one of the compartments or components, a $P_{K|H}$ value is obtained from the appropriate damage correlation data. Thus, a $P_{K|H}$ is obtained for the shotline passing through each grid cell. If more than one such compartment is encountered, the damage is assumed to be independent, and the survivor rule is used to combine the effects. The $P_{K|H}$ for the shotline in a given cell is assumed to be representative of any hit within that cell. Usually, the cellular $P_{K|H}$s are averaged over the presented area or averaged over the presented area and angle.

Figure 4.36 summarizes the approach to the compartment-kill methodology, which was ultimately adopted as a NATO standard (North Atlantic Treaty Organization 1986). Additional details are included in Reed (1992a, 1992b), U.S. Department of Defense (1997), Benjamin and Gholston (1960), and Ege and Harvey (1974).

4. Future Use of the Model

Although the compartment model has been the primary model for production of V/L data for AFVs, it is now being used less and less. Modern AFVs are equipped with more and different mission equipment than vehicles for which the correlation curves were derived. It is extremely difficult to either enhance that

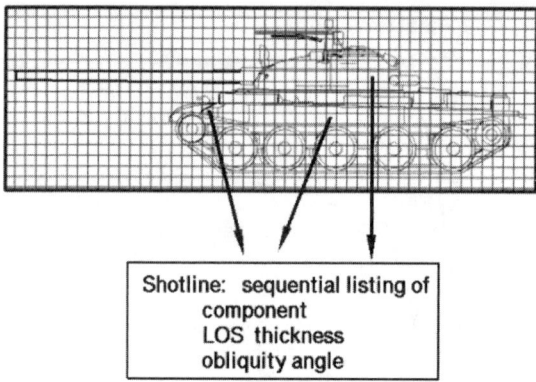

Fig. 4.36 Summary of the approach to the tank compartment-kill methodology.

database or modify extant curves especially for developmental systems that have not been live-fire tested. Further, the compartment model is an empirical model that predicts level 4 metrics that can neither be observed nor tested in the laboratory; thus, the compartment model cannot be validated.

E. Expected-Value Point-Burst Models

Point-burst models are similar to the compartment model up to the point at which the penetrator breaches a compartment. At that point, the various point-burst models use some characterization of the various damage mechanisms (i.e., penetrator, spall, blast, and shock) to determine damage to specific components in the compartment. Then the corresponding subsystem functional degradation can be estimated. Historically, AFV component damage and subsystem LoF have been related to system functional kills by way of the damage assessment list (DAL). Point-burst methodologies have the potential to implement degraded combat states methodologies to produce system level 3 (engineering capability) metrics.

As an example of the type of problem encountered, consider a 1975 TRADOC request for BRL to conduct an analysis of the Swedish S-tank. Because the S-tank had no turret and nothing that really qualified as a crew compartment, the basic premise of the compartment model was inappropriate. Thus, a nascent point-burst model called the Armored Vehicle Vulnerability Analysis Model (AVVAM) had to be used for the analysis, which was performed out by a team led by Rapp (1983).

1. VAST Model

Because funds were inadequate to support a continuing full-scale series of experiments on modern tanks, and because the ability to look at conceptual systems also implied the need for the ability to do *ab initio* analyses, there was a general feeling in the mid-1970s that a new approach to vulnerability analysis was needed. Thus, a cooperative effort among BRL, Watervliet Arsenal, and the Computer Science Corporation created a point-burst model called Vulnerability Analysis for Surface Targets (VAST) (Nail 1982).

In VAST, when a spall-producing armor component was encountered along the shotline, spall rays were traced from the spall burst point through the target to all the target critical components. As the previously mentioned VAMP was the BRL prototypical compartment model, so VAST was its prototypical point-burst model for ground vehicles. VAMP was less demanding on computation and relied heavily on full-scale determination of the damage-correlation functions. VAST was quite computationally intensive and required large amounts of data on spall characteristics and the vulnerability of a myriad of individual components.

In the late 1970s and early 1980s, the advances in armor and SC and KE threats made penetration modeling somewhat frantic. The issue was the need for considering a large number of cases rather than concentrating on limiting conditions. In addition, for the terminal ballistician, the main goal was to determine the levels at which the penetrator could breach the armor or the level at which the armor could defeat the penetrator. The details of under- and over-penetration were of less concern. For the vulnerability analyst, on the other hand, these details were just what were needed to define the level of kills.

Around 1980, canted and high-yaw SC warheads also became matters of concern and were major challenges for the compartment models. The M1 Abrams tank was a case in point. Its modern armor and interior configuration bore little resemblance to the tanks tested at CARDE, and the modern rounds against which the M1 had to defend itself bore little resemblance to those tested. New correlation data were thus needed if the VAMP methodology was to be used. Although there was some success with modifying the CARDE data based on limited M1 tests, it seemed clear that if VAMP were to be the baseline analysis program, a considerable full-scale testing program would be required to define the correlations.

Also at this time, EFPs, with their attractive long-standoff capability and application to top-attack, emerged as an exceedingly important subject for lethality analyses. EFPs depended in large measure on back-face failure for penetration, which resulted in a new distribution of mass and velocity from those of previous penetrators. This implied that the existing correlation functions of VAMP were inappropriate for EFPs; therefore, VAST was viewed as the more appropriate model. However, it also implied that new BAD data would be needed for EFPs. (See Appendix B for a discussion of BAD.)

Attempts have also been made to develop improved models of component damage. These include the work of Robert Shnidman (1988), which is related to earlier (1984) work on EFPs. Shnidman's approach uses direct correlation of the observed data from witness plate firings with component damage, rather than intermediate variables.

Because VAST deals specifically with components, it was quite useful for developing the vulnerable areas of components for use in the sustainability predictions for Sustainability Prediction for Army Requirements for Combat (SPARC), battle damage repair (BDR) analyses, and RAM analyses (see, for example, Chapter 5).

2. COVART Model

The simulation program for COVART is most known for its use in the evaluation of aircraft vulnerability, but it is also used for ground targets. We thus include a discussion of COVART because it does have broad use and because it represents some different perspectives in its use of vulnerable areas and attention to repair times.

One interesting aspect of the COVART model is its use of a vulnerable area to categorize damage to systems and components. The concept of a vulnerable area is particularly useful in considering situations in which the target is attacked by randomly distributed threats such as fragments from artillery rounds.

The vulnerable area (A_V) of a target object is defined as follows (note that the target object might be the system itself, or it might be a component such as an engine):

1) Let the vector $\boldsymbol{n}$ represent the direction from which an attacking penetrator impinges on the target.

2) Call the plane that is perpendicular to $\boldsymbol{n}$ the plane of attack.

3) Any point on the target object can be projected along the direction $\boldsymbol{n}$ to the plane of attack and represented by the coordinates (x, y) in that plane.

4) The projected image of the target object is the set of image points (x, y) that are projected from the points in the target object to the attack plane. The projected area (A_P) of the object is the area of its set of projected points in the attack plane.

5) Let $p(x, y, a, \boldsymbol{n})$ represent the probability that a penetrator produces a particular type of damage given that it passes through the attack plane at the point (x, y) in the direction n and has characteristics a, where the variable a might represent such properties as its speed and mass.

The vulnerable area $A_V(a, n)$ of the target object for such a fragment is then defined as

$$A_V(a, n) = \iint_{A_P} p(x, y, a, \boldsymbol{n}) \, dx \, dy \tag{4.4}$$

where the integration is performed in the attack plane.

The $A_V(a, n)$ values might then be aggregated in a variety of ways, such as summed over a weighted average of fragment characteristics, as might be obtained from the burst of a particular type of artillery round, viz,

$$A_V(n) = \int A_V(a, \boldsymbol{n}) f(a) \, da \tag{4.5}$$

where $f(a)$ is the distribution of fragment characteristics.

One might also average over a weighted set of directions to get an A_V that would represent the vulnerability of the target to a randomly oriented burst from a round.

The concept of a vulnerable area is closely tied to the concept of the Poisson distribution (Hunt 1995). Consider a barrage of penetrators that impact the target plane with an areal density of R. The probability that the target is damaged (given no synergism among penetrator hits) is

$$P_D = 1 - e^{-RA_V} \tag{4.6}$$

A simple related concept is the probability of damage given a random hit on the target ($P_{D|H}$), which is given by

$$P_{D/H} = A_V/A_P \tag{4.7}$$

The individual vulnerable areas of the components can be aggregated to reflect the vulnerability of the system either to a uniform cloud of penetrators or to the pattern of penetrators emerging from a burst point (n and a could be different for each component). In this case, the individual damages can be aggregated by

$$P_D = 1 - \prod_{i=1}^{n}(1 - P_{D,i}) \tag{4.8}$$

where $P_{D,i}$ is the probability of damage for each of the n components that might be damaged.

COVART is quite similar to the other programs discussed in this chapter. It uses a geometric representation of the target, follows shotlines, calculates penetration characteristics, and estimates the effects on components that are hit. For detailed information on the latest COVART release or the latest COVART user manual, visit the Survivability/Vulnerability Information Analysis Center (SURVIAC) website at www.surviac.org.

F. The Stochastic Point-Burst Model

It has long been known that threat–target ballistic interactions involve many random processes. Some of the sources of variability are as follows:
1) Location of the impact point about the aimpoint.
2) Orientation of the munition at the instant of impact.
3) Performance of munition damage mechanisms.
4) BAD fragment distribution and characteristics.
5) Damage caused to vehicle components and the effect of that damage.
6) Geometric representation of the target vehicle.
7) Sample-to-sample variations in actual vehicles because of manufacturing tolerances.

VAMP, VAST, and COVART are expected-value models. That is, they calculate the probability that a given outcome (e.g., an M-Kill) occurs for a given encounter, but they give no information on the variability of that outcome. This limitation of VAMP and VAST became acute when Live-Fire testing began in the mid-1980s. Those tests produced data on the specific damage to the target system for each of a relatively small number of full-scale shots. Furthermore, the tests did not give direct data on the classical measures of effectiveness of M- and F-Kills, although, of course, when K-Kills occurred, they were directly observable.

1. SQuASH Model

There was apparent need for a model that would predict the probabilities that various sets of components would be damaged (component damage vectors) by a given encounter with the threat. Around 1986, BRL significantly reconfigured a point-burst (component-level) vulnerability model to support the many Live-Fire test programs. With the new stochastic point-burst model, SQuASH, cases were modeled probabilistically. As the model was exercised, various discrete damage states were predicted, together with a likelihood of occurrence and the attendant effects on battlefield functions. Basically, the SQuASH program:

1) Picked rays randomly and followed them, much like VAST, until a main compartment was perforated.

2) Traced the main penetrator (possibly deflected) through the compartment, as in VAST.

3) Unlike VAST, selected spall fragments randomly from the spall pattern and traced them in the compartment to find encounters with critical components.

4) Recorded the vehicle's damage state.

5) Repeated the above damage assessment processes many times.

6) Sorted, ranked, and mapped all vehicle damage states into P_K space.

The remainder of this section is based on work reported by Deitz and Ozolins (1989). We use their example of an analysis of the Abrams tank as a way of describing the SQuASH program and as an example of the various steps involved in a stochastic vulnerability analysis. Note that all quantitative results are notional and are included for illustrative purposes only.

2. *Example SQuASH Application and Stochastic Vulnerability Analysis*

SQuASH was designed to accommodate the following threats including the special case of multiple hits (salvo-fired weapons): KE rounds; CE munitions, including SC rounds, EFPs, artillery fragments, and mines.

The following is a simple outline of SQuASH processing:

1) Intersect nine rays with target geometry to simulate a threat munition hit distribution within a small area centered on the aimpoint; from each possible interior spall point, burst 10,000 rays.

2) Randomly pick one of nine rays and fire threat munition.

3) Check for suspension and other exterior damage.

4) Check for perforation.

5) If perforation, randomly deflect residual penetrator (depending on target/munition).

6) Assess components killed because of residual penetrator.

7) Check for K-Kill because of impact on fuel/ammunition.

8) Assess presented area and barrier shielding for critical components in spall domain.

9) For each component, calculate expected number of lethal fragments from spall model and use a Poisson distribution to perform random draw for a specific number (n) of fragments.

10) Play n fragments individually against the component $P_{K|H}$ ($0 \leq P_{K|H} \leq 1.0$); power up individual P_K's using the survivor rule; take a random draw to calculate a kill/no-kill outcome.

11) Repeat spall processing for all remaining critical components.

12) Record vehicle damage state.

13) Repeat the above damage assessment processes 999 times.

14) Sort and rank all vehicle damage states.

15) Map all (weighted) damage states to P_K space to build M, F, and M/F histograms using deactivation diagrams and DAL.

Also, accommodation was made in SQuASH to stochastically vary the following variables.

Hit point Even in ideal conditions, the modeling of a complex target cannot perfectly reflect real vehicles. Actual vehicles vary from copy to copy in details such as wire routing. The geometric interrogation process involves shooting geometric (zero cross section) rays through targets to replicate possible projectile paths. So, rather than a single ray normally used to model a striking projectile, a nine-ray matrix distributed about the aimpoint was used to sample over a 6-in. cross section.

Warhead performance Normally, warhead performance is modeled in terms of its expected (point-value) penetration capability. Repeated warhead/armor experiments using precision components reveal random variations in, for example, depth of penetration. The SQuASH program associates a distribution function with all warhead/armor calculations; in the course of model exercise, random draws are made from this distribution function.

Residual penetrator deflection For KE projectiles incident at oblique angles, a penetrator's residual portion can deflect upon exiting armor. The deflection is greatest near the limit velocity, when the armor is just being overmatched. A distribution function is used to select trajectories near the expected deflection.

Spall production The VAST program uses a spall model based on BAD described in terms of fragment mass, velocity, and shape factor. Because much of this information is lacking for many warhead/armor pairings in the M1A1 program, a spall model based on a notion of lethal fragments was used. Spall collection was standardized by means of a package of thin metallic plates (Corbett et al. 1987). Lethal fragments for these purposes are defined as those fragments that penetrate at least the first plate in this combination pack. (Later versions of SQuASH use the mass/velocity/shape factor BAD characterization.)

The SQuASH spall model is based on a routine that describes the spatial density of lethal fragments as a bivariate Gaussian distribution. The solid angle subtended by any critical component and its location then defines the expected number of lethal fragment impacts. In the exercise of the program for a particular shot, the expected number of fragments is used in a Poisson distribution to draw a specific number of fragments. This particular number of fragments is then evaluated against the given component. (Later versions of SQuASH draw from distributions of BAD fragment characteristics and treat each fragment as a separate penetrator to compute damage to each component it encounters along its path.)

Component $P_{CD|H}$ characterization Each critical component is separately characterized in terms of probability of being killed by main penetrators and single lethal spall fragments. For intermediate threats (e.g., KE penetrator fragments), intermediate kill probabilities are computed using hole size and penetration capability. Multiple hits are assessed using the survivor rule (Rapp 1983).

Secondary kill phenomena Although the primary phenomena are often not adequately characterized, even less is known about the myriad of possible secondary effects. In general, secondary kill phenomena have not been explicitly modeled because they have not appeared to play a consistent and significant role on AFV vulnerability. Nevertheless, particular tests have been conducted in which ballistic shock or blast has been shown to cause critical damage in certain circumstances. For example, a principal goal of the BRL M1A1 assessment program was to gain as much insight as possible into the importance of these secondary mechanisms.

In the context of the Abrams program, there was insufficient time to introduce damage algorithms for these secondary phenomena. However, provision was made in the program structure to support any additional damage algorithms that might be required. In addition, algorithms have continued to be developed to robustly model these damage mechanisms, and rudimentary methodologies have been used in more recent analyses.

Before using SQuASH on the M1A1, the following inputs had to be assembled.

Geometry At the inception of the M1A1 Live Fire program, the existing target description was a moderately detailed version of the M1E1 vehicle. Based on the Bradley experience, it was clear that the target geometry had to be enhanced to an unprecedented level. Using the BRL-CAD solid geometric modeling software, 25 specific subsystems were added to the target description (Muuss et al. 1983; Deitz et al. 1988). These systems were modeled down to the individual wire and hydraulic-line level of detail. The fuel system, including critical fuel lines and filters, as well as the larger fuel tanks, was modeled. Also, the power pack, the turret fire control, and communications gear were modeled. This effort resulted in the largest target-description file assembled up to that time, comprising over 5000 objects.

Criticality analysis Every point-burst analysis program, including SQuASH, requires a criticality analysis. A criticality analysis of a target involves two steps. First, every component of the vehicle that supports the mobility or firepower function must be identified. Second, the logical interconnectivity of each component in its respective system or subsystem must be represented in a deactivation diagram, which is a form of fault-tree analysis. By this process, the effect of the potential loss of a component to a given system function can be assessed so that the SDAL can be invoked in the MMF level 2 to level 4 mapping process. The details of the M1A1 criticality analysis are documented elsewhere (Ploskonka et al. 1988). An example for the fuel system can be found in Fig. 4.37. In this structure, the series layout of components with the lack of redundancy shows that component loss is equivalent to system loss. In contrast, the loss of a single component that operates in parallel with a similar component (e.g., fuel line from left rear fuel cell to tee) does not affect system capability. Code has been written in a program called the Interactive Criticality Estimator (ICE) to assist in the construction of these diagrams and the compilation of the logic structures for the SQuASH input files (Gwyn 1987).

Threat characterization – main penetrator The initial Abrams Live Fire plan called for approximately 50 tests. The M1A1 itself at that time comprised six different armor types. Warhead/armor data were assembled for all possible encounters. In all previous vulnerability models, only the nominal (expected-value) performance parameters were used. However, SQuASH requires that an estimate

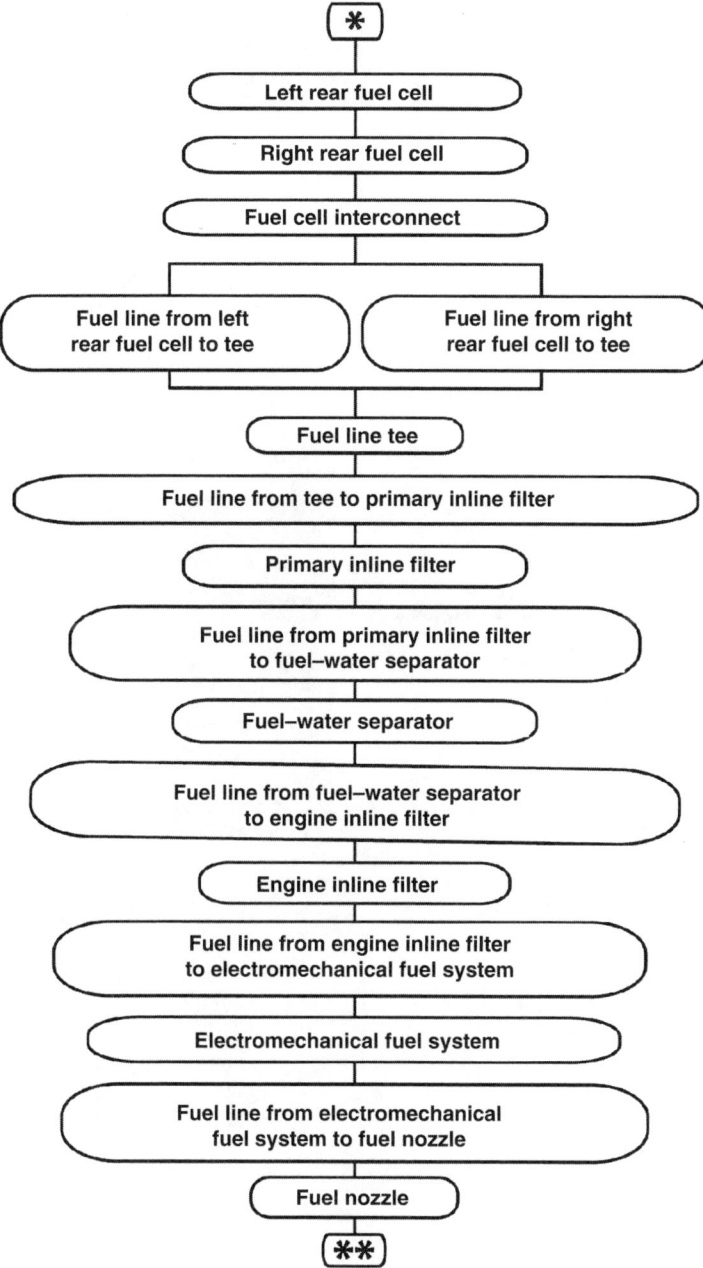

Fig. 4.37 Fuel subsystem.

of the variability of the warhead/armor performance be included. This information is illustrated in Fig. 4.38. At the top of the figure, a test configuration is shown for an SC warhead against a semi-infinite target. This experiment is repeated many times for a series of standoffs. After plotting data from such an experiment, the curve shown in the bottom of the illustration is derived (Dispersio et al. 1965). This solid curve is the relationship normally used in vulnerability models. In the case of SQuASH, data about the variability of penetration depth as a function of standoff were also developed for each round. The error bars on the mean data points imply this. In the course of program execution, nominal penetration values were modulated by random draws from related error functions. For KE rounds, penetration variability is modeled in terms of limit thickness, which is the form of the data provided by terminal ballisticians. Note that data were extremely sparse for many of the threat warheads used in the Live Fire program.

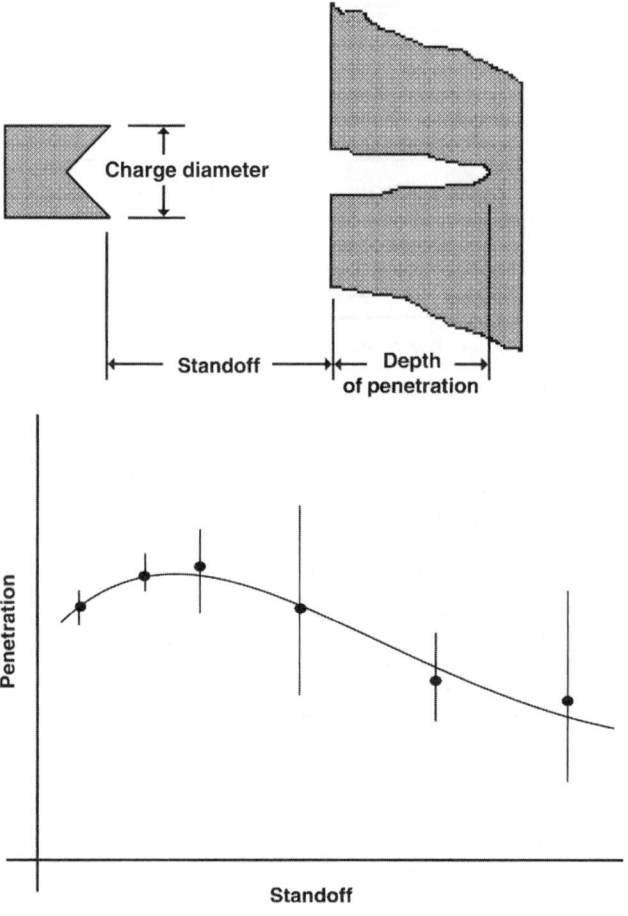

Fig. 4.38 Typical test configuration (above) and data (below) for an SC warhead.

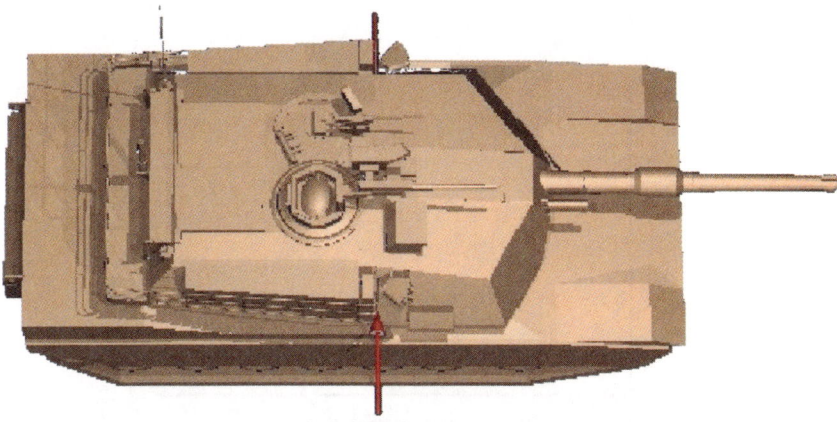

Fig. 4.39 Modeled warhead attack geometry. (Shotline selection is arbitrary and included for illustrative purposes only.)

Threat characterization – BAD Because point-burst modeling involves the explicit interaction of BAD with critical components, behind-armor spall clouds must be described analytically (see Appendix B).

SDAL modifications Minimal modifications in an earlier SDAL were made to include components and systems not present on earlier AFVs. These changes were coordinated with the U.S. Armor School at Fort Knox, Kentucky.

A CE shot into the right turret basket is used to illustrate typical model results. Figures 4.39 and 4.40 show the warhead attack geometry. An arrow normal to the target description illustrates the shotline. In Fig. 4.39, the view is directly along the shot path; Fig. 4.40 gives a perspective view.

Figure 4.41 is a histogram showing the distribution of residual-penetrator overmatch (in inches). Over the course of many similar computations, these curves exhibit complex shapes, sometimes with multi-modal distributions. This is a natural consequence of the randomness of the overmatch, together with the

Fig. 4.40 Modeled warhead attack geometry, perspective view. (Shotline selection is arbitrary and included for illustrative purposes only.)

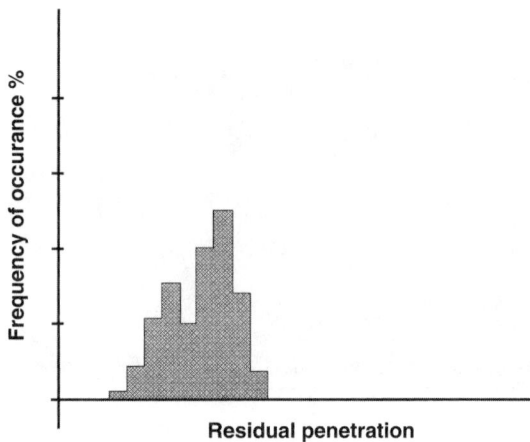

Fig. 4.41 Histogram of residual penetration. (All quantitative results are notional and are included for illustrative purposes only.)

grid ray data derived over nine sample rays. Even though the rays are separated nominally by 3 inches, different combinations of armor are often encountered. The difference in effective protection levels leads to significantly different residual magnitudes.

To support the spall/component interaction, a matrix of divergent rays was fired from each of the nine potential spall points at the grid-ray entry points. To ensure adequate spatial resolution, 10,000 rays were fired from each of the nine spall origins. This hypersampling ensures resolution of individual hydraulic lines and wires at large distances from the spall point. The information from these nine processes was used to calculate the solid angle subtended by each critical component with each spall cone, as well as any intervening (shielding) barriers.

The same ray file used for the vulnerability calculation can be used to form an image. Figure 4.42 shows the view from the center grid ray just after entry into the crew compartment. The gunner is directly in the center shotline, the commander is behind, and the loader is across and away from the spall entry point. Over the course of 1000 program replications, 60 critical components were assessed to have been killed at least once. Table 4.4 lists a subset of these components. The columns list the total P_K, the contribution because of the jet only, and the fragment cloud only.

Table 4.5 shows where SQuASH output departs radically from other point-burst models. Here, the first of five classes of components is listed separately by category. This procedure has been adopted because of the difficulty in interpreting the results of damage states across the complete vehicle. The table lists the category of Crew. For this group, the calculated damage states apply to the personnel located in the turret-basket area. The "0" indicates no calculated damage. An "x" indicates the component has been killed (or, in the case of a crew member, incapacitated). The damage states derived from the 1000 replications were sorted together and then ranked from the most to the least likely in occurrence. Table 4.5 shows that the most likely crew casualty state is for the commander and loader not to be incapacitated and for the gunner to be incapacitated. This outcome occurred 461 times in the

MODELING AND SIMULATION TOOLS AND METHODS

Fig. 4.42 Model view from the center grid ray.

Table 4.4 Partial listing of components killed[a]

Component	Relative frequency of damage		
	Total	Jet	Fragment
Commander	0.399	0.000	0.399
Gunner	0.995	0.683	0.594
Loader	0.201	0.000	0.201
cable 1w100-9	0.018	0.000	0.018
f.line right bow to manifold	0.001	0.000	0.001

[a]All quantitative results are notional and have been included for illustrative purposes only.

Table 4.5 Damage states for the subset Crew[a]

Damage states			Relative occurrence	
Commander	Gunner	Loader	State	Cumulative sum
0	×	0	0.461	0.461
×	×	0	0.237	0.698
×	×	×	0.192	0.890
0	×	×	0.103	0.993
0	0	0	0.005	0.998
×	0	0	0.002	1.000

[a]All quantitative results are notional and have been included for illustrative purposes only.

Table 4.6 Damage states for the subset Propulsion[a]

Damage states		Relative occurrence	
Cable 2w107-9	Cable 2w108	State	Cumulative sum
0	0	0.986	0.986
×	0	0.008	0.994
0	×	0.005	0.999
×	×	0.001	1.000

[a]All quantitative results are notional and have been included for illustrative purposes only.

1000 replications, a net probability of 46 percent. The next most likely crew casualty state is for the commander and gunner to be incapacitated, but not the loader. The likelihood of this outcome is assessed at 24 percent. For this component subset, all outcomes occurred over only six combinations.

The damage states for Propulsion given in Table 4.6 are relatively simple. Only two components from that group were killed. The three damage states in which at least one of these cables was killed occurred in only 14 of the 1000 replications.

Table 4.7 gives the damage states for the category Major Electrical. Six components are involved over 13 specific outcomes. As shown, the most likely damage state involves damage to none of the components, estimated at 87 percent.

The damage states for Armament shown in Table 4.8 reveal the greatest complexity in damage states. This is probably to be expected as nearly half of all the critical components killed during the 1000 replications were part of this group. As seen in other groupings, the most likely damage state assessed for the 29 components in Armament is no damage (28 percent of the outcomes). The most likely state exhibiting damage occurred for five components (numbers 6, 10–12, and 15) on 78 of the 1000 replications, for a 7.8 percent probability. From here on, the 29 components are involved in a slow convergence to the 99th percentile.

The remaining 11 components involved in damage are listed as other and documented in Table 4.9.

At this point in the simulation, we have accumulated a full accounting of the statistics of level 2 in the MMF. The final stage of calculation involves the various categories of kills (see Fig. 4.43). First, K-Kill involves the complete loss of the system. This generally occurs because of encounters with large-caliber ammunition (warhead and/or propellant) or fuel. The probability of this event is shown in the bottom-left corner of Fig. 4.43. For this particular shot, the probability of a catastrophic event is assessed as zero. Note that the histogram associated with K-Kill can be populated only in the first and last bins. This is a consequence of the K-Kill event belonging to the class of Bernoulli trials.

The other kill categories are assessed by mapping each of the thousand damage states via the SDAL over to the appropriate M-Kill and F-Kill values. The category labeled M/F (read M or F), by long-standing agreement with the training and doctrine community, represents the larger of the two values. It is not the "or" of the logical (Boolean) operation.

If we examine the M-Kill plot in the top-left corner of Fig. 4.43, we find the most likely outcome is for about 0.57 mobility LoF (M-LoF), assessed at about 30 percent probability. However, the distribution is extremely broad, with approximately

Table 4.7 **Damage states for the subset Major Electrical**[a]

Damage states component number						Relative occurrence	
1	2	3	4	5	6	State	Cumulative sum
0	0	0	0	0	0	0.866	0.866
0	0	0	0	0	×	0.045	0.911
0	0	×	0	0	0	0.039	0.950
×	0	0	0	0	0	0.014	0.964
0	×	0	0	0	0	0.009	0.973
0	0	0	×	0	0	0.008	0.981
0	0	0	0	×	0	0.008	0.989
×	0	0	0	0	×	0.003	0.992
0	0	×	0	×	0	0.002	0.994
0	0	×	0	0	×	0.002	0.996
0	0	×	×	0	0	0.002	0.998
×	0	0	×	0	0	0.001	0.999
0	×	0	×	0	0	0.001	1.000

Component number	Component
1	cable 1w100-9
2	cable 1w101-9
3	cable 2w105-9
4	cable 2w154-2w155
5	hull network box
6	turret network box

[a] All quantitative results are notional and have been included for illustrative purposes only.

18 percent of the outcomes near the 0.0 bin. The expected M-LoF outcome is 0.36; inspection of the histogram shows that approximately 26 percent of the outcomes are near this value. However, the distribution is broad, and a significant number of occurrences are away from the mean. The corresponding results for firepower LoF (F-LoF) are given in the top-right corner of Fig. 4.43. In this histogram, the mean LoF occurs in a bin with a low population. There is also a significant probability (about 18 percent) that the F-LoF will be zero. The M/F-LoF histogram is given in the bottom-right corner of Fig. 4.43. The M/F value, by definition, is the larger of the M- and F-LoFs on a shot-by-shot basis. The F-LoF tends to dominate in this case.

G. Modeling the Effects of Bursting Indirect-Fire Munitions

Vulnerability of a target to direct-fire munitions generally involves the impact of a single warhead on the target vehicle. Historically (and currently), simulation of a combat engagement in which a particular vehicle is attacked by more than one warhead during an engagement considers each impact to be an independent event, and vehicle vulnerability to each impact is computed as though the other impacts

Table 4.8 (a) Damage states for the subset Armament[a]

	Damage states component number																													Relative occurrence	
	1	2	3	4	5	6	7	8	9	a	b	c	d	e	f	g	h	i	j	k	l	m	n	o	p	q	r	s	t	State	Cumulative sum
	o	o	o	o	o	o	o	o	o	o	o	o	o	o	o	o	o	o	o	o	o	o	o	o	o	o	o	o	o	0.275	0.275
	o	o	o	o	o	x	o	o	o	x	x	x	o	o	x	o	o	o	o	o	o	o	o	o	o	o	o	o	o	0.078	0.353
	o	x	o	o	o	x	o	o	o	o	o	o	o	o	o	o	o	o	o	o	o	o	o	o	o	o	o	o	o	0.077	0.430
	o	o	o	o	o	x	o	o	o	x	o	x	o	o	x	o	o	o	o	o	o	o	o	o	o	o	o	o	o	0.060	0.490
	x	o	o	o	o	o	o	o	o	x	x	x	o	x	o	o	o	o	o	o	o	o	o	o	o	o	o	o	o	0.039	0.529
	x	o	o	o	o	x	o	o	o	o	x	o	o	x	o	o	o	o	o	o	o	o	o	o	o	o	o	o	o	0.026	0.555
	o	o	o	o	o	o	o	o	o	o	o	o	o	o	o	o	o	o	o	o	o	o	o	o	o	o	o	o	o	0.023	0.578
	o	o	o	o	o	x	o	o	o	x	x	x	o	x	x	o	o	o	o	o	o	o	o	o	o	o	o	o	o	0.013	0.591
	o	o	o	o	o	o	o	o	o	x	o	o	o	o	o	o	o	o	o	x	o	o	o	o	o	o	o	o	o	0.011	0.602
	o	x	o	o	o	o	o	o	o	o	x	x	o	x	x	o	o	o	o	o	o	o	o	o	o	o	o	o	o	0.010	0.612
	o	o	o	o	o	x	o	o	o	x	o	o	o	o	x	o	o	o	o	o	o	o	x	o	o	o	o	o	o	0.010	0.622
	o	o	o	o	o	o	o	o	o	x	o	x	o	x	o	o	o	o	o	o	o	o	o	o	o	o	o	o	o	0.010	0.632
	o	x	o	o	o	o	o	o	o	x	x	o	o	o	x	o	o	o	o	o	o	o	o	o	o	o	x	o	o	0.009	0.641
	o	o	o	o	o	x	o	o	o	o	o	x	x	x	o	o	o	o	o	o	o	o	o	o	o	o	o	o	o	0.008	0.649
	o	o	o	o	o	x	x	o	o	x	x	o	o	x	x	o	o	o	o	x	x	x	o	o	o	o	o	o	o	0.007	0.656
	o	o	o	o	o	o	o	o	x	o	o	o	o	o	o	o	o	o	o	o	o	o	o	o	o	o	o	o	o	0.001	0.999
	o	o	o	o	o	x	o	x	o	o	o	o	o	o	o	o	o	o	o	o	o	o	o	o	o	o	o	o	o	0.001	0.999
	o	o	o	o	o	x	x	o	o	x	x	x	o	x	x	o	o	o	o	o	o	o	o	o	o	o	o	o	o	0.001	1.000

[a]All quantitative results are notional and have been included for illustrative purposes only.

Table 4.8 (b) Damage states for the subset Armament

Component number	Component
1	cable 1w104
2	cable 1w104
3	cable 1w105-9 main branch
4	cable 1w107-9
5	cable 1w108-9 to main gun
6	cable 1w200
7	cable 1w201
8	cable 1w202 main branch
9	cable 1w203
a	cable 1w208
b	cable 1w209
c	cable 1w210
d	gunner's primary sight
e	commander's gps extension
f	thermal image control unit
g	filter manifold
h	h.lines filter manifold to
i	h.lines filter manifold to HDM
j	h.lines TDM to azimuth serzo
k	manual azimuth gearbox
l	manual elevation pump
m	gunner's control handle
n	commander's control handle
o	race ring
p	h.line check valve to HDM bypass
q	coaxial ready ammo box
r	azimuth gear box-cws
s	commander's vision block #2
t	loader's sight

Table 4.9 Damage states for the subset Other[a]

Damage states component number											Relative occurrence	
1	2	3	4	5	6	7	8	9	10	11	State	Cumulative sum
0	0	0	0	0	0	0	0	0	0	0	0.700	0.700
×	0	0	0	0	0	0	0	0	0	0	0.064	0.764
0	0	0	0	0	0	0	0	0	×	0	0.022	0.786
0	0	0	×	0	0	0	0	0	0	0	0.019	0.805
×	0	0	0	0	0	0	0	0	×	0	0.016	0.821
0	0	0	0	0	0	0	0	0×	0	0	0.015	0.836
0	×	0	0	0	0	0	0	0	0	0	0.014	0.850
×	0	0	×	0	0	0	0	0	×	0	0.012	0.862
0	0	×	0	0	0	0	0	0	0	0	0.010	0.972
×	0	0	×	0	0	0	0	0	0	0	0.008	0.880
0	0	0	0	0	0	0	×	0	0	0	0.008	0.888
0	0	0	0	0	0	×	0	0	0	0	0.007	0.895
×	0	0	0	0	0	0	0	0	×	×	0.005	0.900
×	0	0	0	×	0	0	0	0	×	0	0.005	0.905
0	0	0	0	×	0	0	0	0	0	0	0.005	0.910
0	0	0	0	0	0	0	0	0	0	×	0.004	0.914
×	0	0	0	0	0	×	0	0	0	0	0.004	0.918
0	0	0	0	0	×	0	0	0	0	0	0.004	0.922
0	×	0	0	0	0	0	0	0	×	0	0.004	0.926
×	0	0	×	0	0	×	0	0	×	0	0.004	0.930
0	0	0	×	×	0	0	0	0	×	0	0.003	0.933
×	0	0	0	0	0	0	0	×	0	0	0.003	0.936
×	×	0	0	0	0	0	0	0	×	0	0.003	0.939
×	0	0	0	0	0	×	0	0	×	0	0.003	0.942
×	0	0	0	0	×	0	0	0	0	0	0.003	0.945
×	×	0	0	0	0	0	0	0	0	0	0.003	0.948
×	0	0	×	×	0	0	0	0	×	0	0.003	0.951

Component number	Component
1	cable 1w301
2	cable 1w304
3	cable 1w306
4	cable 1w309
5	cable 1w310
6	cable 1w311
7	cable 1w312
8	cable 1w316
9	intercom amplifier
10	gunner's intercom control box
11	loader's intercom control box

[a] All quantitative results are notional and have been included for illustrative purposes only.

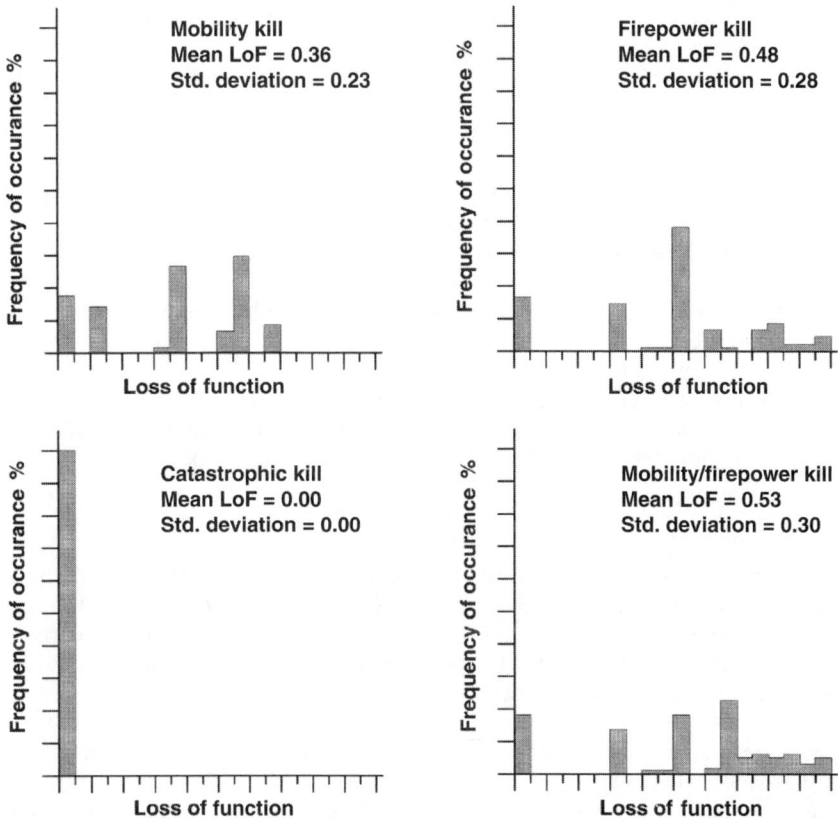

Fig. 4.43 Histograms of four kill categories. (All quantitative results are notional and are included for illustrative purposes only.)

had not occurred. That is, the vehicle is considered to be a pristine undamaged target. [It is worth noting, however, that methodologies are currently being developed to allow a given vehicle to accumulate damage (MMF level 2) and loss of capability (MMF level 3) as it progresses through a mission.] On the other hand, vulnerability to indirect-fire bursting munitions, including artillery and mortar shells as well as fragmenting bomblets, necessarily involves the nearly simultaneous attack by more than one penetrator, namely, fragments.

1. Fragment Characterization

When a fragmenting warhead bursts, fragments of varying masses, speeds, and shapes are dispersed in all directions. The fragment distributions of most bursting warheads are characterized in a series of arena tests in which the warheads are statically detonated. The arena is composed of witness panels arranged in a circular or semi-circular pattern several meters from the warhead. The witness panels, complemented by photography, record or infer individual fragment sizes, speeds, trajectories, and shape factor (a measure of the physical size and shape). The arena data are processed to generate distributions of number, size, and velocity of fragments as a function of angle from the axis of the bursting warhead. This process has been standardized and produces tables of fragmentation characteristics in what is referred to as the JTCG/ME Manual (JMEM) format (Joint Technical Coordinating Group for Munitions Effectiveness 1988). Historically, several warheads were tested and the results averaged to produce the published characteristics. More recently though, characteristics for individual bomblets (U.S. Army Combat System Test Activity 1990) and artillery shells (Klopcic and Lynch 1997, 1998a, and 1998b) have been determined to provide insights into the round-to-round variability. At this time, round-to-round variations have been collected only for a limited set of munitions, including the 155-mm HE, 152-mm HE, 122-mm HE artillery shells, and M74 bomblet.

Fragments from typical warheads have fragments weighing from a fraction of 1 gram to several hundred grams, with speeds of a few hundred meters per seconds to several kilometers per second, and with widely varying shape factors.

2. Historical Analysis Process

In contrast to direct-fire munitions, the vulnerability of a target vehicle to a single fragment is generally meaningless by itself. Of ultimate interest is the vulnerability to all fragments from a single warhead that strike the vehicle and, more often, the vulnerability of the vehicle to a barrage of warheads exploding about the vehicle during a typical encounter.

Historically, two complementary processes were involved. The vulnerability analyst generated tables of vehicle vulnerability data for individual fragments of varying mass, speed, and shape striking the vehicle at various combinations of attack azimuth and elevation. These data were very similar to the data produced in a direct-fire vulnerability analysis except that the results were presented as vulnerable areas rather than probabilities of kill given a hit.

In the complementary process, an operation research analyst combined the vulnerability data with fragment characteristics of the warhead and characterizations of the barrage (number of warheads, distribution about the vehicle, orientation and speed of the warhead when it detonates, and other such characteristics) to generate lethal areas. Lethal areas were determined by the integration of vulnerable areas over the area around the vehicle where individual bursts could occur, which is referred to as the ground plane. Lethal areas were evaluated using a program such as Surface-to-Air Missile Site Mean Area of Effectiveness (SAMSMAE) (Schumacher 1981). The resulting characterization was usually referred to as "effectiveness."

3. Stochastic Processor for Artillery Effectiveness (SPRAE) Model

Effectiveness is useful to estimate the survivability of a vehicle exposed to a defined combat encounter. It does little to provide insight into the vulnerability of the vehicle to a single warhead or into the effectiveness of specific vulnerability reduction measures.

Live fire tests are accompanied by pre-test analyses to determine how well analytical models are able to estimate vehicle vulnerability. These analyses attempt to predict the outcome of specific tests and are essential elements of model validation.

Analyzing the vulnerability of the vehicle to warheads detonating at select locations about the vehicle and understanding the contributions by fragments of various sizes and velocities allow the designer to optimize armor protection and other design features to mitigate the vulnerability to bursting munitions.

To provide a tool for the consideration of the effect of conventional artillery, a single-shot point burst model called SPRAE was developed (Schumacher and Hunt 1992; Hunt 1995). SPRAE combined existing methodologies from the VAST program and the SAMSMAE program. These deterministic models divided the vulnerability process into a calculation of component vulnerable area (A_V) for a set of attack directions followed by burst point and lethal area analysis using these A_V's. SPRAE still relied on a raytracing preprocessor and computed component P_K values for independent fragment impacts. Because rays were produced before generating initial fragment distributions, each ray could contain multiple, partial, or no fragments. Also, a given set of rays was used on each stochastic iteration. This caused geometric sampling errors. Neither ricochet nor tertiary spallation could be computed. Because component P_K values were calculated for each ray and survivor-rule summed over all rays, no multiple hit damage methodology could be used.

4. Stochastic Analysis of Fragment Effects (SAFE) Model

SAFE is an improved version of SPRAE. In addition to having all the capabilities of SPRAE, SAFE enhances fragment analysis by explicitly modeling fragment trajectories, processing multiple fragment impacts on individual components, and adding the potential to perform ricochet calculations.

SAFE needs stochastic distributions of fragment characteristics. Using these distributions, a standard deviation for mass, velocity, shape, and spatial density can be calculated for a Gaussian distribution and made available to assign to appropriate input parameters in SAFE threat files.

SAFE considers the retardation of fragment velocity because of air drag. These calculations are performed for each air or void region the fragment passes through using air drag equations and coefficients for fragments (Dunn and Porter 1955). It must be noted that the target is assumed to be stationary relative to the fragmenting munition, and all fragment trajectories through air, void, and target components are assumed to be linear.

VII. Computer Environments

For our purposes, we consider a computer environment to be a unified package of software that allows a user to perform a variety of tasks. The environment handles all the files necessary for an analysis, controls the use of the programs that carry out the various calculations, and coordinates the functioning of the various parts.

A. MUVES

During the mid-1980s, it was recognized that a computer program was required that would encompass compartment-level programs such as VAMP, along with other vulnerability programs, including point-burst programs such as VAST, SQuASH, and SPRAE. As discussed, these programs share many of the same algorithms. The new computer program would not only share some of this code, but it would provide better configuration management, flexibility of use, and ease of maintenance.

For instance, the SQuASH and SAFE programs required raytracing (or target interrogation) capabilities just like the compartment method. Subsequent to raytracing, penetration and BAD equations (physical interactions), along with the means of controlling or sequencing the interactions, are needed in the assessment process. Subsequent to the calculation of the physical interactions is the evaluation of damage on the target. To some extent, the variations between the compartment and SQuASH vulnerability estimation methods consist of the differences in the level of detail in the algorithms used. The basic estimation process is, however, the same.

The MUVES computer environment was designed and developed to codify and take advantage of the similarities between the various vulnerability assessment methods (Hanes et al. 1991). (Note that developers no longer consider MUVES an acronym; thus, the spelled-out form is not typically used.) MUVES defined computer program libraries for definition of warhead threats, the interaction of those threats with the target (interaction modules), the calculation of the interaction (physics/physical interaction libraries), the evaluation of damage on the target (evaluation modules), target interaction and raytracing libraries, and various computer program "infrastructure" functions (such as memory management, file handling, and user interface operations). With these libraries, and the underlying assessment structure embodied in MUVES, development of new vulnerability estimation methods could be performed much more rapidly and with greater standardization because the existing body of MUVES could be re-used. Only those features specific to the new application (such as indirect-fire effects for SAFE) needed to be developed and added. As such, MUVES became the first vulnerability computation or estimation architecture.

A large quantity of data flows through the MUVES analysis process (see Fig. 4.44). In the most general terms, MUVES sets up the initial geometric and physical parameters of a threat, determines the results of an interaction between the threat and the target, and evaluates damage to the target based on that interaction.

The threat–target interaction is computed using a threat path. A threat path consists of geometric information for the components along a (not necessarily straight) path through the target, plus a set of parameters describing the physical attributes of the threat in question. The geometric information for a single component is called a component trace; a trace is a straight line through a component, containing the coordinates and normals at each intersection. Although each trace through a single component is linear, the threat path may be constructed from nonlinear traces through different components. The parameters describing a particular threat are stored in a data structure called a threat packet. These structures are keyed to specific types of threats (e.g., KE or SC) and sometimes to the algorithms used for evaluating their effects (e.g., Fireman–Pugh vs DSM evaluation for SC munitions); other information, such as geometric values, may also be stored in these packets as needed.

The initial raytracing is determined by the view specified by the analyst. A view specification may consist of a number of individually specified shots, a pattern of shots (e.g., a grid) relative to some defining orientation, or some combination of the two. After each ray in the shot pattern is traced through the target, the initial parameters of the threat are attached to the first component on the path, and this collection of data (now called an initial threat path) is passed to the interaction subroutines.

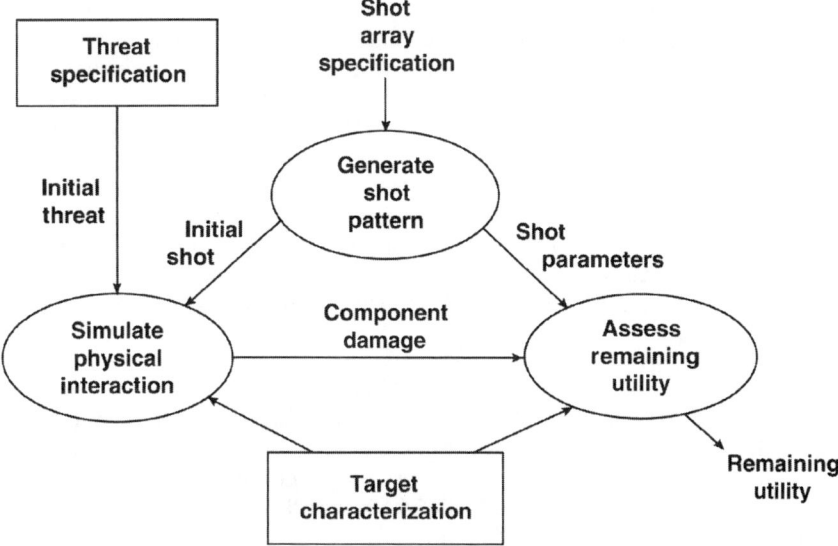

Fig. 4.44 MUVES analysis principal data flows.

MUVES traverses the threat path, computing interactions of the threat with each component in the path as it is encountered. At each component, several things may occur: damage may be recorded for the component; the threat path may be altered, usually by modifying the threat parameters and propagating them to the following component; and secondary threats (e.g., pieces of the main penetrator or material displaced during perforation of components) may be created.

If the interaction indicates that a threat must change direction, this may be performed by raytracing the new direction and attaching the threat parameters to the set of component traces produced along the new raytrace. This process is repeated until the threat no longer retains any damage-producing capability or there are no more components in the threat path. Along the way, secondary threats may be generated and their paths raytraced to produce new threat paths; these are processed in the same fashion until all damage-producing mechanisms have terminated. Damage to components caused by any threat is recorded and stored until all interactions for the shot have been processed. This damage information is recorded in data structures called damage packets, which define the type of resultant damage (e.g., blast, main penetrator impact, or spall fragments) and contain any parameters needed to describe the degree of damage. Damage is recorded only for those components deemed critical for the mission function(s) specified by the analyst. A critical component is any component whose loss affects the target's ability to perform its designated function. To be considered critical, a component must be referred to in an assessment expression (described in the next paragraph) used in the current analysis session.

Finally, some measure of the target's ability to function is evaluated from the damage to the individual components. This is called the target's functionality.

The damage for each component is collated, and a utility value is computed from the damage to that component during the interaction phase. Systems within the target are defined in system definitions, which are Boolean or mathematical combinations of component and subsystem utilities. The utilities of the systems in the target are evaluated based upon the utilities of their constituent components. The functionalities of the target are then computed from the relevant system utilities. These functionalities are determined for one or more contexts (the term used for the combination of mission functions the target is expected to perform and the environment in which it must operate). The contribution of each system to the overall functionality of the target in a given context is defined in a damage evaluation expression, which is also a combination of component or system utilities.

MUVES can easily assess damage for different mission functions (e.g., "mobility," "firepower," and "desert at night"). Although damage-correlation data are currently available only for a few standard mission functions, the eventual need for additional mission functions is foreseen. The VAMP-based (compartment model) programs could have handled this by adding new data and kill evaluations to the program and recompiling. MUVES can handle this through the analyst's input files without the need to recompile code.

Initially, only COMPART, a version of the compartment model, and SQuASH were placed within the MUVES environment. These "approximation methods" were intended to provide a unified environment for analyses of AFVs. Later, SAFE was added for indirect-fire analyses. In addition, to make MUVES generally applicable for all target classes, development of the Modular Aircraft Vulnerability Estimation Network (MAVEN) was started. Before it was completed, however, developers realized that SAFE and the MAVEN were functionally similar in many ways. Both were designed to analyze lightly armored or unarmored vehicles, both were designed to accommodate penetrator types that were small relative to antiarmor munitions, and both needed to analyze bursting warhead munitions. Thus, SAFE and the MAVEN were combined into one approximation method (still called SAFE). Finally, it was recognized that SAFE and SQuASH also had many common features and functional requirements, and SAFE and SQuASH were combined into what is now referred to as S2 (Hunt et al. 2004).

In addition to furthering the original goals of MUVES to reduce duplication of code required by each approximation method, to obtain better configuration management, to improve validation, to better share input files, and to obtain greater flexibility for post-analysis processing, MUVES-S2 offered the distinct advantage of combining features from the individual approximation methods. For example, before MUVES-S2, SQuASH could be used to compute the vulnerability because of an SCJ and BAD. SAFE could be used to compute vulnerability because of SC warhead casing fragments. However, the only way to combine all three of these tasks was to employ some sort of algorithm to add these effects during post-processing of data generated by the individual approximation methods. Accordingly, MUVES-S2 provided the means to integrate all three of these tasks damage mechanisms during the analysis and determine and combine the vulnerability contributions of each method in a logical and nonoverlapping manner. Work is under way to incorporate other damage mechanisms into MUVES-S2 (and/or its planned successor, MUVES 3), including internal and external blast from missiles and large explosive projectiles, ballistic shock, and more robust crew casualty estimation methods.

B. Advanced Joint Effectiveness Model

Commonality of analysis methodologies, processes, and input data among the Services has long been pursued through the JTCG/ME and the JTCG/AS [now part of the Joint Aircraft Survivability Program Office (JASPO)]. Consistency among the Services has also been encouraged by the OSD to standardize live fire test and evaluation (LFT&E) procedures and practices. Over the course of several decades, this type of collaboration, although sometimes difficult to effect, has produced much better understanding of V/L phenomena, algorithms to quantify these effects, and data to support algorithm development and V/L model operation. Some progress was also made in sharing models for system-level V/L analyses. For example, COVART was used by all three Services for aircraft analyses for many years. MUVES-S2 offered the opportunity to collaborate on the development of a stochastic, target-independent V/L analysis model.

In the early 1990s, the JTCG/ME and JTCG/AS recognized the need to replace the Joint Service endgame model. The new model (which is still used today) was the Advanced Joint Effectiveness Model (AJEM) (Hunt et al. 2004). AJEM is a lethality, vulnerability, and endgame computer simulation code capable of analyzing one or more threats attacking a single rotary- or fixed-wing aircraft or ground-mobile target. It combines elements of target and threat modeling, encounter kinematics, weapon burst-point generation, damage mechanisms propagation (e.g., penetration, fire, blast) to and interaction with the target, target system relationships (functionality, redundancies, etc.), and remaining capability or LoF of the target.

AJEM was designed to run in conjunction and integrate closely with the previously discussed BRL-CAD, capitalizing on the extensive modeling and geometric representation details necessary for preparing target descriptions. As an effectiveness model (as opposed to a V/L analysis model), AJEM contains an encounter module and a V/L analysis model. After several years of design and development, MUVES-S2 was designated the V/L analysis module (sometimes referred to as the V/L engine). This designation took advantage of extensive development of MUVES-S2 for stochastic V/L analyses, provisions for major upgrades to implement new or enhanced methodologies and validation based on extensive live fire test data.

Beginning around 2003, a robust and rigid configuration management process was institutionalized for MUVES-S2 maintenance and development. The AJEM model manager was given responsibility for configuration management to ensure tri-Service needs are accommodated. All changes, ranging from bug fixes to methodology enhancements in MUVES-S2 and BRL-CAD, are ported to AJEM through regular managed releases. (For more detail and the latest information about AJEM, visit www.ajem.com.)

C. Modular Effectiveness Vulnerability Analysis (MEVA) Program

Another vulnerability architecture that is available is the MEVA program (Dunn et al. 1998). This program provides assessments of air-to-surface and surface-to-surface threats against buildings and bunkers, as well as the equipment and personnel within. The features include damage propagation within the target, accumulation of damage on and within the target over multiple weapons or threats, multiple target interrogation methods in addition to raytracing, and a novel computer program execution environment.

The MEVA program is a graphical and modular software environment developed for the U.S. Air Force for the purpose of specifying, executing, and visualizing the result of threat–target interaction vulnerability assessment simulations. MEVA is composed of an editor and modules. The MEVA editor is a C++ program that provides a graphical environment for constructing survivability/vulnerability calculations and features multi-process communication among modules. Modules are independently compiled and linked vulnerability analysis algorithms that are executed by the editor to provide input, analysis, output, and post-processing for survivability/vulnerability studies.

MEVA is derived from the Effectiveness/Vulnerability Assessment in 3-D (EVA-3D) program (Maestas et al. 1995). This program was focused primarily on buried hardened targets such as concrete bunkers. For this class of target, it was essential to model the 3-D path that a penetrating bomb would traverse through the target, the potential for weapon failure (bomb case breakup before detonation, because of unfavorable impact conditions with concrete slabs), warhead fuzing, and target damage because of the detonation effects. Although the 3-D penetration paths were important and ultimately became part of the program's name, the more important features in EVA-3D were propagation and accumulation of damage. Previous assessment methodologies assumed that all damage from a bomb detonation inside a hardened target was confined to the room or compartment in which the bomb exploded. The EVA-3D methodology allowed propagation of detonation effects from room to room within the target, contingent upon breaching the walls in between the rooms. Breach events were calculated with algorithms that represented the physical response of walls to the detonation effects. With this feature, the extent of damage calculated was much more accurate for weapons that could breach the interior walls.

Damage accumulation in EVA-3D was accomplished by modifying the target model based on the damage calculated for each weapon. Thus, if the first weapon in an attack penetrated into the target and detonated, breaching walls or floors in the target, the damaged areas of the target geometry were removed from the model. If the next weapon penetrated into the target near the location of the first, it would not find the damaged walls, but instead might fly through the voids or holes in the damaged walls. This feature was essential to accurately assess situations requiring multiple rounds and the accumulation of physical damage on the target to achieve a kill.

The MEVA program embodies these features of EVA-3D and added volumetric target interrogation methods to the standard raytracing techniques, additional structure in the analysis of weapons effects, above-ground buildings in addition to buried bunkers as targets, and a novel computing environment. In MEVA, all target interrogation is performed in a single module, so that all operations on the target and information from the target can be kept consistent. The target interaction module supports raytracing as well as volume intersections to support weapons effects other than ballistic interactions. It also timestamps all damage to the target so that the evolution of damage on the target can be examined after the simulation.

MODELING AND SIMULATION TOOLS AND METHODS 201

Because of the complexity of the physical interaction of the weapon effects with the target structure, the MMF mapping from initial to damaged states (the $O_{1,2}$ mapping) was subdivided into propagation, interaction, load, and response categories. This was done to facilitate verification and validation (V&V) efforts and evaluate the synergism of the weapon effects between environments. Each of the categories is described in further detail in the following text.

The propagation category of algorithms is used to model the manner in which each weapon effect (or environment) propagates through the free field (i.e., space without a target). For instance, air blast or shock is transmitted through the atmosphere in a prescribed manner, and fragments fly through the atmosphere governed by aerodynamic drag and gravity. These effects are well understood, documented, and accepted. The next step involves algorithms that describe the interaction of the weapon effects with the target (or some portion of it). When air blast impinges on a target, it applies pressure against the target surfaces. In addition, the interaction of the target with the air shock causes a reflection, thus increasing the pressure. For fragments, the interaction with the target is characterized with penetration, residual velocity, hole size, and ricochet equations. The interactions in turn develop loads, both on the target structure and the impinging effect. It is then the target's response to these loads in which we are interested.

Although not all weapon effects and target response algorithms map directly into these four categories, it is useful to examine each effect from this point of view. Ultimately, the target structure responds to the loads applied, and to some extent it does not matter how those loads are developed. The loads could develop through the combination of several weapons effects, or as a result of the application of several weapons. The propagation-interaction-load-response approach can be used to great advantage in characterizing the physical response of a target to combined or synergistic effects.

One of MEVA's most important features is execution environment. The program is constructed to provide an environment that is independent of the algorithms used in the survivability/vulnerability analyses. As mentioned previously, MEVA is composed of modules and an editor. Each module is its own independent program that provides input, analysis, output and post-processing functions. The editor is a graphical environment for constructing and executing calculations, and it is a mechanism for facilitating inter-process communication. Because the editor and modules are independent, new modules can be designed, developed, and tested for MEVA without affecting other modules in the system.

Because MEVA calculations may be complex and require numerous user inputs, the graphical user interface (GUI) is another critical feature of MEVA. Each module includes a GUI that provides a straightforward tool for data input. Constructing calculations is accomplished by pulling modules into the editor palette and linking the modules together into a calculation network. Figure 4.45 shows a fully linked MEVA network.

Note that the MEVA network flow naturally follows the MMF taxonomy. The weapons and target are defined at the top, along with the initial conditions between weapons and target (the MEVA delivery module). The penetration, air blast, and other effects modules perform the mapping from level 1 to level 2 by calculating the damage on the target. The mission evaluation module performs the mapping from damage state to a capability or operability state. The flow of execution of the

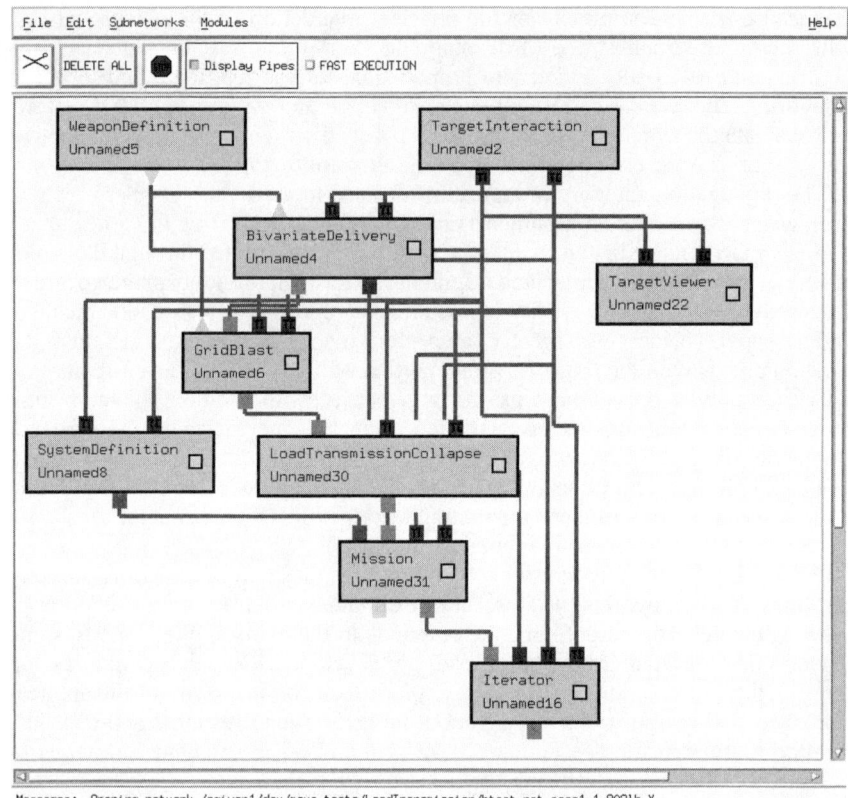

Fig. 4.45 MEVA editor window with a fully linked analysis network.

analysis is downward through the modules, just as in the taxonomy. The request for execution in MEVA, however, comes from the bottom. The iterator module is at the bottom of the network, and it requests the modules above to execute. Each module executes when its inputs have been updated, but those inputs might be outputs of another module further up the network. The request for execution thus propagates up the network, much like system requirements propagate up the taxonomy.

D. ORCA Program

The next-generation soldier survivability or crew casualty assessment methodology is the previously mentioned ORCA model (Neades and Davis 2002). Since 1997, model development has primarily occurred through ARL/SLAD, which today owns and configuration-manages the ORCA model. ORCA embodies state-of-the-art codes, algorithms, and supporting data to calculate the anatomical and physiological damage done to a human body by the insult. The model includes an improved methodology and enhanced capabilities compared with both of the previously mentioned Sperrazza–Kokinakis and ComputerMan methodologies

(see Sec. 4.III.E), and it provides an improvement for all phases of the casualty assessment process, including the following (Davis et al., to be published):
1) More accurate anatomical representation.
2) Mapping of insult(s) to injury evaluation.
3) Mapping injury to medical casualty characterization.
4) Mapping injury to physical and cognitive impairment.
5) Mapping job and task requirements to basic human capabilities.
6) Evaluating basic human capability requirements to post-injury capabilities.
7) Calculating both injury type and severity and operational casualty metrics.

In addition, the ORCA contains (or is planned to contain) modules that permit casualty assessments for all of the following insults: blast overpressure, penetration, thermal, directed energy, toxic substance inhalation/contact, blunt trauma, acceleration.

The ComputerMan digitized anatomical representation and fragment penetration methods are in the ORCA program to analyze injuries from penetration (Clare et al. 1980; Saucier and Kash 1994). Nonauditory blast overpressure injuries are modeled using the injury model, which was jointly developed by the U.S. Army WRAIR and JAYCOR, Inc. (JAYCOR 1995). Blast-overpressure-induced auditory injuries are characterized using a combination of the WRAIR/JAYCOR injury model and an Army model developed by Price and colleagues (Price and Kalb 1991a, 1991b). Injuries associated with rapid acceleration are modeled using the U.S. Air Force's articulated total body (ATB) model (Obergefell et al. 1988) to define the physical forces on the human body and the U.S. Army's Battlefield Laser Acquisition Sensor Test (BLAST) model developed by Miller and Carver (1998). Thermal and burn injuries are characterized using the BURNSIM model developed by U.S. Army and Air Force researchers Knox and colleagues (Knox et al. 1993). Chemical insults and resulting injury are modeled using the U.S. Army Chemical Computer Man – Chemical Response Simulation (CARS) model developed by Davis and Mioduszewski (1988). Blunt trauma and acceleration modeling is still being developed.

In the ORCA program, all injuries are recorded in a standard form via a vector whose elements refer to the state of over 473 body parts that have been defined to constitute the human body. Each element has an associated scale through which the severity of the damage is recorded. For example, in the case of a blast overpressure exposure, the lungs and ears, being the main organs at risk, are the relevant elements of the injury vector. A numerical value ranging from one to five is then used to indicate the severity of damage to the lung or ears as a result of the blast insult.

The determination of medical casualties was not within the charter of the CCWG. However, it was essential that, to the point that medical and operational casualty factors are common, the ORCA model be consistent with the needs of the medical community. DOD Medical Command representatives were involved with the ORCA model development; to this end, significant care was taken to define and record injuries in such a way as to serve future medical analysis needs. In particular, the ORCA model determines and keeps track of each injury's Abbreviated Injury Severity (AIS) score, the previously mentioned anatomical-based injury characterization system common throughout the civilian medical community (American Association for Automotive Medicine 1985).

Significant work was devoted to correlating the various types and levels of injury to the impairment of an individual's capabilities. These capabilities are formally recorded via the elemental capability vector (ECV). The elements of the ECV are representative of the capabilities that humans use to accomplish tasks. The elemental capability set contained in the ORCA model, and shown in Table 4.10, was established as a direct result of a 1994 CCWG Workshop on Human Performance Modeling. Each of these elements is quantified by a set of parameters that permits the consideration of between two and eight levels of capability.

In the assessment process, an anatomical injury is characterized at the time it occurs. However, it is recognized that the effects of physiological injury may change over time. To capture this effect, the ORCA model calculates a number of impairment vectors, one for user-selected post-wounding times ranging from immediate to 72 hours. Currently, the ORCA model can evaluate any selected post-wounding time period in this range. Each vector is in turn compared with the job requirements to determine job performance or casualty status over time.

As previously described, an operational casualty is defined with respect to a particular task or job. Thus, in addition to defining the insult, the task or job of an individual must be input using MOS, NEC, or AFSC job description information. These military jobs have been defined by a list of basic physical, cognitive, and sensory tasks. Each basic task description in the ORCA task library is described using the elemental capability vector. For instance, a pilot's job contains several tasks such as operating controls, reaching above, visual mental processing, communication, and hearing. Each task is described by the ECV, and the job consists of the summation of all the tasks needed by the pilot. At present, ORCA contains task information on the following six military jobs: infantryman, firefinder radar operator, ammunition specialist, light-armor-vehicle crewman, attack-helicopter pilot, air-transport pilot. The addition of other military jobs and specialties that span the Services is planned for the future.

Besides considering the task requirements related to an entire job description (e.g., infantryman), an analyst might want to evaluate only a subset of the tasks. These may be the tasks associated with a predefined or user-specified mission that is related to a specific scenario. Because all jobs are broken down into basic tasks and ultimately into elemental capability requirements, each of these possibilities

Table 4.10 Elements of the ECV

Visual acuity and color discrimination	Somatic senses	Left leg strength
Night vision	Balance	Right leg strength
Visual field of view	Cognitive mental processing	Left arm/hand strength
Visual binocularism and motility	Visual mental processing	Right arm/hand strength
Hearing threshold – low freq.	Auditory mental processing	Left hand/arm dexterity
Hearing threshold – high freq.	Psychomotor mental processing	Right hand/arm dexterity
Binauralism	Speech articulation	Torso support
	Vocal power	Head/neck movement
		Endurance (aerobic)

will be an option. A user-friendly interface allows an ORCA model user to describe his or her own tasks and jobs through common terms and images.

VIII. Force-Level Modeling
A. Combined Arms and Support Task Force Evaluation Model

By force-level models, we mean the simulation of the combat between opposing forces in which interactions can be simulated at levels as low as individual weapon systems. These can involve various sizes of opposing forces and various levels of detail. They usually involve large computer programs and can also involve manned play. For the most part, these models have used rather crude V/L summary data that expressed whether a system was totally functional or totally nonfunctional in its major capabilities (e.g., the abilities to move and fire).

The availability of the more detailed description of system functionality given in level 3 data has important consequences for force-level modeling. It permits more accurate portrayal of the system's remaining capabilities as a result of nonfunctional components, either through combat damage or failure, and allows the system to be modeled as still functional but in a degraded operational mode. This could result in a system remaining in the engagement longer and possibly affecting the outcome of the game. In addition, information from level 2 component damage could provide detailed information on damaged parts, thus providing more accurate data on spare parts requirements and the need for battle damage repair.

Efforts have been made to integrate MMF level 3 metrics into a large-scale combat model called the Combined Arms and Support Task Force Evaluation Model (CASTFOREM) to determine the resulting effect on the war game. A proof-of-principle effort in the mid-1990s demonstrated that level 3 metrics could be integrated into the model and that these metrics had an impact on battle outcome. Significant observations were made during this implementation. These observations varied among modeling shortfalls, computer resources, and the test bed parameters. Some may be described as findings of the study, while others should be taken strictly as observations of problems that were encountered during the course of the effort.

As the proof-of-principle effort showed, CASTFOREM can use specific system capability information that is provided through level 3. Information, such as the probability of acquiring a target and a mean time to acquire, can be provided to CASTFOREM to adjust specific vehicle performance parameters.

Also, because the initial effort was limited to a single type of vehicle and a small scenario, current efforts are aimed at developing level 3 metrics for broad classes of systems and integrating these into CASTFOREM for a combined-arms scenario. This work will provide a clearer picture of the benefits of level 3 metrics in force-level models. A standard set of state definitions should be jointly developed by the vulnerability and the data user communities to make them useful to as many analysts as possible. In addition, model development has been both driven and hampered for many years by data availability. The MMF level 3 provides a mechanism to develop and produce vulnerability data that satisfy the whole spectrum of data needs in the community, but it requires a standard set of states with documented and auditable definitions, produced by both analysts and model developers, so that the data needs for all existing and proposed models and simulations can be satisfied.

Finally, integration of these metrics into CASTFOREM also will involve changes to CASTFOREM's response and decision tables. Along with a standard set of states, development of a corresponding standard set of responses and residual-capability values is needed. The involvement of materiel developers is also needed to provide the user relevant information on what the battlefield effects are for each given state (e.g., if a state is reduced mobility, then how reduced and under what circumstances?).

B. Unit Reconstitution Models

Around 1978, there was a Congressionally inspired program, called Theater Nuclear Force Survivability, to assess how well a theater nuclear force would survive after conventional strikes and possibly after a preemptive nuclear strike. A multiagency team, which included BRL, the Harry Diamond Laboratory, AMSAA, and the TRADOC schools, was created, with BRL being assigned the task of deciding whether the theater nuclear force would survive a conventional attack. Because methodologies for determining what happens on the integrated battlefield were not available, the Residual Combat Capability (RCC) model, and later, the AURA model were developed to address this concern.

The RCC model is an event-sequenced simulation model. If the event is a lethality event (e.g., the arrival of an artillery volley), the effect is calculated. If the event is a reconstitution event, all the time-dependent events are updated, the status of the unit is reported, and the commander reviews the status and does an optimized reallocation of remaining assets. The output shows how well the unit can do its job with the remaining assets.

Around 1982, the RCC became the AURA model to reflect the emphasis on the unit. AURA is a large, interconnected collection of analysis models that provides a detailed evaluation of the ability of an Army unit to accomplish a series of missions in a combat scenario. It is an event-sequenced, one-sided combat-simulation methodology consisting of a number of highly detailed models from the various technical communities interfaced into a large, time-dependent, event-playing, and optimization routine. The interfaces are varied, involving such diverse kill probabilities as lethal footprints for conventional munitions, log-normal kill probabilities for nuclear effects, toxic-chemical dispersions and evaporations, mission-oriented protective posture (MOPP) degradation, reliability, and target-acquisition probabilities. The optimization is a dedicated, nonlinear routine that models the commander's reallocation of the surviving, potentially degraded assets to minimize the choke points in the optimal functional path. The logic process required the development of a general model for the functional structure of a military unit. Such a model was developed and forms an essential part of the AURA methodology (Klopcic 1984).

IX. Verification, Validation, and Accreditation for V/L Assessment

With the DOD increasingly investing in and relying on modeling and simulation (M&S) tools to support V/L analysis (as well as a wide array of other military

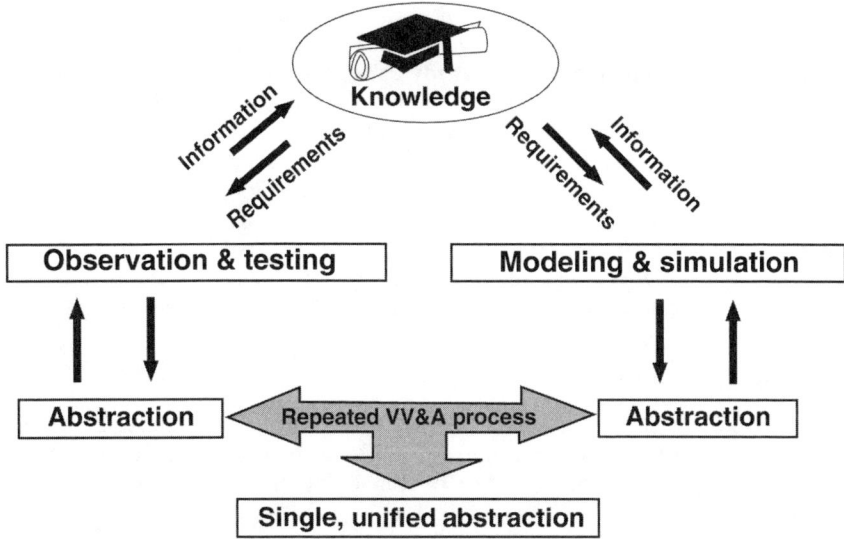

Fig. 4.46 The role of VV&A in unifying testing and simulation.

activities), it is increasingly important that these tools be rigorously checked for accuracy, realism, and suitability for the intended application. The formal process for this checking is known as verification, validation, and accreditation (VV&A).

In general, as shown in Fig. 4.46, effective V/L assessment is primarily based on an accumulation of knowledge that is derived from two sources: 1) observation and testing, and 2) M&S. It is thus the job of the VV&A process to compare the parallel abstractions that these two sources produce, resolve any differences (e.g., in focus, level of granularity, application, semantics, etc.), and achieve a single, unified abstraction that can be a reference source of analytical knowledge for V/L decision-makers.

In addition, just as M&S can provide the V/L community significant benefit in terms of reducing the time and resources needed for extensive (and costly) live fire testing (LFT) and experimentation, VV&A can provide the V/L community significant benefit in terms of reducing the time and resources needed to verify the credibility of an M&S tool. Specifically, VV&A provides a structured framework for internally auditing the M&S development process by comparing computer programs, the mathematical models on which they rely, and the perceived reality that they attempt to simulate. In so doing, VV&A can also serve to facilitate close coordination between users, developers, programmers, subject-matter experts, and any other individuals/organizations involved in the V/L analysis process.

A. VV&A Policy Documents and Resources

There are numerous sources that provide instruction on how, when, and under what circumstances formal VV&A procedures should be executed. They include numerous DOD and Service policy documents, such as DOD Directive 5000.59

(Office of the Secretary of Defense 1994), Instruction 5000.61 (Office of the Secretary of Defense 1996), Army Regulation 5-11 (Headquarters, Department of the Army 1997), Department of the Army Pamphlet 5-11 (Headquarters, Department of the Army 1999), and Developmental Test Command Pamphlet 73-4 (Developmental Test Command 1998). In addition, the Defense Modeling and Simulation Office (DMSO) has developed the "VV&A Recommended Practices Guide" (RPG) (Defense Modeling and Simulation Office 2002), which is accessible via DMSO's web site (http://vva.dmso.mil). Finally, the IEEE Computer Society Press has published *Simulation Validation: A Confidence Assessment Methodology* (Knepell and Arangno 1993), which provides practical concepts and methodologies of simulation assessment, especially as applied to large-scale projects.

B. Definitions of VV&A

Although VV&A is often discussed as a single process, it is made up of elements that focus on an M&S tool from three different perspectives. These elements are as follows (Office of the Secretary of Defense 1996):

Verification   The process of determining that a model implementation and its associated data accurately represent the developer's conceptual description and specifications.

Validation   The process of determining the degree to which a model and its associated data provide an accurate representation of the real world from the perspective of the intended uses of the model.

Accreditation   The official certification that a model, simulation, or federation of models and simulations and associated data are acceptable for use for a specific purpose.

In idiomatic terms, the three VV&A elements can be described as answering a series of simple questions. Verification roughly answers the question "Did I build the thing right?" Validation roughly answers the slightly different question "Did I build the right thing?" And accreditation roughly answers the question "Is the thing appropriate for my use?" (Knepell and Arangno 1993). Figure 4.47 illustrates how V&V are used to compare and connect mathematical and computer models to real-world phenomena (Office of the Secretary of Defense 1994).

It is important to note that success or failure for one of these elements is not necessarily an indicator of success or failure for either of the other two. After all,

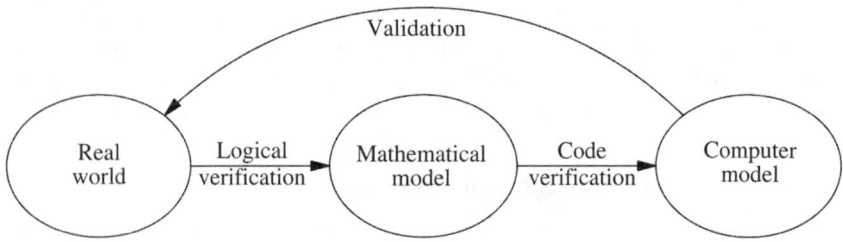

Fig. 4.47   **Verification and validation.**

a model could be determined to be flawless in its design, execution, and compliance with a user's concept and specifications (verification) but still be assessed as unusable when it comes to accurately simulating a specified real-world V/L phenomenon (validation). Likewise, a model could be assessed as accurate in simulating a specified V/L real-world phenomenon for a given purpose or application but still be assessed as inaccurate if used for a different purpose or application (accreditation). For these reasons, VV&A processes must be exercised individually before being used collectively to answer the ultimate idiomatic question about a V/L model or simulation: "Should the thing be used?"

C. Typical Participants of VV&A

Not surprisingly, the implementation of the VV&A process, especially in the complex field of V/L analysis, can be labor intensive. Often, the individual Services and other government agencies employ their respective VV&A processes in different ways, according to their own operational structure, doctrine, policies, procedures, and priorities. In addition, each of the Services has its own set of names to describe the roles of the key people/groups/organizations that participate in its particular VV&A processes [see Table 4.11, which is taken from Defense Modeling and Simulation Office (2002)].

But regardless of the way in which participants are named or individual duties are assigned, the overall VV&A process is roughly equivalent for all Services and can be seen to fall under the following six common basic roles of responsibility.

User   The individual/organization that needs to solve a problem or make a decision using a simulation. The user defines the requirements, establishes the criteria by which simulation fitness will be assessed, determines what method(s) to use, makes the accreditation decision, and ultimately accepts the results.

M&S program manager   The individual/organization responsible for planning and managing resources for simulation development/modification, directing

Table 4.11   Comparison of names of typical VV&A participants

DOD instruction 5000.61	Army	Navy	Air Force
Accreditation authority	Application sponsor	Accreditation authority	Accreditation authority
M&S application sponsor	Application sponsor	Accreditation authority	Accreditation authority
M&S user	Application sponsor	User	User
—	Developer	Developer	Model developer
M&S proponent	M&S proponent	Proponent	Model manager
Verification agent	V&V agent	V&V agent	V&V agent
Validation agent	V&V agent	V&V agent	V&V agent
Accreditation agent	Accreditation agent	Accreditation agent	Accreditation agent
—	Subject-matter expert	Subject-matter expert	Subject-matter expert

the overall simulation effort, and overseeing configuration management and model maintenance.

Developer The individual/organization responsible for actually constructing or modifying the simulation, preparing the data for use in the simulation, and providing technical expertise regarding simulation capabilities as needed by the other roles.

V&V agent The individual/organization responsible for providing evidence of the simulation's fitness for the intended use by ensuring that all the V&V tasks are properly carried out.

Accreditation agent The individual/organization responsible for conducting the accreditation assessment, providing guidance to the V&V agent to ensure that all the necessary evidence regarding simulation fitness for use is obtained, collecting and assessing the evidence, and providing the results to the user to make the decision.

Subject-matter expert (SME) An individual/organization that is recognized as an authority in a specific area and that can contribute, as a secondary player, to a VV&A effort by providing expert opinion in a variety of areas (e.g., the problem domain being simulated, required input data, and required computing technology) as well as assisting the user in various activities (e.g., establishing requirements and acceptability criteria and assessing validation and accreditation results).

The basic VV&A responsibilities for these roles are illustrated in Table 4.12, which is taken from Defense Modeling and Simulation Office (2002). Note that a single person, group, or organization could perform one or several roles depending on the size of the project. Detailed discussions of the participants and their responsibilities are provided in DMSO's RPG as well as in the other previously mentioned VV&A resources.

D. The VV&A Process

VV&A is sometimes thought of as an external step in the M&S development/modification process; however, for it to have maximum benefit for V/L analyses, it must be an integral part of all phases of the M&S life cycle, with all participants working cooperatively to assess the strengths and weaknesses of the M&S tool, methodologies, and data [see Fig. 4.48, also taken from the Defense Modeling and Simulation Office (2002)]. VV&A plans and processes should begin on the first day of development and continue throughout the process. In addition, the same documentation used for M&S requirements, design, development, and configuration control should be used to support VV&A activities.

1. Verification

Verification not only determines whether or not an M&S tool accurately represents the developer's conceptual description and specifications (as designated in the requirements documentation), but it also evaluates the extent to which the model or simulation incorporates sound software engineering techniques and methodologies. For large-scale M&S development projects (which are common in V/L analysis), verification is often executed at each stage of the M&S life cycle

Table 4.12 Typical roles and responsibilities of VV&A

Activity	User	M&S PM	Developer	V&V agent	Accreditation agent	SME
Define requirements						
Define measures						
Define acceptability criteria						
Plan M&S development or modification[a]						
Develop V&V plans						
Develop accreditation plan						
Verify requirements						
Develop conceptual model[b]						
Validate conceptual model						
Develop design[c]						
Verify design						
Implement design						

(*Continued*)

Table 4.12 Typical roles and responsibilities of VV&A (continued)

Activity	User	M&S PM	Developer	V&V agent	Accreditation agent	SME
V&V data	Approve	Monitor	Assist	Lead	—	Perform
Verify implementation	Approve	Monitor	Assist	Lead	—	Assist
Test implementation	Approve	Monitor	Lead	Assist	—	Assist
Validate results	Assist	Monitor	Assist	Perform	—	Assist
Prepare V&V report	—	Monitor	—	Lead	—	—
Configure for use	Monitor	Lead	Assist	—	—	—
Gather additional accreditation info	—	—	—	Assist	Lead	—
Conduct accreditation assessment	Monitor	—	—	—	Lead	Assist
Prepare accreditation assessment report	Review	—	—	—	Perform	—
Determine accreditation	—	—	—	—	Perform	—
Prepare accreditation report	—	—	—	—	Perform	—

Legend

- **Lead** (yellow): Leads the task. Normally involves active participation from others.
- **Perform** (red): Actually does the task. Normally involves little active participation from others.
- **Assist** (blue): Actively participates in task (e.g., conducting tests, providing information).
- **Review** (brown): Participation normally limited to reviewing results of task and providing recommendations.
- **Monitor** (dark blue): Oversees task to ensure it is done appropriately but does not normally participate.
- **Approve** (green): Determines when an activity is satisfactorily completed and another can begin. Determines what activity should be pursued next (e.g., whether to continue on to the next scheduled activity or to return to a previous activity).

[a] This activity refers to planning and scheduling of any M&S development, modification, or preparation.
[b] This activity refers to development of new, as well as modification of existing, conceptual models.
[c] This activity refers to development of new M&S designs as well as modification of existing M&S designs.

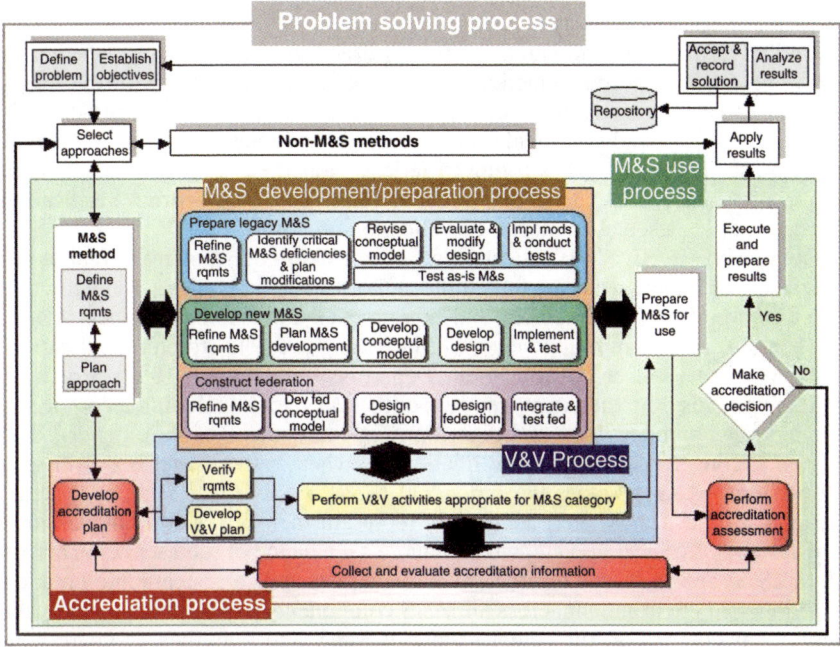

Fig. 4.48 VV&A in the M&S life cycle.

to verify that the "products" of each stage accurately implement the output from the previous stage.

There are two main components to M&S verification: 1) logical verification, and 2) code verification (Headquarters, Department of the Army 1999). Logical verification is the process that ensures that the implemented algorithms function as intended (e.g., checking the independence of certain events or comparing pseudocode logic with the actual code implementation). Methods to execute logical verification include management and peer reviews, algorithm checks, design "walk-throughs," submodel-to-submodel interface (or model-to-model interface in a federation) analysis, and traceability assessments. There are also computer-aided software engineering (CASE) tools available to help the verifier analyze and troubleshoot logical implementations. Code verification, on the other hand, is the process that ensures that all compilable code properly implements the verified logic in an M&S tool (e.g., checking that there are no division-by-zero errors occurring at boundary conditions or ensuring the stability of given mathematical properties in a specific computer hardware/software environment). Methods to execute code verification include peer review, sensitivity analyses, code walk-throughs, mathematical stability testing, unit checking, statistical testing for repeatable and nonrepeatable stochasticism, and rule-based system checking. In addition, automated test and CASE tools (e.g., variable name spell checkers, memory maps, subroutine call trees, and static code analyzers) can greatly assist in the code verification process.

2. Validation

Unlike verification, validation is not concerned with determining whether or not M&S logic and code function are accurate and consistent with the user's design and specifications. Instead, validation is concerned with ensuring that this logic and code together provide an accurate representation of the real-world phenomena the M&S tool is trying to reflect. Nonetheless, the complementary nature of verification and validation means that some results from verification tests can also be used as input for the validation process.

Obviously, effective V/L validation requires clear understanding of real-world processes, as identified through credible authorities such as SMEs, scientifically accepted algorithms and theories, and experimental and test measurements/data.

There are two main components to M&S validation: 1) structural validation, and 2) output validation (Headquarters, Department of the Army 1999). Structural validation focuses on the internal portion of the M&S tool, including incorporated assumptions, architectures, and algorithms in the context of the intended use. Common concerns in structural validation are whether the model is sensitive to proper input data, whether the individual M&S pieces (e.g., functional areas, weapon systems, units, etc.) adequately represent their real-world counterparts, whether the M&S is complete, whether there's a balance of representation across all M&S components, and whether there is adequate and consistent representation of terrain and environment across all M&S components. By contrast, output validation concentrates on the actual results of M&S and how they compare with perceptions of the real world. Concerns commonly addressed in output validation are whether the M&S output is "feasible"; whether it is reasonable relative to the input; whether differences in input produce proportional changes in output; whether M&S output compares well with historical, test, laboratory, and/or exercise data; and whether graphical outputs and visualizations are realistic.

There are several technical methods commonly employed for both structural and output validation. The first is face validation, a process that involves SMEs performing a high-level "surface" check to ensure that a model's performance is reasonable. Other methods include comparison to previously accredited M&S internal algorithms, code, or documentation; functional decomposition (or "piecewise validation"); stress tests and sensitivity analysis; turing tests; and graphics and visualization playback.

In addition, a particularly effective validation method in the field of V/L analysis is the model-test-model (MTM) approach. This approach requires close coordination between the test and evaluation and M&S communities and involves a series of pretest modeling (or prediction) of a particular scenario, followed by actual scenario execution (e.g., through field/laboratory tests or experiments), followed by comparison of results and refinement of the M&S tool, followed by post-test modeling.

If needed, the series can be repeated multiple times; however, as is discussed in succeeding sections, the resource-intensive nature of typical V/L testing is such that multiple replications of the series is often not possible.

3. Accreditation

The final element in the VV&A process is accreditation, a process by which the user formally issues a "stamp of approval" for an M&S tool to be used for a specific application or purpose. At this point, questions regarding the correct functionality of the M&S tool and its comparability to real-world phenomena should have been answered. All that is left to confirm is whether or not the tool will meet the objectives of the user.

There are two basic types of accreditation: 1) class accreditation and 2) application-specific accreditation. The basic procedure is relatively equivalent for both types. As its name implies, class accreditation is an endorsement of an M&S tool for a generic set or class of applications. In addition, it produces a core accreditation report that can serve as a starting point for application-specific accreditation. But application-specific accreditation is the endorsement for each specific use of an M&S tool. Note too that tools that have been class accredited do not necessarily need to be entirely accredited for a new application. Rather, as long as the application falls within the parameters of the class, only those parts of the tool that exclusively address the new application need to undergo another VV&A procedure (Headquarters, Department of the Army 1999).

The first step in accreditation is to formally establish the accreditation acceptability criteria for the intended application. These criteria, which are usually defined in the accreditation plan, help define what "success" is for the M&S tool as they identify the essential issues that the tool must be able to address to be considered acceptable for use. The second step is a high-level M&S review. This step includes reviewing the audit trail for the development and use of the tool, the V&V documentation, configuration management procedures and records, assumptions, and previous successful uses. Other factors that can influence the accreditation decision include the developer's past history, the hardware configuration and software support required, and personnel and security issues (Headquarters, Department of the Army 1999).

E. Common Challenges to the VV&A of V/L Processes

As today's increasingly advanced computer technologies facilitate the design and development of increasingly realistic M&S tools, VV&A can be a highly complex process to execute. Moreover, the field of ballistic V/L assessment inherently involves a number of characteristics that can make VV&A even more challenging. These characteristics, which are discussed in turn, include the following: small sample sizes; variability within materials, conditions, and processes; stochastic and unpredictable outcomes; and nonlinear thresholds.

It has been said that the "art" of VV&A lies in the sampling of the data. Unfortunately, the nature of ballistic V/L testing and assessment of combat systems is such that VV&A personnel must typically work from few samples (especially as one moves up the MMF structure toward full-up system-level and force-level analysis at level 4). In addition, this characteristic of V/L analysis presents VV&A personnel with a particularly formidable challenge in that it serves to exacerbate the other previously listed challenges.

There are numerous reasons for the typically small sample size in V/L assessment. First and foremost is cost. Combat materiel assets are often expensive and difficult to acquire (especially for the purpose of destructive testing), and live fire testing and experimentation often require specialized manpower, equipment, and facilities (e.g., DU testing). Thus, fully implemented VV&A in the V/L arena is relatively rare, and assessors are often challenged by users to focus on only a few elements of the overall process. Another reason behind small sample sizes is the type and frequency of events that V/L models attempt to simulate. Battlefield conditions and phenomena are relatively rare occurrences and, therefore, are often difficult to measure, simulate, and predict. In addition, the catastrophic damage that commonly results from physical testing often precludes a sufficient number of test replications from being performed.

In addition, even if reduced to their most basic forms, ballistic V/L testing and analysis can involve many variables. There are different structures/compositions of the materials themselves (i.e., no two plates are alike), different configurations and conditions of those materials (i.e., no two tanks are alike), different complex physical processes occurring in target-threat events (i.e., no two bullet strikes are alike), and different environmental and operational conditions surrounding the event (i.e., no two missions are alike).

Also, the multivariate nature of many V/L processes can be significantly worsened by tests and analyses that attempt to address too many variables all at once, a problem known as a product of parameter growth. To avoid this situation, V/L assessors try to design tests and analyses with variables that are mutually exclusive and exhaustive.

Because of their multivariate nature, and thus an inability to simulate all the contributing factors, V/L events are also highly stochastic (i.e., random) and unpredictable. They are like a baseball player stepping up to the plate: even though he may have gotten a hit the last time at bat (or the last 10 times at bat), one cannot predict with any certainty the outcome he will achieve this time. There are too many variables and too much randomness to consider and calculate. One could, of course, make predictions based on probabilities (e.g., considering the hitter's batting average in this type of situation), but a reasonable amount of confidence would require many samples/replication, which, as mentioned, is not a normal convenience in V/L studies.

Finally, V/L processes do not often behave linearly, and the math in M&S tools is notoriously weak in making nonlinear predictions. The effect of this problem is that several observations/measurements in a nonlinear configuration may reflect one outcome, an outcome that may be in total disagreement with reality. For example, in characterizing BAD fragmentation and flight, mathematical averaging of the BAD pattern can lead to false beliefs about lethal penetration locations. One way to deal with nonlinearity is to sample as much as possible from a "nonbiased" population to introduce some randomness (which, ironically, is itself a challenge for VV&A processes).

For more information about specific issues actually encountered in the V/L VV&A process, see Appendix D, which is a case study of the VV&A of MUVES–SQuASH for the BFV.

Chapter 5

Applications

I. System Acquisition

FIGURE 5.1, adapted from DOD 5000.2 (U.S. Department of Defense 2003), shows the framework for acquiring major combat systems.

In an ideal system acquisition cycle, a continuum of analyses, experiments, and tests would be conducted throughout the cycle. These activities would build on preceding activities to provide V/L data to define and refine design requirements; support technology development, selection, and integration; foster innovative design concepts to achieve efficient vulnerability reduction or lethality enhancements; and support meaningful system evaluations.

A. Requirements Documents

An additional application of fault trees is the determination of performance criteria for requirements documents. These documents contain performance and related operational parameters for a proposed concept or system.

One of the major objectives of the requirements documents is to define the required performance capabilities and requirements. These definitions are to include a performance objective that represents a measurable, beneficial increase in capability or operations and that supports the minimally acceptable level specified in the document. However, these performance objectives historically have been defined in terms that are both physically immeasurable and sufficiently vague in quantification (e.g., P_K's). Because of these inherent problems, definition of the requirements in terms of the system's required capabilities would provide well-defined, measurable performance objectives. It should be noted that

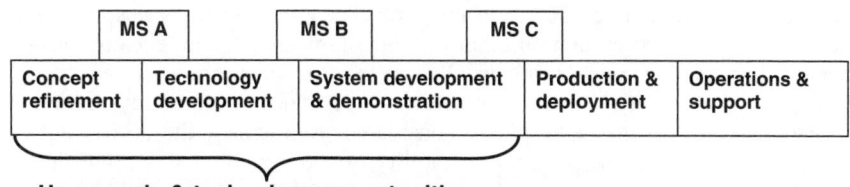

Fig. 5.1   Defense acquisition management framework.

Table 5.1   Example firepower requirements

Acceptable levels	Rate of fire
Upper bound	At least 12 rounds/min for 5 min
Lower bound	No fewer than 6 rounds/min for 5 min

this approach works just as well for the lethality as it does for the vulnerability/survivability of combat systems. Firepower, for example, could be expressed in terms of lower and upper bounds of an acceptable measurable level, as shown in Table 5.1.

The same type of physical, or engineering, metric could be applied to the mobility requirement. For example, speed could be defined in terms of the required (and desired) level for different environments (e.g., at least 40 mph in European rolling hills or 25 mph in a southwest Asian desert).

The quantification of the requirement in terms of engineering metrics permits easier evaluation as to whether or not the requirement has been met. The benefits of this approach are numerous. First, the requirements are expressed in terms of capabilities that can be explicitly measured and that are separated into the different capability categories (i.e., mobility, firepower, and acquisition). The approach provides a means for the user to prioritize the capabilities most worth preserving as well as a greater insight into the military utility of the system. Most importantly, it provides greater clarity as the user and developer of the system discuss trade-offs between what capability is desired vs what can affordably be built.

B.   Concept Refinement and Technology Development

In the early stages of the system acquisition process, the primary focus with regard to vulnerability reduction, analysis, and experimentation is on low-resolution modeling to evaluate candidate technologies that may be implemented in the system design. Because only concepts are being considered at this stage, there is not enough system design detail to permit high-resolution modeling. In fact, the need is for analyses to define the potential of competing concepts to reduce vulnerability and to define the limits or bounds of vulnerability reduction that may be achieved if a given concept is selected for system development. Ancillary analyses may be performed to assist in the refinement of design requirements. It is essential that the acquisition strategy motivate vulnerability reduction requirements that are appropriate given the projected threat environment and use of the objective system but that, at the same time, are achievable with acceptable technical risk and without putting an undue burden on other design parameters. Again, because design details of competing concepts are usually sparse at this point, the analyses attempt to capture the potential (or lack of potential) of major vulnerability reduction technologies known to be seriously considered. Remember, the aim is not to quantify, in absolute terms, the vulnerability of the system at this point but to ensure the proper rank ordering of competing concepts.

This point is an optimum time in the process to conduct a high-resolution analysis of the system being replaced (if that is the case) or of the system being supplemented with the new system. The analysis will establish a baseline assessment of vulnerability that will be used later in the acquisition cycle to gauge the improvement achieved in the new system. Care must be taken, however, to resist comparing low-resolution model analyses of concepts with the high-resolution analysis of existing systems because the basis of comparison (i.e., the level of detail of the system design) will be faulty. Experiments should always be conducted to get a handle on the vulnerability reduction characteristics of emerging technologies. In some cases, experiments must be conducted because there are no relevant data from which to base a technology assessment. Examples include characterizing the ballistic performance of composite or other new materials, determining the impact sensitivity of stowed ammunition energetic components, and identifying likely failure modes of sensitive electronic components. It is seldom possible for these experimental investigations to be extensive because there is rarely adequate hardware. In fact, it is seldom necessary for them to be extensive because there is no assurance that investigated technologies will be pursued in development. For example, composite armor might be ultimately used, but the specific armor recipe may not be known at this point.

It is, however, often necessary to conduct adequate experimentation to generate rules of thumb that will define the bounds of ballistic performance, provide insights into potential failure modes, and guide technology integration. For example, it is not necessary to generate new penetration equations for all new types of armor materials, but it is necessary to conduct enough experiments to calibrate existing equations for similar materials to be able to predict approximate penetration resistance.

There may also be a need to improve or even extend analysis methodologies at this point if it appears that existing methodologies will be deficient later in the program. Whether this is desirable or even possible, however, depends on the likelihood of various technologies being pursued later in system development. Thus, the importance of a comprehensive and continuing experimental program beginning at the earliest stages of the program cannot be overstated.

As design concepts mature, analyses and experiments become much more system specific than they were earlier. However, there may still be many gaps in system design details, and tradeoffs will continue to be necessary to define the eventual system configuration. Low-resolution analyses, and even high-resolution analyses of existing prototypes in many cases, should begin to focus on specific questions and issues related to system design and to address specific system design requirements. One important task at this stage is to recast system vulnerability reduction requirements, typically stated in terms of MMF level 4 metrics (i.e., loss of combat utility), into a suite of level 3 metrics (i.e., system performance). These level 3 metrics, which are observable, testable, and quantifiable, are valuable for characterizing the vulnerability of the system in terms that can, with follow-on effectiveness modeling, truly address survivability in specific mission scenarios.

C. System Development and Demonstration

As prototypes mature, it is possible to critique emerging designs and apply engineering design principles that foster vulnerability reduction. Examples include using redundancy for hydraulic, fuel, and electrical circuits; locating critical components to use protection afforded by heavy structures; and determining optimum armor coverage.

In the same way, experiments should be conducted to characterize component vulnerability, including identifying significant component failure modes, and thereby determining those damage mechanisms that must be addressed in high-resolution modeling. Experimental efforts will focus on level 2 metrics and the mapping algorithms between levels 1 and 2, although consideration of level 3 metrics and consequent level 2–3 mapping should also begin. For armor systems, BAD characterizations are usually necessary. The scope and complexity of these experiments will vary, depending on hardware availability and design maturity. Nonetheless, it is desirable for these experiments to be as thorough as possible at this point because the data will be required for subsequent analyses and will also be useful for supporting detailed design. It may also be desirable to conduct limited subsystem-level tests at this time.

Vulnerability analysis model development should now focus on continuing development of algorithms to account for new damage mechanisms, addressing all pertinent failure modes, and modeling new system technologies. The definition of degraded states begun previously should be refined to reflect maturity in both system design and operational deployment concepts.

As the definition of the system matures, it is both possible and necessary to conduct more definitive analyses and more comprehensive experiments and tests to help perfect system design. Both high-resolution modeling and the continued application of vulnerability reduction design principles are necessary to accomplish this objective. Even though system design is now focused on meeting specific design requirements, including those for vulnerability reduction, high-resolution modeling affords an excellent opportunity to analyze the vulnerability of the system to the spectrum of battlefield threats. It is usually apparent that some threats can be easily defeated or tolerated if the design requirements are met. Others will be shown to be overmatches that may be impractical or impossible to defeat totally. Nevertheless, it may be possible to mitigate the effects of these threats under at least some combat conditions through the judicious application of vulnerability reduction engineering principles. Often, significant gains in protection can be achieved at little additional cost. It is not the purpose of modeling to design the system, but rather to generate data to motivate robust design decisions.

As with analyses, this point is the opportune time to conduct experiments to gain insights into the threat damage mechanisms to which the design is particularly susceptible. The experiments conducted to this point will have produced the basis for continued experiments that can focus more on subsystem and system attributes. Subsystem-level and, later, system-level experiments and field tests should be conducted to identify and investigate synergisms that cannot be thoroughly addressed at the component level. Although many of these synergisms can be anticipated quite early, even in preceding phases, experiments and tests at this point can be conducted in a manner that will foster the identification of unanticipated synergisms. This is accomplished by including as many components in the test assembly as possible and by installing instrumentation to collect data that otherwise might not be deemed necessary. Subsystem tests are also necessary to fully establish selected vulnerability reduction performance limits. For example, a full suite of production armor mounted on a complete chassis may behave far differently than laboratory coupons in a laboratory fixture. Development and operational tests should be exploited when possible, both to add to the test database and to reduce the amount of full-up testing to meet vulnerability requirements. In this way, dedicated vulnerability tests can be reduced, thus reducing costs.

D. Full-Up LFT

A major problem in the conduct of V/L analyses is the need for an extensive database that includes data from firing at complete targets. It was difficult to generate the necessary enthusiasm to allow a significant number of complete targets such as M1 tanks to be sacrificed in a comprehensive firing program, yet new developments in armor concepts and in munitions such as long-rod penetrators and EFPs made the older database inadequate.

The JLF test series and the Congressional requirement for LFT provided the impetus for reviving the databases. The Director of Defense Test and Evaluation, Office of the Secretary of Defense, chartered the JLF program in March 1984. The series was established to develop data on the lethality of U.S. weapons against Soviet systems and the vulnerability of U.S. systems to Soviet weapons. The responsibility for the program was assigned to the previously mentioned JTCG/ME for tests against armored systems and to the JTCG/AS for tests against aircraft.

As noted, there was a dearth of full-scale testing of the vulnerability of armored vehicles from the CARDE tests from the late 1950s until the mid-1980s. From 1980 to 1983, a firing program of SC warheads against armor found the blast effect to be less severe than previous predictions (based on spherical charges). The SC warheads from the front were found to have little blast effect, but they were highly effective from the sides, especially for rounds with a good bit of metal that could provide fragments. Moreover, if the warhead got into the running gear, the blast effect was greater.

The side spray of the warhead did produce big holes in light armor, so a box was built to simulate the crew compartment. The pressure and temperature rises observed in that compartment were generally found not to be life-threatening.

In 1983, the popular press played up the vaporific effect that was alleged to make the BFV a "death trap." A catastrophic increase in fire and pressure was alleged to occur because of jet penetration of the aluminum armor. Some basic data supported this contention, but no full-scale compartment data were available.

Accordingly, BRL saw the need for full-scale testing for all major combat vehicles, such as the Abrams tank, BFV, and M109 howitzer. At that time, Air Force LtCol. James Burton, a representative from DOD, was visiting AMSAA in search of proposals for programs to gather vulnerability data. He received a BRL proposal to perform full-scale testing of the vaporific effect that involved adding aluminum to the box mentioned previously. Burton supported the idea, and the tests were carried out in 1984. The results showed some enhancement of lethality against aluminum over that for steel, but the effect was not enough to cause a significant increase in crew casualties.

These events helped lead to the Live Fire Testing Law. On September 19, 1984, LTC Burton testified before Congress on the need for realistic, full-scale testing of the vulnerability or lethality of all major U.S. combat systems. As a result, the requirement for live-fire testing was included in the FY86 DOD Authorization Act. Subsequently, the requirement was codified in Title 10, U.S. Code Section 2366: "Major System and Munition Program: Survivability and Lethality Testing Required Before Full Scale Production." This law is still in effect.

The Bradley Live Fire tests started in March 1985 and were completed in mid-1987. As a result of those tests, ARL (then BRL) developed spall liners, redesigned stowage of ammunition, and enhanced armor protection. It also developed a reactive-armor package for protection against hand-held SC weapons. Many of these items were tested during the live-fire testing effort. Similarly, the Abrams tank was also the subject of live-fire testing from which important lessons were learned.

Department of Defense Instruction (DODI) 5000.2 (U.S. Department of Defense 2003) requires test and evaluation programs be structured to integrate all developmental, operational, and live-fire tests and evaluations as well as M&S activities into an effective continuum throughout the acquisition cycle for applicable combat systems. Even though full-up, system-level (FUSL) testing satisfies the requirements of Title 10, U.S. Code Section 2366, a common misconception is that LFT&E involves only the FUSL test series that precedes the full-rate production decision. In fact, DODI 5000.2 requires test and evaluation planning to begin in the presystems acquisition phase of the acquisition cycle (Fig. 5.1). Thus, the LFT&E strategy must be incorporated into the Test and Evaluation Master Plan (TEMP) as early as Milestone A, the time of the first program milestone review. [Note that implementation of the instruction is Service-specific. Army implementation is described in Department of the Army Pamphlet 73-1 (U.S. Department of the Army 2003). In addition, because changes to the regulations occur at variable intervals, the interested reader should avail himself of copies of the latest issue before making any decision concerning matters under regulatory control.]

Figure 5.2 depicts the relatively high emphasis on understanding and evaluating candidate vulnerability reduction technologies early in the acquisition cycle. As the system design matures, less emphasis is placed on technology issues and increasing emphasis is placed on system implementation of chosen technologies.

Conventional ballistic vulnerability reduction concerns the response of the target vehicle to an interaction with a threat munition, and it is a major component of survivability. The success and effectiveness of the vulnerability reduction effort are obviously related to the choices of technical solutions to mission needs, the maturation of these technologies during system development, and their incorporation into system design. A comprehensive vulnerability reduction, analysis, and experimentation program, though, is required to ensure a successful transition from concept to production in terms of the selection, development, optimization, and integration of vulnerability reduction alternatives into the final system design. Systems engineering, for example, is a management process espoused in DODI 5000.2 to ensure operational needs are met through a systematic integration of technical inputs from the entire development community. Risks are characterized

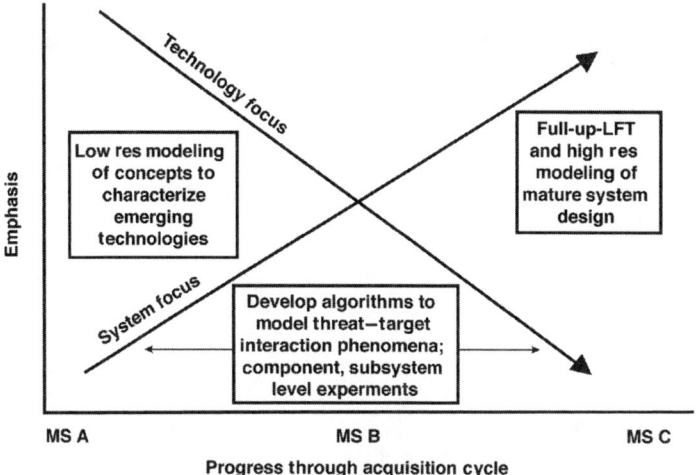

Fig. 5.2 Relative emphasis on technology and system issues during the acquisition cycle.

through early testing and demonstrations of system elements. The ultimate objective of the systems engineering process is to ensure and verify that the system design meets the operational needs with acceptable technical risk.

The only specific vulnerability or lethality test program mandated (for covered systems) is the LFT. Because this test requires production hardware, it generally occurs immediately before full-rate production.

Under Title 10, U.S. Code Section 2366, Congress has mandated in that Aquisition Category (ACAT) I and II programs for covered systems (vehicles, weapon platforms, or conventional weapon systems designed to provide some degree of protection to the user in combat) must include provisions for vulnerability LFT. Such testing is intended to demonstrate and, in conjunction with modeling, determine the vulnerability characteristics of the candidate system to the spectrum of threats likely to be encountered in combat. The statute has provisions for a waiver if one can demonstrate that LFT would be prohibitively expensive or impractical and if one can show that the survivability of the system can be evaluated in other ways. Such a waiver, however, must be made before Milestone B, as previously referenced.

Because the LFT is the only vulnerability or lethality testing specifically identified in DODI 5000.2, it may be appealing to defer experiments, testing, and analysis until the time the LFT is to be conducted. However, the LFT is the culmination of a much more comprehensive and long-term vulnerability reduction, analysis, and experimentation program that includes not only the formal LFT but all other vulnerability analyses, experiments, and testing conducted from concept exploration through low-rate production. Relying on only the LFT to address system vulnerability not only places undue risk on the program but practically ensures the system will exhibit less than optimum vulnerability characteristics. A well-formulated vulnerability reduction, analysis, and experimentation program initiated early and structured in a building-block manner will ensure more robust system survivability and, if planned, conducted, and documented properly, will reduce the scope of the full-up LFT.

The vulnerability reduction, analysis, and experimentation program must address system survivability from three points of view. First, it must ensure requirements for vulnerability reduction are reasonable, achievable, and practicable. The whole battlefield threat likely to be encountered must be addressed. System survivability must not depend on technologies that cannot be reasonably expected to mature during the acquisition process or that have an unacceptable risk of doing so. On the other hand, requirements should be sufficiently aggressive to take advantage of, or even challenge, extant applicable technologies. Incorporation of vulnerability reduction features into system design must not unduly burden other design parameters. Even more important, these features must not sacrifice the ability of the crew to perform all combat requirements.

Second, the vulnerability reduction, analysis, and experimentation program must be formulated to help the materiel developer ensure that design requirements are met as the acquisition cycle progresses. But, even beyond requirements, the program must stimulate measures that can and should be taken to optimize vulnerability reduction against threats or conditions beyond the actual design requirements. The focus here is on applying sound engineering principles to reduce vulnerability and to ensure the design is relatively insensitive to at least minor variations in threat characteristics with minimal additional design cost. Within cost and design constraints, design solutions that protect against a wide range of threats are preferable to those that protect against only a few specific threats.

Third, the vulnerability reduction, analysis, and experimentation program should be structured to reduce program risk by ensuring successful completion of the LFT. Because this test is conducted on production hardware, it is conducted late in the development cycle. Clearly, problems identified during the LFT will be costly to remedy and will most certainly delay the full-rate production decision or require a follow-on product improvement program, often with its own LFT requirement. In an aggressive vulnerability reduction, analysis, and experimentation program, problem areas will be identified and addressed long before the design is finalized and in time to effect appropriate corrective measures.

Additionally, efforts by Nelson (2000) have established a foundation for the development of a methodology to identify, measure, and categorize the costs and benefits of FUSL LFT&E as an element of a cost-effective vulnerability assessment plan and to compare competing vulnerability assessment plans for a system, considering the costs, benefits, and risks or uncertainties associated with the individual and collective elements of each plan. A vulnerability assessment plan is defined by its elements (e.g., LFT of components, FUSL LFT&E, controlled damage experiments, etc.), and competing plans would differ in elements included.

To make a vulnerability assessment not only effective but cost-effective, budgetary constraints must be considered. LFT&E strategies must be efficient and designed with an eye toward the final objective, the assessment of the ability to prosecute the mission at the joint forces command level. To conduct a cost-effective vulnerability assessment, test planners must ascertain what data are required by system evaluators for assessment decisions (i.e., required data set) and compare that required data set to the subset of reliable and relevant data available. The subset of data required for assessment that are unavailable or unable to be relied upon is identified as the data voids. With consideration given to the limitations in available resources, the data voids must be prioritized and addressed in the design of the vulnerability assessment plan.

Accordingly, the methodology developed by Nelson to design a cost-effective assessment plan incorporates many of the characteristics of the Cost As an Independent Variable (CAIV) strategy (U.S. Department of Defense 2004). The core of the CAIV strategy is the tradeoff process, which in the case of selection of the optimal vulnerability assessment plan includes the plan's performance (capacity to assess prioritized data voids and critical issues of personnel/system vulnerability) against the budget for plan implementation, with consideration given to availability of hardware, production schedule, and risks or uncertainties identified in achieving the plan's objectives.

To make a valid comparison of assessment strategies, decision-makers must be provided with cost data that are identified and measured according to a consistent methodology employed across assessment plans. Activity-Based Costing (ABC), a methodology that focuses on activities as the basic cost objectives, is suggested as appropriate for costing elements of a vulnerability assessment plan, such as FUSL LFT&E.

Employing the ABC methodology, the costs of FUSL LFT&E are determined by summing the direct material costs of test assets, spare parts, targets, and munitions and the indirect costs of resources consumed in the cross-functional FUSL LFT&E activities of planning, M&S, test performance, documentation, and evaluation. Questions that must be addressed include the following:

1) How are the costs of tested items measured if the items are damaged but salvageable? How are the costs measured if the items are completely destroyed?

2) How are the costs of spare parts used in repairs measured?

3) How are the costs of targets used in lethality testing or munitions used in vulnerability testing measured?

4) How are the labor costs of individuals participating in the planning, M&S, testing, documentation, or evaluation activities determined?

5) How are the costs of the test ranges and instrumentation measured and allocated across the many systems tested at the test center or across the multiple time periods over which the center is in operation?

Data supplied by the implementation of the ABC methodology to the costing of elements of both a LFT&E strategy and a vulnerability assessment plan provide input to decision-makers addressing the strategic question (which is similar to the validation question in Chapter 4), "Are we doing the right things?" and the operational question (which is similar to the verification question in Chapter 4), "Are we doing things right?" (Cooper and Kaplan 1991). A complete description of Nelson's work is described in Nelson (2000).

The vulnerability reduction, analysis, and experimentation program should be structured to progress in a manner that allows all efforts to take advantage of, and build on, previous efforts. Procedurally, a combination of analytical modeling, experiments, and analysis methodology development is necessary.

Both modeling and component/subsystem-level experiments support the LFT. Modeling is used to help select shotlines and also (particularly in high-resolution modeling) to predict test outcomes (preshot predictions). The experiments leading up to the LFT provide input data for modeling (e.g., component kill probabilities, failure modes, limits of redundancies) and also reduce the likelihood of unpleasant surprises during the full-up LFT. To the extent that analysis models are accredited for the system and threat munitions tested, modeling can actually replace selected LFT firings.

Just because a system has undergone LFT, has been produced, and has been fielded, survivability cannot be neglected. Seldom will a system be fielded and retired without requiring design changes to address design deficiencies identified after production or to reflect changes in operational use or threat escalation. Advances in materials, design, and manufacturing technology also lead to design changes to reduce costs, improve performance, or add capabilities to the system.

Design changes for any of these reasons can affect system survivability. In fact, the Live Fire legislation requires LFT for product improvements if the vulnerability of the system is potentially affected. The roles of analysis and experimentation in supporting these LFT requirements are essentially the same as they are during system design except that it is usually possible to conduct more high-resolution modeling since the baseline system is well-defined and, hopefully, well-characterized.

II. System Life Cycle

The usefulness of the structure of the V/L analysis process can also be extended to other areas of interest that are concerned with combat-system performance. All events, as defined in the structure, through the determination of remaining capabilities, are engineering observables or measurables; that is, one could physically observe or measure these phenomena in the field. The starting point is the definition of a common set of combat-system fault trees that can then be used in analyses across the spectrum of Army concerns. This definition increases clarity about which capabilities are important and provides a tool for communication among different areas of the analysis community, some of which are discussed in the following subsections.

A. SPARC/BDR Methodology

For many years, spare parts for military systems were stocked according to life expectancies and mean-time-between-failures in a peacetime environment. Little consideration was given to parts likely to be damaged in battle. In the early 1980s, the Sustainability Prediction for Army Requirements for Combat (SPARC) program was initiated to address this oversight. The SPARC code was developed to quantify spare parts requirements in support of combat damage repair (Saccenti and Schumacher 1984). The SPARC methodology is really an adaptation of the point-burst methodology and therefore accounts for damage from both the main penetrator and spall debris. The major difference, however, is the introduction of the concept of the mission-essential component, a more inclusive classification than the standard V/L critical component. Any component whose loss would require replacement, either to prevent damage to other components or to maintain long-term operational readiness reliability, is considered mission essential.

APPLICATIONS

All components considered critical in a point-burst analysis are also considered mission-essential, however, noncritical components may become mission essential if they must be replaced when damaged to maintain long-term operational status. For example, consider the third roadwheel of a tracked vehicle. This roadwheel is not considered V/L critical because the vehicle is still functional when the roadwheel is damaged. However, this damaged roadwheel could lead to problems with other suspension components, such as torsion bars and the track, and is, therefore, considered a mission-essential component for a SPARC analysis. All mission-essential components, even redundant components, must be uniquely identified. It is necessary, therefore, to ensure that the regions used to model a component are uniquely identified, and any other occurrence of the component must be distinguished from the others.

Fault trees and DSVM analysis not only help clarify the capabilities of concern for the combat system but also provide the same starting point for both vulnerability and analyses related to logistics questions. For example, although the damage mechanisms may be different, for reliability failure and combat damage, the effect on the combat system's functionality should be the same for a given set of lost components.

Figure 5.3 depicts the structure behind the BDR methodology, laid out in terms of the MMF. BDR, or any kind of repair, can be modeled using the MMF approach in the following manner. Given an initial set of component damage at level 2, mapping can be performed (using DSVM) to determine the remaining capabilities of the system at level 3 (Roach 1996). This represents the capability of the system

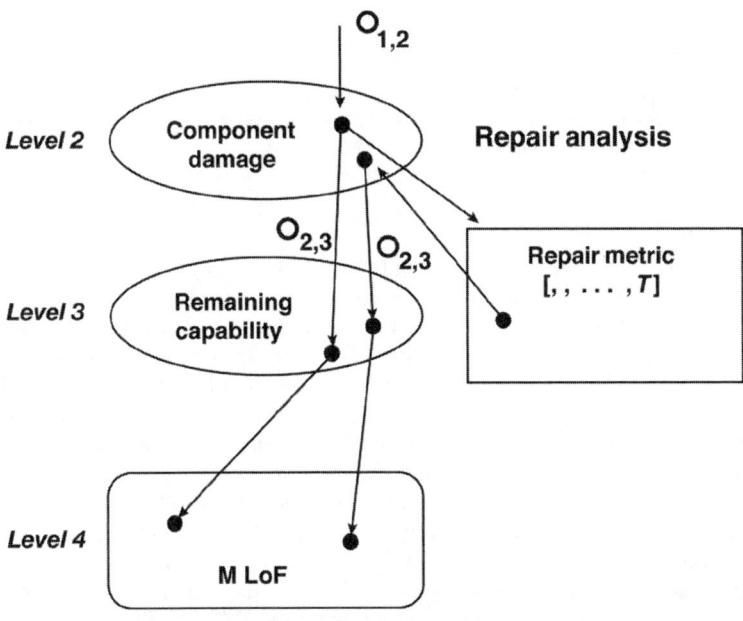

Fig. 5.3 The BDR methodology.

assuming no repair is performed. If one can establish repair priorities and required repair times, one can perform a sensitivity analysis to determine the usefulness of repairs by attempting to do whatever repairs are possible in the allotted time. This provides a second set of damaged components, one that is (possibly) a subset of the original set. Using this new damage-component vector (level 2), mapping is again performed to determine the remaining capabilities of the system given the effected repairs. After a comparison is made between the original set of remaining capabilities and the new set resulting from repair, an assessment can be made of the usefulness of the repairs (i.e., what was gained in terms of system capabilities).

A series of repairs can be identified, and sensitivity analyses performed, to determine what capabilities the system gains as a result of varying amounts of repair time and parts stockage. These analyses can indicate what repairs are necessary in terms of system performance, what types of spare parts should be stocked, and what the critical path is in terms of needed repair. It should be noted that when the repairs are attempted, a system may remain at the same damage point as before repairs (i.e., not enough time was allocated). Conversely, a system may be returned to fully functional if all damaged components are repaired. In addition, the sensitivity analyses may indicate whether or not the system can continue a certain mission, given the capabilities available.

Another way to view the BDR methodology is to start at level 4 and ask what is the system's mission and what capabilities are needed to accomplish it. This approach returns the system to level 3 to determine whether or not the capabilities are available. If not, the path is traced back to level 2 to determine what components are needed to permit the functioning of the capabilities required to accomplish the mission, giving an indication of what components must be repaired.

Of course, because the level 2 analysis gives data on which components are damaged, those data can be used for resupply planning. The older traditional metrics (kill probabilities) merely told that a vehicle was rendered inoperable in a major area such as firepower or mobility; they did not report which components were dysfunctional and in need of immediate replacement, or possibly damaged and in need of future repair or replacement.

B. RAM Analyses

Studies have been performed to identify the commonalities between the vulnerability-analysis and reliability-analysis techniques. It was found that one of the principal commonalities was the criticality or fault-tree analysis (Roach 1993). The use of fault-tree analysis not only clarifies the capabilities of concern for the combat system but provides the same starting point for both vulnerability and RAM analyses. Although the damage mechanisms may be different, the effect on the combat system's functionality should be the same for a given set of lost components.

The initial premise of the study was that the methods for assessing the damage may take different approaches but should yield the same results. Both analyses are concerned with determining remaining combat system capability after a component becomes nonfunctional. That is, both must assess combat system damage. The V/L analyst determines the lethality of a weapon or the vulnerability of a combat system in terms of functional damage; the RAM analyst determines the reliability of the combat system by investigating functional failures. To do this, both analysts must develop an understanding of component interrelationships and relate these components to the system's required capabilities. Although the V/L analyst is concerned with critical components, that is, components required to allow mission performance, the RAM analyst is concerned with all components, regardless of the effect on mission performance. For example, the commander's seat is noncritical in terms of mission performance for a V/L analysis and, therefore, is not considered during the analysis, however, a RAM analysis needs to assess the commander's seat to determine whether or not it meets its reliability criteria. Thus, the components of concern for the V/L analyst are a subset of those of concern to the RAM analyst. After the capability levels are identified and the fault trees developed, each analyst could select those fault trees applying to a particular analysis.

The objective of the initial investigation was to determine if the two methods yielded the same results and to investigate the differences and similarities to determine where the two processes could be combined. An initial effort established the functional loss of the M1A1 for a given set of killed components, selected from a previous vulnerability analysis. This set of killed components was used by both reliability and vulnerability analysts to assess which functions on the vehicle were affected. Each used his normal procedures. For the vulnerability analyst, the criticality analysis of the M1A1 was consulted. The reliability analyst made use of a failure criteria document for the M1E1 tank as well as conversations with the M1A1 project manager. The RAM failure criteria uses block diagrams which group components by function (e.g., mobility, fire control). This allows the analyst to show the relationship between the function and the components that make up the function. A description is then developed for the block diagram, which contains narrative representations of the basic functions, followed by failure modes of the hardware associated with the function. This description identifies which components cause functional loss, cause functional degradation, or have no effect at all.

After the individual analyses were completed, the results were compared. Although there appeared to be, in general, no disagreement in assessed functional loss, the different processes and nomenclature made it difficult to be certain. The reliability assessment provided more specific information on functional loss but required more effort than the vulnerability approach. Generally, the comparisons were fairly straightforward; however, some difficulties did arise. Examples are displayed in Table 5.2 with a discussion provided in succeeding text. The first example lists a specific cable lost. According to the M1A1 criticality analysis, the loss of this cable results in the loss of the gunner's primary sight (GPS)/thermal imaging sight (TIS). The RAM functional loss description reads "1) opens cable disconnection at the thermal receiver, thermal power unit, fire control malfunction will come on; 2) disables power between thermal power control unit and thermal receiver." This description represents a fairly good comparison, unfortunately, not all were this good. In some instances, components contributed to more than one function and thus were described in more than one narrative section, depending on the function being described. Unless the RAM analyst knew this a priori, it is possible that not all lost functions were identified. For example, one component identified as nonfunctional was the GPS, which is composed of both optics and electronics. Because these components are handled

Table 5.2 Example problems from V/L and RAM comparisons[a]

Nonfunctional component	V/L functional loss	RAM functional loss
Cable 1wxxx	Gunner's primary sight (GPS-day) and (GPS-TIS)	1) Opens cable disconnection at the thermal receiver, thermal power unit, fire-control malfunction will come on 2) Disables power between thermal power control unit and thermal receiver
GPS	All power traverse lost All power elevation lost GPS-day and TIS target range	Loss of day and night target sighting capability for accurate main/coax weapon laying using the GPS
Computer electronics unit	All power traverse All power elevation	Loss of ability to calculate ballistic corrections

[a]All results are notional.

separately, the RAM analyst identified functional loss relating to the optics (i.e., "loss of day and night target sighting capability for accurate main/coax weapon laying using the GPS"); the electronics were not included. However, the V/L analyst, using the criticality analysis, identified four lost systems: all power traverse, all power elevation, GPS-day and GPS-TIS, and target range. A second, more common, problem dealt with matching functions selected by the analyst when developing criteria. For example, the loss of the computer electronics unit resulted in the loss of all power elevation and all power traverse in the criticality analysis. The description from the failure criteria document read "loss of ability to calculate ballistic corrections." It is unclear if the two analysts were describing the same functions.

During the conduct of this comparison, it became clear that one set of combat-system capabilities should be identified early in the developmental cycle as identification of required combat-system capabilities is conducted for all facets of system analyses, including V/L, RAM, and logistics. The techniques employed have not been consistent across organizations, most likely because of a lack of realization by the various analysts and project managers of the inherent similarities, resulting in the development of different capability requirements for different applications. Developing the list of required capabilities early in the cycle would avoid the aforementioned comparison problems as well as provide the basis for all subsequent analyses of the combat system.

These examples indicate that the use of fault-tree analysis could make the RAM process easier and faster. Additionally, if V/L and RAM analysts, early in the analytical process, developed fault trees for a given combat system jointly, some of the differences in answers could be avoided, or more easily discovered. Common standards and practices for the identification of required combat-system capabilities, the components that contribute to each capability, and the interrelationship of these components to overall system functionality and impact when the

Table 5.3 Damaged components effects in terms of DSVM capability levels[a]

Component	Capability level
Throttle-steering housing	M2 – significant reduction in speed
	M3 – total immobilization
Parking-brake lock	M2 – significant reduction in speed
Laser rangefinder	F1 – loss of main armament
	F2 – unable to fire on the move
	F3 – increased time to fire
	F4 – reduced delivery accuracy
	A2 – unable to acquire while moving
Breech-mechanism assembly	F1 – loss of main armament

[a]All results are notional.

component (or subassembly) is lost or partially lost, enhance the Army's ability to provide detailed and consistent system-capability requirements across the spectrum of analyses. In addition, savings can be identified in terms of both time and money by the reduction of duplication of effort.

C. BDR Example

A simple example is provided to illustrate the basic BDR analysis process, using information available from previous DSVM analyses of a tank (Abell et al. 1989; Abell, Burdeshaw, and Rickter 1990). Assume the following components on the vehicle have been rendered nonfunctional, either because of combat damage or reliability failure: throttle-steering housing, parking-brake lock, laser rangefinder, and the breech-mechanism assembly. The effect of these components, in terms of DSVM capability levels, is shown in Table 5.3, whereas the associated repair time for each component is listed in Table 5.4. Also note that more than one capability within the mobility and firepower categories are affected by these components. In these cases, either a combination of levels occurs (as with firepower capability levels), or the more serious damage level takes precedence (as with the mobility-capability levels).

Table 5.4 Component repair times (in man-hours)[a]

Component	Repair time
Throttle-steering housing	0.7
Parking-brake lock	3.0
Laser rangefinder	0.4
Breech-mechanism assembly	1.2

[a]All results are notional.

For the purposes of this example, it is assumed that all required spare parts are available and that the tank's organizational repair unit is nearby and can begin repairs immediately (thus, no time is lost awaiting repair personnel and spare parts). It is also assumed that the repairs will be performed sequentially, from shortest to longest time period. Because the laser rangefinder only requires 0.4 man-hours, it is repaired first. Recall from Table 5.3 the loss of this component causes the loss of most firepower capabilities and some acquisition capabilities. With repair, the F2, F3, F4, and A2 capabilities are restored. However, the main armament (F1) is not, as the breech mechanism assembly has yet to be repaired. After approximately 24 min, the tank has most of its firepower and acquisition capabilities restored, but it still does not have the use of its main armament or its mobility capability.

Repair of the throttle-steering housing is next. After 0.7 man-hours, the tank is no longer totally immobilized, but it still has significant speed loss until the parking-brake lock is repaired. Based on the repair times, the breech-mechanism assembly is repaired next, restoring full firepower capability to the system. Finally, requiring 3.0 man-hours of repair time, the parking brake-lock is repaired, restoring the vehicle to full capability. Thus, if sequential repair is performed, the tank is fully functional after 5.3 man-hours.

If the assumptions are changed, then additional analyses may be conducted to give further insight into the problem. For example, if the repairs are performed concurrently, 3.0 man-hours would be required to restore the vehicle to full capability, though some capabilities (such as the firepower and acquisition) would be restored earlier. One may also wish to examine only a subset of the needed repairs, those that would enable the system to continue its current mission. For example, suppose our example tank was in an overwatch position, where mobility was not required. In this case, the repair of the parking-brake lock and the throttle-steering housing could be bypassed in favor of repairing the breech-mechanism assembly and the laser rangefinder. These components would enable the tank to continue with its assigned mission in less than 1.5 hours, assuming the time until mission continuance was reasonable. Of course, those components not repaired may affect the tank's ability to do other missions. If, for example, the tank must subsequently move, the parking-brake lock and the throttle-steering housing must first be repaired. This factor could also be considered.

Another area that lends itself to these types of analyses is spare-parts requirements and stockage. With the MMF mapping from level 1 to level 2, a series of damage vectors can be computed to infer a distribution for the likelihood of component damage. This information can then be used to determine wartime spare-parts stockage for the different components. Additional analyses can be conducted, comparing these wartime requirements with those currently computed for peacetime. Insights can be gained as to whether or not the wartime and peacetime requirements are similar and what types of components, and in what quantities, should be stocked. If the system's capabilities are considered, one can concentrate on the components most likely needed for anticipated missions.

III. Vulnerability Reduction

We now introduce the general principles that lead to good non-nuclear vulnerability reduction (VR) design practices for armored vehicles and discuss current and emerging technologies that can reduce vulnerability.

It is first necessary, however, to define VR. Usually, vulnerability is associated with the probability of kill given a hit ($P_{K|H}$) term in the definition of survivability, where the equation for survivability is as follows:

$$P_{SURV} = 1 - (P_{DET})(P_{H|DET})(P_{KILL|HIT}) \tag{5.1}$$

where

P_{SURV} = probability of survival
P_{DET} = probability of detection
$P_{H|DET}$ = probability of a hit given detected
$P_{KILL|HIT}$ = probability of a kill if hit

This equation shows that vulnerability is concerned with what happens to the vehicle once it encounters a threat. Thus, VR is concerned with negating or mitigating the effects of such an encounter. Whereas the term *vulnerability reduction* is used to denote processes to eliminate or mitigate ballistic damage or the effect of that damage on component or subsystem function, the term *survivability enhancement* refers to actions taken to improve survivability either by reducing the likelihood of being hit or by reducing vulnerability [i.e., reducing any of the probabilities in equation (5.1)].

VR is an engineering discipline similar to maintainability and reliability, and it is based on sound engineering principles. VR must be afforded the same level of importance as other engineering disciplines to ensure battlefield survivability, and as such it must be considered in trade-off studies to optimize total-system design. In considering VR, the analyst and designer must consider such factors as threat, scenario, and design constraints to select or develop component designs and materials intelligently and to locate components where they are least likely to be damaged. For example, if a vehicle is to be used in mine-clearing operations, the designer should protect components by keeping them off the floor or by adding an anti-mine appliqué to the bottom of the vehicle.

In addition, it is not only important to consider VR for friendly vehicles. Vulnerability/VR studies of foreign equipment may identify weaknesses that may be exploited. In this way, VR techniques used in threat systems can impact combat strategies for choosing aimpoints and assessing whether the threat system has been killed.

One way of looking at the processes involved in VR is to consider, once again, its relation to the MMF:

Level 1 The initial configuration includes the details of the environment of a warhead and the target at the instant of interaction. The things that can be done to enhance survivability of the target are those things that reduce the likelihood that the threat will interact with the target. Clearly, such actions are not the concern of the vulnerability analyst, but they are an important part of the trade-offs in the total-design process of which VR is a part.

Level 2 The resulting physical state of the components is of concern in VR. One can visualize this level of VR as comprising some sublevels, as follows: a) keeping the threat from gaining access to sensitive areas of the target; armor is commonly used to protect sensitive areas; b) keeping a threat that has gained access to a compartment from interacting with components that are critical to the mission; locating, hardening, and burying components are common ways of reducing such interactions; c) keeping interactions of a threat with components from seriously affecting their performance; physical hardening of the components is a way of making it more difficult to break them, but one also must be concerned with making them able to function with some capability in spite of damage (e.g., self-sealing tanks).

Level 3 The (resulting) system capability state of the target is often best described by a fault tree. Building redundancy and alternate paths in the fault tree is an important VR tool.

In addition, there are four notable myths about the VR process.

Myth 1 VR is always expensive. This statement is simply not true. Considerable benefits can be realized from low-cost efforts such as designing stowage schemes to reduce the likelihood of damage to ammunition and routing cables and fluid lines through areas of the vehicle in which they are least likely to be hit. Furthermore, the most substantial payoffs occur early in the design process before hardware has been produced. An additional benefit that VR sometimes provides is improved reliability.

Myth 2 What works for one system will work for another system. Even if this were true for a given case, the vehicle designer should not assume it is true. The designer must always consider the complete set of design criteria and constraints for the specific vehicle before determining that something that worked for another vehicle should be implemented in a new vehicle. Consideration should be given to threat, intended use of the vehicle, tactics, mobility requirements, fabrication, and cost, to name a few. As is the case with Myth 1, the best time to make these decisions is early in the design phase, when the impact on other design considerations can be minimized.

Myth 3 A system should be designed only to its threat requirements. Threats change throughout the design life of a vehicle, therefore, the designer must plan ahead for such changes or try to incorporate designs that are not threat-dependent. Designs with less threat dependency are the most desirable. For example, redundancy and separation of critical components can provide protection against a wide range of threat munitions. Armor, on the other hand, often provides protection against a rather narrow range of threats. Once a threat is capable of defeating the armor, any threat with more penetration capability can also defeat the armor. Regardless of the design threat, however, the designer must always give consideration to overmatching threats simply because they do (or will) exist and because they will cause severe, possibly catastrophic, damage to the system. Depending on the system, a known overmatching threat may be identified as primary. This means that consideration must be given to the vulnerability of interior components and vehicle crew, and the designer must be given a requirement to design vehicles in such a way as to minimize the effect of overmatching threats.

Myth 4 VR can be achieved only through the use of armor. There are many problems with this single-minded approach to system design. There is always a threat that can overmatch whatever armor protection is present. Also, armor generally has an associated weight penalty and can result in shock problems as well. Armor and its supporting structure prevent perforation at the expense of

APPLICATIONS   235

absorbing much of the KE of the attacking penetrator. This energy can create and propagate lethal shock waves to sensitive electronic and optical components. Finally, many exotic protections have drawbacks in that they quite often are threat-dependent (i.e., can be optimized only for a narrow class of threats) and may provide for only single-engagement protection.

Obviously, it is necessary to have an understanding of the inherent vulnerability of a system before its vulnerability can be reduced. With knowledge of what kill types are important and likely to occur in a particular system, it is possible to identify subsystems and components critical to maximum system performance. Moreover, this criticality analysis must be detailed and thorough because it forms the basis for a comprehensive and integrated VR effort. Critical components (including personnel) and items that can cause catastrophic damage (primarily fuel and ammunition) must be given priority for vulnerability reduction.

In addition, before determining how to protect components or design subsystems, there must be an understanding of what component damage mechanisms exist and how certain components fail when subjected to these mechanisms. Predicting component failure can be a highly complex engineering problem. Different components receiving identical damage may fail in different ways, and any given component may have multiple failure modes. Understanding the failure modes of each critical component and subsystem is thus essential to reducing vulnerability. After all of these factors are known, VR reduces to a matter of applying the general principles that follow to the specific design problem at hand.

A. Current Vulnerability Reduction Technology

In the course of performing VR studies, a set of general VR principles has evolved and remains applicable for a wide range of systems. These principles are referred to as the ABCDEs of VR:

A – Add protection (e.g., armor, spall liners)   Note that although this approach can reduce vulnerability, there is a cost that must be paid, and that cost is weight. If the system is already in the hardware stage, this may be the only available alternative, however. For personnel, armor is usually the only alternative as there is not much that can be done to re-engineer a human being.

B – Bury critical components   The initial phase of a design is the best time to implement this principle. Judicious placement of critical components so that they are masked by noncritical components or heavy structural members greatly reduces vulnerability.

C – Concentrate   Another alternative VR method is to group several critical components together to reduce the total vulnerable area. This decreases the probability of hitting critical components, thus reducing vulnerability. This also allows for more efficient use of armor (selective armoring), thus decreasing the weight penalty.

D – Duplicate and separate   This principle may appear to be in conflict with concentrate, but actually it is not. Many times systems have redundancy built in for safety and performance reasons. In certain cases, redundancy is good for VR reasons as well. For example, if the redundant features are separated so that a threat cannot defeat both paths, then one has achieved VR and enabled the system to survive.

E – Eliminate   A critical review of components will sometimes show that what was initially thought to be mission critical may, in fact, be optional. By making the simplest machine consistent with all requirements, savings are experienced in vulnerability, reliability, and maintainability.

F – Failsafe   For vulnerable subsystems, auxiliary or back-up modes of operation should be provided in the event of damage to the primary system. In this case, failsafe could be considered a subset of duplicate and separate.

Upon examination of current vehicles, many technologies and guidelines that are effective in reducing vulnerability become apparent. Vehicle and crew gain protection through the use of various types of armor, spall liners of various materials and thickness, and ammunition compartments that keep ammunition events out of the crew area.

For example, the M1 Abrams tank has an ammunition compartment that keeps the crew separate from the ammunition. When an ammunition compartment works well, it protects the crew from reactions in stowed ammunition and leaves the vehicle in a repairable condition, even though an F-Kill may result from the loss of ammunition or damage to ammunition-handling mechanisms. On the other hand, if the stowed ammunition reacts too violently, the compartment, vehicle, and crew will be destroyed. To be survivable, a compartment must be designed to withstand the detonation of at least one warhead and contain devices to prevent the sympathetic detonation of neighboring warheads. However, extremely violent nondetonative reactions from stowed propellant can also destroy a compartment. Two types of propellant reactions must be considered. As described previously, some propellants react very rapidly (in tens or hundreds of microseconds) as the projectile passes through. If the compartment survives this phase, flame spreads through the bed and a larger volume of propellant reacts more slowly (in milliseconds). Accordingly, venting of the compartment is the key to success because even relatively nonviolent reactions can destroy the compartment if the vent is not large enough to release combustion products fast enough or if it doesn't open because pressure builds slowly (relative to violent reactions).

To assess the situation, a vulnerability analyst must know what pressures the compartment can withstand. This must be determined by experiment or structural analysis. For prompt reactions, the TNT equivalency derived from the deposited KE relation can be used in a simulation code such as ConWep or BLASTX (which are codes distributed by the Army Corps of Engineers) to determine the pressure delivered to the walls (U.S. Army Corps of Engineers 2005). It is harder to deal with the slower phase of the reactions. A code called AMMOBOX treats these reactions, but it requires knowledge of many parameters that can normally only be determined by considerable experimentation (Trott and White 1991). The vulnerability analyst should focus on the prompt reactions and be aware that slower reactions can be destructive. If the slower reactions are believed to be substantial risks, subject-matter experts should be consulted to perform engineering analyses and experimentation to determine adequacy of the compartment design.

Personnel vulnerability is further reduced by use of protective clothing (e.g., vest, helmet, goggles, and gloves) and by eliminating toxic-fume hazards by increasing ventilation, minimizing plastics in design, and using ammunition compartments.

In an armored vehicle, BAD is a major threat to soft components and personnel. Spall liners are effective for reducing both the number of fragments and the cone angle. Space is always at a premium inside combat vehicles. Spall liners mounted directly on the outside wall of the vehicle mitigate the spall cloud, but the greater the space between liner and wall, the more effective the liners will be. Spall liner material and thickness can also affect liner effectiveness. Consequently, a design trade comparing liner material, thickness (and weight), cost, spacing from the vehicle wall, and ease of attaching the liner to the vehicle wall with effectiveness is necessary.

Understandably, the vulnerability of fuel systems is always of concern. Some of the methods that can be used to reduce fuel system vulnerability include the following:

1) External cells: To reduce internal fires. Note that external fuel cells must have at least as much armor protection as the rest of the vehicle.

2) Multiple cells and redundant supply paths: To prevent loss of all fuel from the damaged fuel cell or line and, thus, preserve mobility.

3) Fuel lines routed off the floor: To prevent rupture from mines and behind heavy structural elements to minimize perforation by penetrators and BAD.

4) Fuel management system: To effectively use remaining resources of a damaged system (e.g., select which cell supplies fuel to the engine, use fuel from a damaged cell or a cell in harm's way as soon as possible, and shut off pumps transferring fuel from an undamaged cell into a damaged cell).

5) Get-home fuel cells: To preserve and isolate a small quantity of fuel to be used as a last resort to exit the battlefield.

Generally speaking, vulnerability can be reduced further by using an automated fire extinguishing system (especially in crew-occupied areas), shock-mounting fragile electronic and optical components, incorporating component or even subsystem redundancies, using ductile materials in lieu of brittle materials, and incorporating intelligent stowage designs. Note that pyrolysis products formed from the reaction of the fire extinguishant with the fire can be hazardous to onboard personnel unless the vehicle has collective toxic gas protection or soldiers have readily accessible individual protection (e.g., gas masks).

Stowed ammunition should be placed low and away from the natural aimpoint (center of presented area). Missiles should be stowed with the warheads facing rearward. An exception is small-caliber ammunition, which can be used to augment vehicle armor providing the ammunition is in sturdy metal containers and not located adjacent to crew members or passengers. Although it may seem contradictory to principles regarding large-caliber ammunition, reactions within containers of small-arms ammunition are generally not severe enough to breach the stowage container or produce enough toxic gases to be injurious to anyone not close to the reaction.

B. Future Trends and Technology

It should now be apparent that there are many ways that armored vehicle vulnerability can be reduced, and there will be even more ways in the future because of the emergence of new materials and technologies that will reduce vulnerability to current threats as well as new technology designed to mitigate the effects of future threats.

Undoubtedly, armor will continue to be extremely important, and considerable research is being devoted to developing armors that are lighter, require less space, and are more affordable than current designs.

Personnel vulnerability reduction is also of great importance. One method for achieving a reduction in casualties in armored vehicles is to reduce the number of crew members through the use of automatic ammunition loaders and robotics. Personnel vulnerability can also be reduced through selective armoring of the vehicle and, of course, improvements in protective clothing.

There has also been considerable discussion about diagnostic systems. These systems can reduce vulnerability if they provide self-healing and instructional capabilities. Combat override of safety features such as hatch interlocks should be included in the design of such systems. An example is a system that could detect that a critical electrical cable was damaged and then either directly reroute the electrical power or the control signal over another wire or notify the crew that the signal path is interrupted.

In addition, the power plant is the single most important system to the mobility of a vehicle and is often quite vulnerable because of its size and location. These systems can be especially vulnerable to fuel fires when supply hoses are cut and fuel continues to be pumped into the fire. However, a suction fuel-feed system can eliminate this hazard. Another fuel-fire hazard is created by the recirculation of excess fuel used to cool injectors in a diesel engine. After long periods of operation, the fuel temperature in the fuel cells of a vehicle can approach the flash point of the fuel because of the heat added by the recirculated fuel. Accordingly, one alternative would be to recirculate the heated fuel, or part of it, immediately back into the fuel-feed path within the engine compartment. This would achieve two things: 1) the fuel would be preheated to a higher temperature, thus increasing combustion efficiency, and 2) the temperature of the fuel in the fuel cells (large presented area) would remain considerably lower than the fuel flash point.

Power plants can also be less vulnerable to small damage mechanisms if fuel and oil lines and filters are routed internally (in the engine) or masked by other components. Fuel and oil filters become less important to the vulnerability of a vehicle if combat override capabilities exist that allow these filters to be bypassed if they are damaged.

Another trend pertains to positioning of components, people, and explosives. Not surprisingly, the future premises to give much more consideration to the location of items (from a VR point of view) than in past efforts. One method that may be incorporated to achieve this is the use of CAD/CAM systems. Trade-off (optimization) studies could be performed before design finalization, thus saving time and money while reducing vulnerability.

Finally, there are five points that must be made about vulnerability reduction:

1) VR requires the application of leading-edge technology to outpace the ever-changing threat environment; this is especially true for armor.

2) The ABCDEs of VR should always apply.

3) Current technology exists that can further reduce vulnerability (e.g., redundant controls, drive-by-wire, and shock-mounted crew seats).

4) High-technology advances must not increase vulnerability. For example, sophisticated electronics may be shock-sensitive whereas mechanical systems may be more robust.

5) VR must be considered early in the design process to reduce the impact on cost and other design considerations.

Note that these basic methods are general and, in some instances, they can conflict with one another as well as with other design parameters. However, they illustrate the fundamental design principles that must be used to reduce vulnerability. The designer, thus, applies these principles on a case-by-case basis, choosing the best principle for the given design problem. The availability of certain technology will also influence which principles are applied.

IV. Tactics and Doctrine

The MMF described in Chapter 2 provides a ready method to allow the vulnerability of systems (and networks, as described in Chapter 3) to be used in tradeoff studies involving tactics and doctrine. The introduction of force-level simulations to the vulnerability analyst's toolkit, due largely to the mission dependence of networked systems analysis, has expanded the role and utility of vulnerability analysis.

There are three levels at which V/L analyses can be useful to tactical planners: 1) defining performance requirements; 2) accommodating strengths and weaknesses of systems in tactics; and 3) performing trade studies on materiel and procedures.

A. Performance Requirements

DSVM can be used to address the problems with defining requirements discussed in Sec. 5.I.A. For example, rather than probability of kill given a shot ($P_{\text{KILLISHOT}}$), speed could be defined in terms of the required (and desired) different environments (e.g., at least 40 mph in European rolling hills or 25 mph in the southwest Asian desert).

This quantification of the requirement in terms of engineering metrics permits easier evaluation as to whether or not the requirement has been met, and the benefits of this approach are numerous. First, the requirements are expressed in terms of capabilities that can be explicitly measured and separated into the different capability categories (i.e., mobility, firepower, and acquisition). The requirements also provide a means for the user to prioritize the capabilities most worth preserving, as well as providing greater insight into the military utility of the system. Most importantly, the requirements provide greater clarity as the user and developer of the system discuss tradeoffs between desired capabilities vs affordable capabilities vs achievable capabilities.

Understandably, there is a continual competition between VR and lethality enhancement both within and among nations. Each improvement in one area brings (usually successful) efforts to mitigate it by opposing areas. For example, in the 1960s, the development of infantry anti-tank guided weapons threatened the role of tanks on the battlefield. However, shortly thereafter, new armor concepts as exemplified in the M1 Abrams tank re-established the tank as a highly survivable weapon system. More recently, the great strength of the frontal and side armor of tanks has led to the consideration of top-attack weapons, which in turn may lead to the strengthening of the top of armored vehicles.

B. Tactics and Doctrine Tradeoff Studies

Tactics might be modified to accommodate strengths and weaknesses of combat systems. If anti-armor weapons pose a threat to an armored attack, they might have to be neutralized by preparatory fire. If a system has weak sides, it might be used in formation to prevent that weakness from being exploited. If a weapon has an unexpectedly high lethality, it might be used more aggressively. In any event, it is necessary for close coordination between the user and the developer so that nascent problems can be recognized early and either corrected or accommodated in planning. The MMF allows changes in tactics to be tied directly to mission success or failure, injecting vulnerability analysis results into the process in a meaningful way (Sheehan et al. 2003). For more detailed discussions about the MMF, see Appendix E.

The process for analyzing the survivability of networked combat systems, described in Chapter 3 [see Walbert (2004b)], also includes the ability to examine the combat utility of various survivability technologies in a mission context. This provides military planners, requirements developers, and system designers with a means to decide which components or systems are best suited for optimizing combat utility and mission success.

References

Note: References cited in an appendix are listed at the end of that appendix.

Abell, J. M., Burdeshaw, M. D., and Rickter, B. A. (Oct. 1990), "Degraded States Vulnerability Analysis: Phase II," U.S. Army Ballistic Research Lab., Rept. BRL-TR-3161, Aberdeen Proving Ground, MD.

Abell, J. M., Roach, L. K., and Starks, M. W. (June 1989), "Degraded States Vulnerability Analysis," U.S. Army Ballistic Research Lab., Rept. BRL-TR-3010, Aberdeen Proving Ground, MD.

Allen, F., and Sperrazza, J. (Oct. 1956), "New Casualty Criteria For Wounding By Fragments," U.S. Army Ballistic Research Lab., BRL Rept. No. 996, Aberdeen Proving Ground, MD.

American Association for Automotive Medicine (1985), "The Abbreviated Injury Scale (AIS) – 1985 Revision," Des Plaines, IL.

Anderson, J. R., and Edwards, E. W. (May 2004), "BRL-CAD Tutorial Series: Volume IV – Converting Geometry Between BRL-CAD and Other Formats," U.S. Army Research Lab., Rept. ARL-SR-121, Aberdeen Proving Ground, MD.

Armendt, B. F., Kinsler, R. E., and Wilson, R. D. (Aug. 1972), "The Vulnerability of the Soviet ZIL-157 Truck to Interdiction Kills by Compact Steel Fragments," U.S. Army Ballistic Research Lab., BRL Rept. No. 1601, Aberdeen Proving Ground, MD.

Baker, P., and Stegall, S. (1997), "The Response of Solid Gun Propellants to Impacts by Kinetic Energy Rods," U.S. Army Research Lab., Rept. ARL-TR-1294, Aberdeen Proving Ground, MD.

Baker, S. P., and O'Neill, B. (1976), "The Injury Severity Score: An Update," *Journal of Trauma*, Vol.16, No. 11, pp. 882–885.

Baker, S. P., O'Neill, B., Haddon, W., and Long, W. B. (1974), "The Injury Severity Score: A Method for Describing Patients with Multiple Injuries and Evaluating Emergency Care," *Journal of Trauma*, Vol. 14, No. 3, pp. 187–196.

Baker, W. E. (1973), *Explosions in Air*, Univ. of Texas Press, Austin, TX.

Ball, R. E. (2003), *The Fundamentals of Aircraft Combat Survivability Analysis and Design*, 2nd ed., American Inst. of Aeronautics and Astronautics, New York.

Beichler, G. P. (Jan. 1956), "The Vulnerability of Armor Protected Diesel Fuel and Gasoline to the 3.5" M28A2 Heat Round," U.S. Army Ballistic Research Lab., Rept. BRL-MR-963, Aberdeen Proving Ground, MD.

Bely, D. C., Bodt, B. A., and Schumacher, R. N. (Sept. 1992), "M74 Bomblet Penetration Evaluation," U.S. Army Ballistic Research Lab., Rept. BRL-TR-3400, Aberdeen Proving Ground, MD.

Bender, D., and Carleone, J. (March 1990), "Explosively Formed Penetrator Design and Analysis," Fundamentals of Shaped Charges Course, Drexel Univ., Philadelphia, PA.

Benjamin, W. C., Jr. (Aug. 1960), "Compilation of Results Obtained from Some US Firings of Kinetic Energy Projectiles Against Tanks," U.S. Army Ballistic Research Lab., Rept. BRL-MR-1295, Aberdeen Proving Ground, MD.

Benjamin, W. C., Jr., and Gholston, W. (Sept. 1960), "Compilation of Results Obtained from Some U.S. Firings of HEAT Warheads Against Tanks," U.S. Army Ballistic Research Lab., Rept. BRL-MR-1301, Aberdeen Proving Ground, MD.

Bergin, T. J. (Sept. 2000), "50 Years of Army Computing: From ENIAC to MSRC," U.S. Army Research Lab., Rept. ARL-SR-93, Aberdeen Proving Ground, MD.

Biggs, J. M. (1964), *Introduction to Structural Dynamics*, McGraw-Hill, New York.

Bingham, B. L., Dunn, P. E., Hacker, W. L., and Whitehouse, S. (Aug. 1996), "Blast Effectiveness Against Mobile Systems (BEAMS): A UNIX-Based Blast Vulnerability Code – Methodology Development," Applied Research Associates, Inc., Rept. No. 5244, Albuquerque, NM.

Boggs, T. L., Alexander, M. D., and Richter, H. P. (1987), "Small Scale Hazard Testing of Solid Propellants," Chemical Propulsion Information Agency, Publ. 464, Vol. 1, 1987 JANNAF Propulsion Systems Hazards Subcommittee Meeting, Huntsville, AL.

Boyle, V., Frey, R., Watson, J., Stegall, S., Reeves, H., Osowski, J., and Ryzyi, M. (1996), "Vulnerability and Sympathetic Detonation Data on the M483A1 and M107 Projectiles," U.S. Army Research Lab., Rept. ARL-TR-993, Aberdeen Proving Ground, MD.

Bruchey, W., Jr., and Sturdivan, L. (1968), "An Instrumented Range Meeting the Requirements of a Wound Ballistics Small Arms Program," U.S. Army Ballistic Research Lab., Rept. BRL-TN-1703, Aberdeen Proving Ground, MD.

Butler, L. A., and Edwards, E. W. (Feb. 2002), "BRL-CAD Tutorial Series: Volume I – Overview and Installation," U.S. Army Research Lab., Rept. ARL-SR-113, Aberdeen Proving Ground, MD.

Butler, L. A., Edwards, E. W., and Kregel, D. L. (Sept. 2003), "BRL-CAD Tutorial Series: Volume III – Principles of Effective Modeling," U.S. Army Research Lab., Rept. ARL-SR-119, Aberdeen Proving Ground, MD.

Butler, L. A., Edwards, E. W., Schueler, B. J., Parker, R. G., and Anderson, J. R. (April 2001), "BRL-CAD Tutorial Series: Volume II – Introduction to MGED," U.S. Army Research Lab., Rept. ARL-SR-102, Aberdeen Proving Ground, MD.

Canadian Armament Research and Development Establishment (Nov. 1959), "Tripartite Anti-Tank Trials and Lethality Evaluation, Final Report Part II," Vols 2, 5, and 6, Valcartier, Quebec, Canada.

Champion, H. R., Copes, W. S., Sacco, W. J., Lawnick, M. M., Keast, S. L., Bain, L. W., Jr., Flanagan, M. E., and Frey, C. F. (1990), "The Major Trauma Outcome Study: Establishing National Norms for Trauma Care," *Journal of Trauma*, Vol. 30, No. 11, pp. 1356–1365.

Chick, M., Bussell, T., Frey, R., and Bines, A. (1989), "Jet Initiation Mechanisms and Sensitivities of Covered Explosives," *Proceedings of the Ninth Symposium (International) on Detonation*, Office of the Chief of Naval Research Publication OCNR 113291–7.

Clare, V., Ashman, W., Broome, P., Jameson, J., Lewis, J., Merkler, J., Mickiewicz, A., Sacco, W., Sturdivan, L., Lamb, D., and Sylvanus, F. (Nov. 1980), "The ARRADCOM

REFERENCES

ComputerMan – An Automated Approach to Wound Ballistics," U.S. Army Chemical Systems Lab., Rept. ARCSL-TR-80021, Aberdeen Proving Ground, MD.

Cook, M. D., Haskins, P. J., and James, H. R. (1989), "Projectile Impact Initiation of Explosive Charges," *Proceedings of the Ninth Symposium (International) on Detonation*, Office of the Chief of Naval Research Publication OCNR 113291-7.

Cooper, R., and Kaplan, R. S. (1991), *The Design of Cost Management Systems*, Prentice Hall, Englewoods Cliffs, NJ.

Copes, W. S., Champion, H. R., Sacco, W. J., Lawnick, M. M., Gann, D. S., Gennarelli, T., MacKenzie, E., and Schwartzberg, S. (1990), "Progress in Characterizing Anatomic Injury," *Journal of Trauma*, Vol. 30, No. 11, pp. 1200–1207.

Corbett, S., Suckling, J., Chick, M., and Helleur, C. (July 1987), "Development of Improved Techniques for the Evaluation of Behind Armour Effects," Report of the Key Technical Areas 9 and 12, The Technical Coordination Program (TTCP), Panel W-1.

Courant, R., and Friedricks, K. O. (1948), *Supersonic Flow and Shock Waves*, Interscience, New York.

Davis, E. G. (1998), "Characterization of Trauma: Evaluation and Refinement of Indices of Severity and Survival Probability Models," Doctoral dissertation, Johns Hopkins Univ., Baltimore, MD.

Davis, E. G. (2005), Personnel Communication, U.S. Army Research Lab., Aberdeen Proving Ground, MD.

Davis, E. G., Kennedy, C. M., Mermagen, W. H., Jr., et al. (to be published), "Operational Requirement-based Casualty Assessment (ORCA) Improves U.S. Army Operational Casualty Assessments Over Both Sperrazza-Kokinakis and ComputerMan Methodologies," U.S. Army Research Lab., ARL TR, Aberdeen Proving Ground, MD.

Davis, E. G., and Mioduszewski, R. J. (March 1988), "Chemical Computer Man: Chemical Agent Response Simulation (CARS)," U.S. Army Chemical Research, Development, and Engineering Center, Rept. CRDEC-TR-88067, Aberdeen Proving Ground, MD.

Davis, E. G., and Neades, D. N. (23–27 Sept. 2002a), "Injury Severity Scoring for Military Wound Ballisticians: Overview and New Directions for Future Research," *20th International Symposium on Ballistics*.

Davis, E. G., and Neades, D. N. (18–22 Nov. 2002b), "Novel Application of Trauma Severity Scoring in the Design, Development, and Evaluation of U.S. Army Body Armor Against Ballistic Threats," *Personal Armour Systems Symposium 2002* (PASS2002).

Defense Modeling and Simulation Office (2002), "VV&A Recommended Practices Guide," http://www.msiac.dmso.mil/vva/default.htm [retreived Aug. 2002].

Dehn, J. (Jan. 1979), "A Fuel Fire Model for Combat Vehicles," U.S. Army Ballistic Research Lab., BRL Memo Rept. No. 02892, Aberdeen Proving Ground, MD.

Dehn, J. T. (April 1975), "Armored Vehicle Combat Losses Accompanied by Fire, The M48 Tank Series in Vietnam, FY68 to FY70," U.S. Army Ballistic Research Lab., BRL Rept. No. 1777, Aberdeen Proving Ground, MD.

Deitz, P. H. (Dec. 1996), "A V/L Taxonomy for Analyzing Ballistic Live-Fire Events," *Proceedings of the 46th Annual Bomb & Warhead Technical Symposium*, 13–15 May 1996, Monterey, CA; also "Modeling Ballistic Live-Fire Events Trilogy," U.S. Army Ballistic Research Lab., Rept. ARL-TR-1274, Aberdeen Proving Ground, MD.

Deitz, P. H., Mermagen, W. H. Jr., and Stay, P. R. (10–12 May 1988), "An Integrated Environment for Army, Navy and Air Force Target Description Support," *Proceedings of the Tenth Annual Symposium on Survivability and Vulnerability of the American*

Defense Preparedness Association, Military Operations Research Society, Alexandria, VA.

Deitz, P. H., and Ozolins, A. (May 1989), "Computer Simulations of the Abrams Live-Fire Testing," U.S. Army Ballistic Research Lab., Rept. BRL-MR-3755, Aberdeen Proving Ground, MD.

Deitz, P. H., Smith, J. H., and Suckling, J. H. (March 1990), "Comparison of Field Tests with Simulations: Abrams Program Lessons Learned," U.S. Army Ballistic Research Lab., Rept. BRL-MR-3814, Aberdeen Proving Ground, MD.

Deitz, P. H., and Starks, M. W. (1995), "The Generation, Use, and Misuse, of 'PKs' in Vulnerability/Lethality Analyses," *Proceedings of the 8th Annual TARDEC Symposium*, 25–27 March 1997, Monterey, CA; also U.S. Army Ballistic Research Lab., Rept. ARL-TR-1640, Aberdeen Proving Ground, MD, March 1998; also *The Journal of Military Operations Research*, Vol. 4, No. 1, pp. 19–33.

Deitz, P. H., Starks, M. W., Smith, J. H., and Ozolins, A. (Nov. 1990), "Current Simulation Methods in Military Systems Vulnerability Assessment," U.S. Army Ballistic Research Lab., Rept. BRL-MR-3880, Aberdeen Proving Ground, MD.

Developmental Test Command (7 May 1998), "Modeling and Simulation Verification, Validation, and Accreditation Methodology," DTC Pamphlet 73-4, Aberdeen Proving Ground, MD.

Dickinson, D. L., Yatteau, J. D., and Recht, R. F. (1987), "Fragment Breakup," *International Journal of Impact Engineering*, Vol. 5, pp. 249–260.

Dispersio, R., Simon, J., and Merendino, A. (1965), "Penetration of Shaped Charge Jets into Metallic Targets," U.S. Army Ballistic Research Lab., Rept. BRL-TR-1296, Aberdeen Proving Ground, MD.

Dobratz, B. M., and Crawford, P. C. (1985), "LLNL Explosives Handbook," Lawrence Livermore National Lab., Rept. UCRL-52997, Change 2, Livermore, CA.

Drimmer, B. E. (June 1983), "Navy Bank of Explosives Data," Naval Surface Weapons Center, Rept. NSWC-MP 83-230, Albuquerque, NM.

Dunn, D. J., Jr., and Porter, W. R. (Aug. 1955), "Air Drag Measurements of Fragments," U.S. Army Ballistic Research Lab., Memorandum Rept. No. 915, Aberdeen Proving Ground, MD.

Dunn, P. E., et al. (Jan. 1998), "MEVA Software User's Manual," Applied Research Associates, Inc., Albuquerque, NM, for Contract No. F08630–95-C-0023, Wright Lab., Armament Directorate, Eglin Air Force Base, FL.

Dziemian, A. J. (May 1960), "A Provisional Casualty Criterion for Fragments and Projectiles," U.S. Army Chemical Warfare Lab., Rept. CWLR 2391, Edgewood Arsenal, MD.

Ege, H. W. (Feb. 1972), "A Method for Estimating Vulnerability of Armored Vehicles to Attacking Projectiles." Interim Memorandum Rept., Aberdeen Proving Ground, MD.

Ege, H. W., and Harvey, S. M. (June 1974), "Computer Program for Determining Vulnerability Volume II: Analyst Manual," U.S. Army Ballistic Research Lab., Rept. BRL-IMR-238 and -239, Aberdeen Proving Ground, MD.

Eycleshymer, A., and Schoemaker, D. (1911), *A Cross-Section Anatomy*, D. Appleton, New York.

Ezell, E. C. (1981), *Handguns of the World*, Stackpole Books, Mechanicsburg, Pennsylvania, p. 282.

Finnerty, A. E. (Nov. 1987), "Predictions of Probabilities of Sustained Fires for Combat Damaged Vehicles," U.S. Army Ballistic Research Lab., Rept. BRL-TR-2875, Aberdeen Proving Ground, MD.

REFERENCES

Finnerty, A. E., and Dehn, J. T. (April 1994), "Alternative Approaches to Fuel Fire Protection for Combat Vehicles," U.S. Army Research Lab., Rept. ARL-TR-377, Aberdeen Proving Ground, MD.

Finnerty, A. E., Meissner, R. R., and Copland, A. (July 1985), "Fragment Attack on Ground Vehicle Hydraulic Lines," U.S. Army Ballistic Research Lab., Rept. BRL-TR-2661, Aberdeen Proving Ground, MD.

Finnegan, S. A., Pringle, J. K., and Heimdahl, O. E. R. (1995), "Further Studies of Impact-Induced Delayed Detonation Behavior of Solid Rocket Motors Using a Planar Model," Chemical Propulsion Information Agency, Publ. 628, 1995 JANNAF Propulsion Systems Hazards Subcommittee Meeting, Huntsville, Alabama.

Frey, R., Howe, P., Trimble, J., and Melani, G. (1979), "Initiation of Explosive Charges by Projectile Impact," U.S. Army Ballistics Research Lab., Rept. ARBRL-TR-02176, Aberdeen Proving Ground, MD.

Frey, R., Watson, J., Gibbons, G., Boyle, V., Finnerty, A., Lawrence, W., Leveritt, C., Peregino, P., Pilarski, D., Blake, O., Bines, A., and Canami, A. (1996), "Compartmentation Technology for Liquid Propellant," U.S. Army Research Lab., Rept. ARL-TR-956, Aberdeen Proving Ground, MD.

Fritz, W. B. (1994), "The ENIAC – A Problem Solver," *IEEE Annals of the History of Computing 16*, No. 1, pp. 25–45.

Gillich, W., Wilson, J., Giglio-Tos, L., and McLaughlin, R. (July 1977), "Trends in Development and Performance of Kinetic Energy Ammunition," U.S. Army Ballistic Research Lab., Rept. BRL-MR-2771, Aberdeen Proving Ground, MD.

Goldstine, H. H. (June 1993), *The Computer from Pascal to von Neumann*, Princeton Univ. Press, Princeton, NJ.

Grote, R. L., Moss, L. L. C., and Davisson, E. O. (Nov. 1996), "Tire Penetration and Deflation Models for Wheeled Vehicles Subjected to Fragment Attack," U.S. Army Research Lab., Rept. ARL-TR-1233, Aberdeen Proving Ground, MD.

Groves, A. D. (Feb. 1986), "A Multiple Burst Vehicle Kill Probability Model, Where the Probability of Igniting Spilled Flammable Fluids Is Time Dependent," U.S. Army Materiel Systems Analysis Agency, Rept. USAMSAA-TR-418, Aberdeen Proving Ground, MD.

Gwyn, D. A. (ed.) (Aug. 1987), "The BVLD/VMB UNIX Supplementary Manual," U.S. Army Ballistic Research Lab., Aberdeen Proving Ground, MD.

Haddix, F. (May 2003), "The Functional Descriptions of the Mission Space (FDMS) Data Interchange Format (DIF)," Ver. 2.0, Defense Modeling and Simulation.

Halsey, C. C., and Roquemore, J. C. (March 1987), "Live Mk 82 Firings Against Radar Vans on Mounds at Cactus Flat Test Range," U.S. Naval Weapons Center, Rept. NWC-TM-5999, China Lake, CA.

Hanes, P. J., Henry, S. L., Moss, G. S., Murray, K. R., and Winner, W. A. (Dec. 1991), "Modular UNIX-Based Vulnerability Estimation Suite (MUVES)," U.S. Army Ballistic Lab., Rept. BRL-MR-3954, Aberdeen Proving Ground, MD.

Hanna, J. W., and Goodwin, B. S. (Jan. 1955), "The Ignition of Diesel Fuel by Statically Detonated 3.5" HEAT Rocket Heads," Armor Test Rept. No. AD-1188, Aberdeen Proving Ground, MD.

Headquarters, Department of the Army (10 July 1997), "Management of Army Models and Simulations," Army Regulation 5-11, Washington, DC.

Headquarters, Department of the Army (20 Sept. 1999), "Verification, Validation, and Accreditation (VV&A) of Army Models and Simulations," Army Pamphlet 5-11, Washington, DC.

Holloway, G., Danish, M., and Matts, J. (1978), "Penetration Relations for Tungsten Alloy Fragments vs. Selected Target Range Materials," U.S. Army Ballistic Research Lab., Rept. BRL-TR-02087, Aberdeen Proving Ground, MD.

Hoyt, R. C. (Feb. 1969), "The Magic-Sam C Target Analysis Technique: Volume I, Combinatorial Geometry and Its Application to Armored Combat Vehicles for Both Nuclear and Conventional Vulnerability," U.S. Army Materiel Systems Analysis Agency, TR 10, Aberdeen Proving Ground, MD.

Hunt, J., Scungio, R., and Burdeshaw, M. (Oct. 2004), "Advanced Joint Effectiveness Model Accreditation Support Package, Volume I," Applied Research Associates, Inc., Belcamp, MD.

Hunt, J. E. (28–30 March 1995), "An Indirect-Fire MUVES Approximation Method," *Proceedings of the 6th Annual Combat Vehicle Survivability Symposium*, Monterey, CA.

JAYCOR, Inc. (May 1995), "INJURY User's Manual (Version 4.00 Beta)," Silver Spring, MD.

Johns Hopkins Univ. (1961), "The Resistance of Various Metallic Materials to Perforation by Steel Fragments; Empirical Relationships for Fragment Residual Velocity and Residual Weight," Ballistic Analysis Lab., Project Thor TR No. 47, Baltimore, MD.

Joint Technical Coordinating Group for Munitions Effectiveness (Feb. 1971), "VAREA Computer Program, Volume I – User Manual," 61JTCG/ME-71-6-1; also "Volume II – Analyst Manual," 61JTCG/ME-71-6-2.

Joint Technical Coordinating Group for Munitions Effectiveness (Sept. 1985), "COVART II – A Simulation Program for Computation of Vulnerable Areas and Repair Times, User Manual," 61JTCG/ME-84-3.

Joint Technical Coordinating Group for Munitions Effectiveness (1 July 1988), "Fragmentation Characteristics and Terminal Effects Data for Surface-to-Surface Weapons," Joint Munitions Effectiveness Manual 61S-3-4.

Kennedy, D. R. (Sept. 1973), "Review of the Early History of the Vaporific Effect," from FMC Corporation, Defense Technology Lab., to U.S. Air Force Armament Lab., Rept. No. AFATLTR-73-198, Eglin Air Force Base, FL.

Kirby, R. L. (Unpublished notebooks), U.S. Army Ballistic Research Lab., Aberdeen Proving Ground, MD.

Klopcic, J. T. (Sept. 1984), "An Introduction to the User of the Army Unit Resiliency Analysis (AURA) Methodology: Volume I," U.S. Army Ballistic Research Lab., Rept. BRL-MR-3384, Aberdeen Proving Ground, MD.

Klopcic, J. T. (May 1999), "The Vulnerability/Lethality Taxonomy as a General Analytical Procedure," U.S. Army Research Laboratory, Rept. ARL-TR-1944, Aberdeen Proving Ground, MD.

Klopcic, J. T., and Lynch, D. D. (Sept. 1997), "Static and Dynamic Characterization of the M107 (Composition B Filled) Artillery Projectile," U.S. Army Research Lab., Rept. ARL-TR-1496, Aberdeen Proving Ground, MD.

Klopcic, J. T., and Lynch, D. D. (Sept. 1998a), "Static Characterization of the OF-462 122-mm TNT-Filled Artillery Projectile," U.S. Army Research Lab., Rept. ARL-TR-1762, Aberdeen Proving Ground, MD.

Klopcic, J. T., and Lynch, D. D. (Sept. 1998b) "Static Characterization of the OF-540 152-mm TNT Filled Artillery Projectile," U.S. Army Research Lab., Rept. ARL-TR-1763, Aberdeen Proving Ground, MD.

REFERENCES

Klopic, J. T., and Reed, H. L. (April 1999), "Historical Perspectives on Vulnerability/Lethality Analysis." U.S. Army Research Laboratory, Rept. ARL-SR-90, Aberdeen Proving Ground, MD.

Klopcic, J. T., Starks, M. W., and Walbert, J. N. (May 1992), "A Taxonomy for the Vulnerability/Lethality Analysis Process," U.S. Army Ballistic Research Lab., Rept. BRL-MR-3972, Aberdeen Proving Ground, MD.

Knepell, P. L., and Arangno, D. C. (1993), *Simulation Validation: A Confidence Assessment Methodology*, IEEE Computer Society Press, Los Alamitos, CA.

Knox, F. S., III, Bonetti, D., and Perry, C. (Feb. 1993), "User's Manual for BRNSIM/BURNSIM: A Burn Hazard Assessment Model," U.S. Army Aeromedical Research Lab., USAARL Rept. No. 93–13, Fort Rucker, AL.

Kokinakis, W., and Sperrazza, J. (Jan. 1965), "Criteria for Incapacitating Soldiers with Fragments and Flechettes," U.S. Army Ballistic Research Lab., Rept. No. 1269, Aberdeen Proving Ground, MD.

Korba, A. (1974), "Vulnerability of Soviet Nuclear Warheads to Fragment Impacts and Blast," U.S. Army Ballistic Research Lab., Rept. BRL-MR-2363, Aberdeen Proving Ground, MD.

Kruse, L. R., and Brizzolara, P. L. (Dec. 1971), "An Analytical Method for Deriving Conditional Probabilities of Kill for Target Components," U.S. Army Ballistic Research Laboratory, Rept. BRL No. 1563, Aberdeen Proving Ground, MD.

Lee, E. L., and Tarver, C. M. (1980), "Phenomenological Model for Shock Initiation in Heterogeneous Explosives," Lawrence Livermore Lab., Rept. UCRL-83618, Livermore, CA.

Libersky, L. D, Petschek, A. G., Carney, T. C., Hipp, J. R., and Allahdadi, F. A. (Nov. 1993), "High Strain Lagrangian Hydrodynamics, A Three-Dimensional SPH Code for Dynamic Material Response," *Journal of Computational Physics*, Vol. 109, No. 1.

Liddiard, T. P., and Roslund, L. A. (1993), "Projectile Impact Sensitivity of Explosives," Naval Surface Weapons Center, Rept. NSWC TR-89-184, White Oak, MD.

Maestas, F. A., Young, L. A., Streit, B. K., and Peterson, K. J. (July 1995), "Effectiveness/Vulnerability Assessment in Three-Dimensions (EVA-3D)," User's Manual, Versions 4.1F and 4.1C, Applied Research Associates, Inc., Contract No. F08630-91-C-0005, Albuquerque, NM.

Mahaffey, M. E. (to be published), "An Engineering Foundation for Modeling Combat Vehicle Mobility in Degraded States," U.S. Army Research Lab. Report, Aberdeen Proving Ground, MD.

Malick, D. (1968), "The Resistance of Steel Targets to Perforation by Small-Caliber Armor-Piercing Projectiles," Ballistic Analysis Lab., TR No. 66, Johns Hopkins Univ., Baltimore, MD.

Malick, D. (1969), "The Resistance of Aluminum Alloy Targets to Perforation by Small-Caliber Armor-Piercing Projectiles," Ballistic Analysis Lab., TR No. 70, Johns Hopkins Univ., Baltimore, MD.

Mathematical Applications Group Inc. (Aug. 1967), "A Geometric Description Technique Suitable for Computer Analysis of Both Nuclear and Conventional Vulnerability of Armored Military Vehicles," Report MAGI-6701, AD847576.

McGlaun, J. M., and Thompson, S. L. (1990), "CTH: A Three-Dimensional Shock Wave Physics Code," *International Journal of Impact Engineering*, Vol. 10, pp. 351–360.

Miller, R. E, and Carver, B. (June 1998), "A Submodel for Combat Casualty Assessment of Ocular Injury from Lasers – Final Report," VERIDIAN, Inc., Brooks Air Force Base, TX.

Muuss, M. (compiler) (Oct. 1991), "A Solid Modeling System and Ray-Tracing Benchmark Distribution Package, Release 1.2, June 1987, Release 3.0, October 1988, and Release 4.0, October 1991," U.S. Army Ballistic Research Lab., Aberdeen Proving Ground, MD.

Muuss, M., Applin, K., Suckling, K., R., Moss, G., Weaver, E., Stanley, C. (March 1983), "GED: An Interactive Solid Modeling System For Vulnerability Assessments," U.S. Army Ballistic Research Lab., Rept. BRL-TR-02480, Aberdeen Proving Ground, MD.

Nail, C. L. (Aug. 1982), "Vulnerability Analysis for Surface Targets (VAST) An Internal Pointburst vulnerability Assessment Model Revision 1," Computer Sciences Corporation Technical Manual, CSC TR-82-5740, Madison, AL.

Nail, C. L., Bearden, T. E., Jackson, E. (Oct. 1979), "Vulnerability Analysis Methodology Program (VAMP): A Combined Compartment Kill Vulnerability Model," Computer Sciences Corporation Technical Manual, CSC TR-79-5585, developed under BRL Contract DAAK4078D0005 (Order No. 0029), Madison, AL.

Neades, D. N., and Davis. E. G. (23–27 Sept. 2002), "A New Approach to the Simulation of Battlefield Injuries and Their Effect on the Performance of Military Tasks," *20th International Symposium on Ballistics*, Orlando, FL.

Neades, D. N., and Prather, R. N. (Aug. 1991), "The Modeling and Applications of Small Arms Wound Ballistics," U.S. Army Ballistic Research Lab., Rept. BRL-MR-3929, Aberdeen Proving Ground, MD.

Nelson, M. K. (April 2000), "Vulnerability and Lethality Assessment: The Role of Full-Up System Level Live-Fire Testing and Evaluation," U.S. Army Research Lab., Rept. ARL-CR-450, Aberdeen Proving Ground, MD.

North Atlantic Treaty Organization (March–April 1986), "NATO Armaments Committee Proceedings AC/225," Panel III, Subpanel 3, D-3, Chap. 2, Brussels, Belgium.

Obergefell, L. A., Gardner, T. R., Kaleps, I., and Fleck, J. T. (1988), "Articulated Total Body Model Enhancements User's Guide," U.S. Air Force Research Lab., Rept. AAMRL-TR-88-043, Wright-Patterson Air Force Base, OH.

Office of the Director for Live Fire, Undersecretary of Defense for Research and Engineering (18–19 October 1988), "Proceedings of the Live Fire Test Crew Casualty Assessment Workshop," Naval Submarine Base, Groton, CT.

Office of the Secretary of Defense (4 January 1994), "DOD Modeling and Simulation (M&S) Management," DOD Directive 5000.59, Washington, DC.

Office of the Secretary of Defense (29 April 1996), "DOD Modeling and Simulation (M&S) Verification, Validation, and Accreditation (VV&A)," DOD Instruction 5000.61, Washington, DC.

Osler, T., Baker, S. P., and Long, W. B. (1977), "NISS: The New Injury Severity Score." Tenth Annual Meeting of the Eastern Association for the Surgery of Trauma, Sanibel, FL.

Piekutowski, A. J. (1990), "A Simple Dynamic Model for the Formation of Debris Clouds," *International Journal of Impact Engineering*, Vol. 10, Nos. 1–4, pp. 453–571.

Platt, J. (April 1973), "Pressure and Temperature Enhancement Inside a Simulated Aluminum Armoured Fighting Vehicle Attacked by Hollow Charge Weapons," Military Vehicles Engineering Establishment, MVEE Report 72047, Chertsey, U.K.

REFERENCES

Ploskonka, J. J., Muehl, T. M., and Dively, C. J. (June 1988), "Criticality Analysis of the M1A1 Tank," U.S. Army Ballistic Research Lab., Rept. BRL-MR-3671, Aberdeen Proving Ground, MD.

Price, R. G., and Kalb, J. T. (1991a), "A New Approach to a Damage Risk Criterion for Weapons Impulses," *Scandinavian Audiology*, Supplementum, Vol. 34, pp. 21–37.

Price, R. G., and Kalb, J. T. (July 1991b), "Insights into Hazard from Intense Impulses from a Mathematical Model of the Ear," *Journal of Acoustical Society of America*, Vol. 90, No.1, pp. 219–227.

Price, S. K., Boyd, P., and Burdeshaw, M. D. (Sept. 1993), "Degraded States Analysis of an M2A1 Bradley Infantry Fighting Vehicle," U.S. Army Research Lab., Rept. ARL-TR-212, Aberdeen Proving Ground, MD.

Randers-Pehrson, G., and Juriaco, P. (June 1978), "Computer-Aided Self-Forging Fragment Design," Large Caliber Weapon Systems Lab., U.S. Army Armament Research and Development Command, Dover, NJ.

Rapp, J. R. (July 1983), "An Investigation of Alternative Methods for Estimating Armored Vehicle Vulnerability," U.S. Army Ballistic Research Lab. (AD B076394LO), Rept. BRL-MR-03290, Aberdeen Proving Ground, MD.

Reed, H. L., Jr. (ed.) (1992a), "Ballisticians in War and Peace, Volume III: A History of the United States Army Ballistic Research Laboratories, 1977–1992," U.S. Army Ballistic Research Lab., Aberdeen Proving Ground, MD.

Reed, H. L., Jr., (ed.) (Sept. 25, 1992b), "Vulnerability Day," Proceedings, Special U.S. Army Research Lab. Publication, Aberdeen Proving Ground, MD.

Repa, J. V., "Existing Computational Combustion Models at Los Alamos," Department of Defense Programs, Slide Presentation, Los Alamos National Lab., Los Alamos, NM.

Ripple, R., and Mundie, T. (eds.) (Sept. 1989), "Medical Evaluation of Nonfragment Injury Effects in Armored Vehicles," Walter Reed Army Inst. of Research, Washington, DC.

Roach, L. (June 1993), "Fault Tree Analysis and Extensions of the V/L Process Structure," U.S. Army Research Lab., Rept. ARL-TR-149, Aberdeen Proving Ground, MD.

Roach, L. (Nov. 1996), "The New Degraded States Vulnerability Methodology: A Change in Philosophy and Approach," U.S. Army Research Lab., Rept. ARL-TR-1223, Aberdeen Proving Ground, MD.

Roach, L. K. (Sept. 1990), "BRL Implements Methodology Breakthrough," *BRL Update*, Vol. 2, No. 5, U.S. Army Ballistic Research Lab., Aberdeen Proving Ground, MD.

Saccenti, J. C., and Schumacher, R. N. (April 1984), "SPARC Analysts' Methodology Handbook," U.S. Army Ballistic Research Lab., Rept. BRL-TR-02562, Aberdeen Proving Ground, MD.

Saucier, R., and Gilman, A. W. D. (Dec. 1996), "The Concept of Ballistic Dose and Its Use as a Predictor of Personnel Incapacitation and Survivability," U.S. Army Research Lab., Rept. ARL-TR-1242, Aberdeen Proving Ground, MD.

Saucier, R., and Kash, H. (Aug. 1994), "ComputerMan Model Description," U.S. Army Research Lab., Rept. ARL-TR-500, Aberdeen Proving Ground, MD.

Saucier, R., Shnidman, R., and Collins J. C., III, (March 1995), "A Stochastic Behind-Armor Debris Model," *6th Annual TARDEC Combat Vehicle Survivability Symposium*, Monterey, CA.

Schmidt, E. M., Heaps, C. W. and Jacobson, J. R. (Sept. 1986), "Blast Overpressure Internal to Vehicles Perforated by Shaped Charges," U.S. Army Ballistic Research Lab., Rept. BRL-TR-2757, Aberdeen Proving Ground, MD.

Schmidt, J. G. (ed.) (a), "Ballisticians in War and Peace, Volume I: A History of the United States Army Ballistic Research Laboratories, 1914–1956," U.S. Army Ballistic Research Lab., Aberdeen Proving Ground, MD.

Schmidt, J. G. (ed.) (b), "Ballisticians in War and Peace, Volume II: A History of the United States Army Ballistic Research Laboratories, 1957–1976," U.S. Army Ballistic Research Lab., Aberdeen Proving Ground, MD.

Schumacher, R. N. (Feb. 1981), "SAMSMAE, A Computer Code for Determining Mean Areas of Effectiveness for Multiple Component Targets, Technical Report," U.S. Army Ballistic Research Lab., Rept. ARBRL-TR-02295, Aberdeen Proving Ground, MD.

Schumacher, R. N., and Hunt, J. E. (Sept. 1992), "SPRAE Users Guide," U.S. Army Research Lab., Aberdeen Proving Ground, MD.

Sheehan, J. H., Deitz, P. H., Bray, B. E., Harris, B. A., and Wong, A. (1–4 December 2003), "The Military Missions and Means Framework," *Proceedings of the Interservice/Industry Training, Simulation, and Education Conference* (ITSEC), Orlando, FL.

Shnidman, R. (1988), "Direct Fragment Lethality Inference from Witness Plate Array Data," *Proceedings of the 25th ADPA Survivability and Vulnerability Symposium*, San Diego, CA.

Skinner, G. W. (May 1959), "Effectiveness of Antitank Projectiles Against Tanks and Armor Protected Diesel Fuel," Test Report AD-1266.

Starks, M. W. (May 1991), "Improved Metrics for Personnel Vulnerability Analysis," U.S. Army Ballistic Research Lab., Rept. BRL-MR-3908, Aberdeen Proving Ground, MD.

Stein, A. (1946), "Optimum Caliber Program," U.S. Army Ballistic Research Lab., Rept. M437, Aberdeen Proving Ground, MD.

Sturdivan, L. M. (Oct. 1981), "Spheres, Cubes, and Fragments," *Handbook of Human Vulnerability Criteria*, Chemical Systems Lab., U.S. Army Armament Research and Development Command, Rept. ARCSL-SP-81005, Aberdeen Proving Ground, MD.

Title 10, U.S. Code Section 2366 (1987), Chap. 139.

Tri-Analytics and the SURVICE Engineering Company (Aug. 1995), "Development of Improved Personnel Serious and Lethal Models for Fragments and Burns," U.S. Army Research Lab., Rept. SURVICE-TR-95-010, Aberdeen Proving Ground, MD.

Trott, B. D., and White, J. J. (1991), "AMMOBOX Code Revision," Battelle Columbus Lab., Rept. BCD 91–03, Columbus, OH.

United Nations (1990), *Recommendations on the Transport of Dangerous Goods; Tests and Criteria*, 2nd ed., United Nations Economic Commission for Europe, Geneva, Switzerland.

U.S. Army Combat System Test Activity (July 1990), "Follow-on Production Test of Army Tactical Missile System (ATACMS), M74 Tungsten Bomblet/Fragment Penetration Test," Firing Record No. LS-00077 and LS-00079.

U.S. Army Corps of Engineers, Protective Design Center, https://pdc.usace.army.mil/software [retreived Sept. 2005].

U.S. Army Materiel Command (Nov. 1962), *Elements of Terminal Ballistics, Part One: Introduction, Kill Mechanism, and Vulnerability*, AMCP 706-160, Engineering Design Handbook.

U.S. Army Materiel Command (Jan. 1971), *Properties of Explosives of Military Interest*, AMCP 706-177, Engineering Design Handbook, Explosive Series.

U.S. Army Materiel Command (15 July 1974), *Explosions in Air, Part One*, AMCP 706-181, Engineering Design Handbook.

REFERENCES

U.S. Department of Defense (June 1977), "Procedures for Performing a Failure Mode, Effects, and Criticality Analysis," MIL-STD-1629A.

U.S. Department of Defense (May 2003), Operation of the Defense Acquisition System, DOD Instruction 5000.2.

U.S. Department of Defense (31 May 1997), "Design of Projectiles for Terminal Ballistic Effects," MILHDBK-1226(AR), Department of Defense Handbook.

U.S. Department of Defense (Nov. 2004), *The Defense Acquisition Deskbook*, Office of the Secretary of Defense, Washington, DC.

U.S. Department of Defense (9 May 2005), "Department of Defense Dictionary of Military and Associated Terms," Joint Publication 1-02.

U.S. Department of the Army (July 1977), "Engineering Design Handbook: Basic Target Vulnerability," DARCOM-P-706-163.

U.S. Department of the Army (Feb. 1995), "Design of Combat Vehicles for Fire Survivability," MILHDBK-684.

U.S. Department of the Army (May 2003), *Test and Evaluation Support of Systems Acquisition*, Pamphlet 73-1.

U.S. Department of the Army, Navy, Air Force, and Defense Logistics Agency (Jan. 1998), "Department of Defense Ammunition and Explosives Hazard Classification Procedures," Joint Technical Bulletin TB 700-2, Washington, DC.

Walbert, J. N. (Nov. 1994), "The Mathematical Structure of the Vulnerability Spaces," U.S. Army Research Lab., Rept. ARL-TR-634, Aberdeen Proving Ground, MD.

Walbert, J. N. (5–7 October 2004a), "Survivability of Networked Combat Systems," *International Test and Evaluation Association Conference*, Plenary Address, Baltimore, MD.

Walbert, J. N. (19–20 October 2004b), "A Methodology for Analyzing the Survivability of Networked Combat Systems," *Proceedings of the XLIII Annual Meeting of the Army Operations Research Symposium*, Fort Lee, VA.

Waldon, D. J., Dalton, R. L., Kokinakis, W., and Johnson, W. P. (Sept. 1969), "A Parametric Analysis of Body Armor for Ground Troops," Aberdeen Research and Development Center Report No. 2, Aberdeen Proving Ground, MD.

Walters, W. P., and Zukas, J. A. (1989), *Fundamentals of Shaped Charges*, Wiley, New York.

Watson, J. L., Serrano, D. F., and Pilarski, D. L. (1991), "Propellant Response to Shaped Charge Jet Impacts," Chemical Propulsion Information Agency, Publ. 562, 1991 JANNAF Propulsion Systems Hazards Meeting.

Wise, S., Rocchio, J. J., and Reeves, H. (1980), "Propellant Binder Chemistry and Sensitivity to Thermal Threats," Chemical Propulsion Information Agency, Publ. 330, 1980 JANNAF Propulsion Systems Hazards Subcommittee Meeting.

Wolfowitz, P. (Aug. 2003), "DOD Architecture Framework. Version 1.0," DOD Instruction 8800.x.

Wright, W. P., and Slack, W. A. (March 1980), "Diesel Fuel Fire Test Data and Model for Armored Vehicles," U.S. Army Ballistic Research Lab., BRL Memorandum Rept. No. 03004, Aberdeen Proving Ground, MD.

Yatteau, J. D., Dunn, J. A. and Dickinson, D. L. (March 1994), "Terminal Ballistic Impact Fracture of Steel Cubes," Applied Research Associates for the Naval Surface Warfare Center, Rept. NSWC TR 91-397, Dahlgren, VA.

Yatteau, J. D., Zernow, R. H., Ford, S. R., and Dickinson, D. L. (21 May 1995), "Debris Cloud Constituents and Plate Damage Throughout a Spaced Plate Array Impacted at Speeds Between 1.5-4.0 km/sec," *15th International Symposium on Ballistics*, Jerusalem, Israel.

Yatteau, J. D., Zernow, R. H., and Recht, R. F. (Jan. 1991), "Compact Fragment Multiple Plate Penetration Model (FATEPEN 2)," Volumes I and II, Naval Surface Warfare Center, NAVSWC TR 91-399, Dahlgren, VA.

Zeller, G. A. (Aug. 1952), "Vulnerability Firings Against Tank Fluid Containers," U.S. Army Ballistic Research Laboratory, BRL TN No. 735, Aberdeen Proving Ground, MD.

Zeller, G. A. (April 1961), "Methods of Analysis of Terminal Effects of Projectiles Against Tanks," U.S. Army Ballistic Research Lab., Rept. BRLMR-1342, Aberdeen Proving Ground, MD.

Zeller, G. A. (Sept. 1962), "HEAT Warheads vs. Thin Aluminum and Steel Armors – Behind Armor Effects," U.S. Army Ballistic Research Lab., Rept. BRL-MR-1431, Aberdeen Proving Ground, MD.

Zeller, G. A., and Armendt, B. F. (Nov. 1987), "Update of the Standard Damage Assessment List for Tanks: Underlying Philosophy and Final Results," Volume X: Vulnerability Models; Part 1A, ASI Systems International, AD-TR-87-65.

Appendix A

Penetration from Fragmentation Munitions

FRAGMENTATION weapons pose a unique set of challenges for the vulnerability analyst, because they present a number of KE penetrators (fragments) to the target nearly simultaneously, with impacts distributed over an area substantially larger than the size of each penetrator. Fragmentation weapons encompass the class of munitions, such as artillery rounds and mortar shells, which carry HEs contained within a metal case. Upon fuzing of the weapon, the explosive detonates and the rapid expansion of the detonation products accelerates the case outward from the axis of the warhead. At some point in the expansion, the case breaks into fragments that have a distribution of masses and velocities. The fragments travel through the intervening space between the detonation location and the target, suffering velocity decay because of aerodynamic drag. Any fragments that hit the target cause damage primarily by KE penetration.

Other fragmenting warheads use scored cases, or pre-formed fragments (generally cubes or rectangular parallelepipeds) to control the distribution of fragment mass and improve the warhead lethality. Controlled fragmentation is typically used with anti-aircraft and antimissile warheads and is also found in bomblets and grenades.

The Thor model, a simple analytical model with experimental parameters, was the fragment-penetration model for many years, but it does not capture high-velocity penetration phenomena, such as fragment shatter. Engagement kinematics of air-to-air missiles carrying modern warheads dictate extremely high-velocity fragment impacts. In some analysis organizations, the more elaborate FATEPEN (Fast Air Target Encounter Penetration) model has now largely replaced Thor.

The following description of FATEPEN is taken largely from the "MUVES/FATEPEN Analyst's Guide," written by Yatteau et al. (1995b). In addition, the reader is referred to Yatteau et al. (2005) for more current information on FATEPEN.

The FATEPEN model and computer code were developed to simulate compact fragment penetration of thin to moderately thick spaced plates at impact velocities up to about 16,000 ft/s (Yatteau et al. 1991). The model was developed primarily for V/L assessments involving air targets or lightly armored surface targets. As a result, the terminal ballistic models in FATEPEN pertain to penetration where the ratio of plate thickness to fragment diameter (T/D) is about 2:0 or less. FATEPEN has been developed primarily as an interactive PC-based penetration code. A batch-processing version of the code is maintained for the Naval Surface Warfare

Center Dahlgren Division (NSWCDD), and the primary FATEPEN penetration algorithms have been extracted for inclusion in other Joint Technical Coordinating Group (JTCG) vulnerability assessment codes (HISVART, COVART 3 and 4).

Impact velocities between missile warhead fragments and high-altitude missile targets are expected to reach or exceed 16,000 ft/s (Dickinson 1979). This velocity is well above the typical threshold fracture speed for steel cubes impacting an aluminum plate. The fragment penetration equations, which have generally been used for air target vulnerability assessments, are taken from the JTCG Penetration Equations Handbook for Kinetic Energy Penetrators (Joint Technical Coordinating Group for Munitions effectiveness 1977). These equations recognize fragment mass loss because of shock erosion, extrusion shear, and impact fracture; but they consider only the major residual fragment particle subsequent to perforation. When impact speeds are below or just above threshold fracture speed, the primary ballistic threat to air targets by warhead fragments remains well represented by a single penetrator fragment throughout the target. However, when impact speeds exceed about twice the threshold fracture speed, impact fracture is extensive and the major fragment particle represents only a small fraction of the damage potential of the post-impact debris cloud. Thus, it was recognized that fragment penetration and plate damage predictions for impact speeds above 6500 ft/s would, in general, require new models to predict the transformation of single fragments into high-speed, multi-particle debris clouds.

Beginning in 1978, the Navy funded a series of experimental and analytical efforts to expand the available penetration database for impact speeds between about 5000 and 16,000 ft/s. An initial test program of 18 shots provided preliminary observations regarding the effects of fragment shape, size, and impact speed on penetration of a spaced plate array representing a typical air target (Dickinson 1979). These and other tests provided the experimental basis behind development of the multiple plate fragment penetration model and a computer code called Fast Air Target Encounter (FATE) (Yatteau 1982). FATE was developed for applications involving steel fragments penetrating thin aluminum and/or steel plates at high velocities. The model development focused on formulas to predict the formation, dispersion, penetration (residual masses and velocities), and plate damage associated with a multi-particle debris cloud.

From 1984 through 1990, additional experiments and analyses were performed at the Naval Research Laboratory (NRL), Denver Research Institute (DRI), and Applied Research Associates, Inc. (ARA) to expand the high-speed impact database and to improve the penetration model. Approximately 750 impact fracture tests were conducted during this period, and the results were used to develop a comprehensive impact fracture model for steel, tungsten alloy, aluminum, and depleted uranium cubes impacting a variety of metallic plate materials (Yatteau et al. 1991; Yatteau et al. 1994a).

Parallel efforts under sponsorship by the Raytheon Company, the Naval Weapons Center, and the Naval Surface Warfare Center led to development of a penetration code called Fragment Penetration and Impact Initiation Model (FPIIM), which simulated multiple-spaced plate penetration by steel, tungsten alloy, or depleted uranium compact fragments (Recht and Finnegan 1986; Recht 1988). FPIIM recognized fragment mass loss because of shock-erosion, extrusion shear, lateral erosion, and impact fracture, and it monitored the mass, velocity, and orientation of the primary residual fragment throughout the plate array (but not secondary particles).

In 1988, versions of FPIIM and FATE were combined to form the original FATEPEN code. In 1989–1990, FATEPEN was completely restructured and streamlined and the user interface was substantially improved. The generalized cube fracture model was also installed along with an additional secondary penetration particle size and improvements to the plate damage calculations. The new version of FATEPEN was named FATEPEN2 (Yatteau et al. 1991). The naming convention for new releases of the code has since been revised, and FATEPEN2 is now called simply FATEPEN. As with most analysis codes, improvements are always being made and new versions are released periodically. In this section, code details refer to Version 2.x.

Since its original release in 1991, several experimental and analytical efforts have been accomplished to improve FATEPEN penetration models. That work was sponsored by the Naval Surface Warfare Center, the Air Force Armament Test Laboratory (AFATL), and the Raytheon Company. These later model enhancements include new impact fracture models for bar fragments (Yatteau and Dunn 1991), cylinders and spheres (Yatteau 1991), multi-particle, synergistic effects on plate damage (Yatteau and Dickinson 1993), impact fracture and penetration models for tantalum fragments, an improved lateral erosion mass loss model, new residual mass and velocity models for laminated plates, and 3-D fragment tumbling and trajectory deflection models (Yatteau et al. 1994b). Also, portions of the FATEPEN code were modified further to improve computational flow. All of these improvements are incorporated in Revision 2.5.0, which is the version of FATEPEN installed in MUVES.

I. Simple Methods

The distribution of individual weights for natural fragmentation follows the Mott law (Weiss 1949), and initial velocities follow the Gurney law (Gurney 1954). For cylindrical casings, the initial fragment velocity, V, can be estimated from

$$V = (2E)^{1/2}[(C/M)/(1 + (1/2)(C/M))]^{1/2} \tag{A1}$$

where M is total weight of fragmenting metal, C is total weight of explosive, and E is the explosive energy per unit mass. Values of $(2E)^{1/2}$ for some typical explosives are given in Table A1.

Table A1 Values of the square root of $2E$

Explosive	Values of $(2E)^{1/2}$
TNT	2440 m/s
Composition B	2680 m/s
Composition A3	2740 m/s

The decay of fragment velocity with distance from the exploding munition is

$$(V/V_0) = \exp[-(C_D \rho_a x)/(2K^{2/3} m^{1/3})] \quad (A2)$$

where V_0 is the initial velocity, V is the velocity along the trajectory, C_D is the coefficient of drag, ρ_a is the air density, x is the distance traveled, K is a coefficient associated with fragment shape, and m is the weight.

Critical to the estimation of damage are the weights, densities, shapes, angles of attack, and striking velocities of each fragment. Brown (1977) presents the results of hundreds of experiments carried out to develop the basic data. Table A2 provides an example of the significance of the investigation, where ballistic limits (V_{50}) of steel and tungsten fragments against steel, aluminum, and titanium armor can be estimated from the following formula, where the angle of attack is 90°:

$$V_{50} = a\, t^b\, m^{-c} \quad (A3)$$

where m = fragment weight; t = plate thickness; and a, b, and c are coefficients determined from least-square fits to the data. Typical values for a, b, and c are given in Table A2.

A different, but simplistic, approach is to assume that during the penetration process, the resisting force on the penetrator is directly proportional to the penetrator's velocity and mass and is inversely proportional to its presented area. Integration of the equation of motion shows that penetration is then a function of momentum per unit area,

$$p \sim m\, v/A \quad (A4)$$

For a homologous set of parameters,

$$M \sim \rho L^3 \quad (A5)$$

where L is a characteristic length and

$$m \sim \rho A\, A^{1/2} = \rho A^{3/2} \quad (A6)$$

Table A2 Constants for estimating ballistic limits

Fragment material	Steel plate			Al plate			Ti plate		
	a	b	c	a	b	c	a	b	c
	English units: feet/second, grains, inches								
Steel	40,600	0.906	0.359	18,900	0.903	0.339	56,600	1.325	0.431
Tungsten	24,700	0.906	0.359	11,800	0.903	0.339	31,100	1.325	0.431
	Metric units: meters/second, grams, centimeters								
Steel	1990	0.906	0.359	980	0.903	0.339	1540	1.325	0.431
Tungsten	1210	0.906	0.359	613	0.903	0.339	848	1.325	0.431

Table A3 Units for equation (A8)

	Metric units	USA common usage units
Weight of fragment m	grams	grains
Density of fragment ρ	grams/cm^3	lbs/in^3
Velocity of fragment v	meters/s	ft/s
Thickness of plate t	cm	inches

Substituting equation (A6) into equation (A4) yields

$$p \sim m^{1/3} \rho^{2/3} v \qquad (A7)$$

The exponents of this equation are close to those computed from Brown's experiments.

From an analysis of available data on the penetration of mild steel and aluminum targets by cubical fragments of steel and sintered tungsten, we can develop a generalized rule for the thickness t of a target that could be penetrated. For normal incidence (90 deg), the rule is

$$t = (1/K) \, m^{1/3} \rho^{2/3} v \qquad (A8)$$

where K is a constant.

This equation is valid for two systems of units, given in Table A3.

If we subsume the value of ρ into a constant (now lowercase k), we can write

$$t = k \, m^{1/3} \, v \times 10^5 \qquad (A9)$$

where k is given by Table A4.

II. Compact Fragment Penetration Characteristics

At low speeds, fragments perforate thin plates without deformation or loss of mass. As impact velocity increases, impact interface pressures become more intense and fragments begin to mushroom (extrude radially). Against harder and/or heavier plates, penetrator material extruded beyond a certain radius will be sheared from the fragment as it passes through the plate (extrusion-shear mass

Table A4 Values of k for equation (A9)

Target material	Metric units Fragment		English units Fragment	
	Steel	Tungsten	Steel	Tungsten
Steel	49.9	83.3	2.40	4.02
Aluminum	102.0	170.0	4.93	8.22

loss). Above a certain impact velocity, the relative velocity between the undeformed portion of the penetrator and the impact interface will exceed the speed at which plastic deformation (under uniaxial stress conditions) can propagate into the penetrator. When this occurs, a shock wave will form in the penetrator just upstream of the impact interface, and penetrator material passing through the shock wave will be ejected radially outward (shock-erosion mass loss). The relative velocity between the undeformed portion of the penetrator and the impact interface continuously decreases during the penetration process. When the relative velocity falls below the plastic wave speed for uniaxial stress, the relative motion can be accommodated by plastic deformation in the penetrator, and shock erosion gives way to extrusion-shear mass loss. Steel cylinders (BHN 285) perforating thin mild steel plates begin to deform when impact velocity exceeds about 1500 ft/s. The onset of extrusion-shear mass loss occurs at a velocity near 2000 ft/s, and shock-erosion mass loss is predicted to begin at a speed of about 2500 ft/s (Recht 1978, 1990).

Above a threshold impact velocity, compact fragments fracture on impact as a result of tensile stresses associated with the interaction of relief waves from the free surfaces of the fragment and plate (fracture mass loss). The threshold fracture speed depends on fragment and plate material properties and the impact geometry. For a particular set of fragment and plate characteristics, the severity of fracture or shatter is determined primarily by the initial impact pressure and thus by the ratio of the impact speed to the threshold fracture speed. Mild steel cubes begin to fracture on impact with mild steel plates at speeds near 2400 ft/s and on impact with aluminum plates at speeds near 3000 ft/s (Yatteau et al. 1994a). The plate material driven out the back of a plate will also be fractured when impact speed exceeds a certain threshold fracture speed. Plate plugs produced by steel cylinders impacting mild steel plates begin to break up at speeds above 3000 ft/s (Recht et al. 1969).

Mass loss from shock erosion, extrusion shear, and impact fracture during the perforation of thin plates at obliquities less than about 45 deg is determined principally by the component of the impact velocity normal to the plate surface. (Obliquity is defined as the angle between the shotline and the inward normal to the target surface.) Impact pressures associated with the lateral component of the impact velocity are lower (because of the predominantly plane stress state in the plate) and do not generally produce mass loss until impact obliquities exceed about 45 deg. At impact obliquities above 45 deg, an erosion-type mass loss associated with the lateral component of the impact velocity becomes noticeable (lateral erosion mass loss). Lateral erosion becomes the principal mass loss mode at impact obliquities above 60 deg (Recht 1988; Yatteau et al. 1994b).

Figure A1 illustrates the general features of high-velocity, multiple-plate penetration by a compact fragment. Upon impact with the first plate, the fragment loses a portion of its mass near the impact side of the plate to shock-erosion and extrusion-shear mass loss mechanisms (and/or to lateral erosion at high impact obliquities). The remainder of the fragment mass and most of the plate plug removed by the penetration appear behind the plate. Conservation of momentum and energy dictate that the residual velocity of the combined plate/penetrator particle mass center (including any particles sprayed out the front side of the plate) be less than the impact velocity.

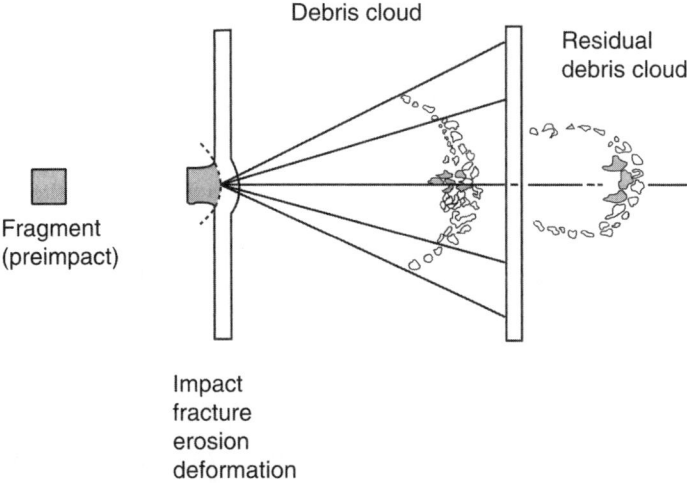

Fig. A1   High-velocity, multiple-plate penetration by a compact fragment.

At low-impact velocities, the residual fragment and plate plug will remain intact. At impact speeds above the penetrator or plate plug fracture threshold velocities, the material behind the plate will form a multi-particle debris cloud, as shown in Fig. A1. The numbers of penetrator and plate debris particles in the cloud increase with increasing impact speed along the threshold fracture speeds. For example, steel cubes fracture into three to five particles when impacting aluminum plates at speeds near 4000 ft/s and will produce about 100 debris particles (not counting tiny dust-like particles) at speeds near 8000 ft/s (Yatteau et al. 1994a). In debris collection experiments involving steel spheres impacting copper plates at speeds near 13,000 ft/s, approximately 800 steel and copper particles were individually counted, along with an additional 18,000 particles estimated from a sieve analysis of the smaller particles (Yatteau et al. 1995a). The shape of the debris cloud reflects the spatial variations in particle velocity and depends on fragment material and shape, impact orientation, velocity, obliquity, and plate material and thickness. For compact steel fragments perforating steel or aluminum plates at impact speeds up to about 16,000 ft/s, the debris clouds are nearly ellipsoidal in shape, with most of the plate and penetrator debris particles residing on or near the leading surface of the hollow, expanding cloud (Dickinson et al. 1987). Debris clouds with penetrator particles extending back into their interiors have been observed for aluminum and copper spheres and cylinders impacting thin aluminum plates at speeds between 11,000 and 21,000 ft/s (Piekutowski 1990). Debris particles disperse radially behind the plate, and the cone angle bounding the penetrator particle trajectories is usually smaller than the cone angle bounding the plate particle trajectories.

In general, upon impact with the next plate, some of the particles will perforate and some will be stopped. The perforating particles can lose mass and velocity and will drive new plate material into the debris cloud. Thus, the debris cloud behind the second plate will generally contain penetrator fragments along with plate

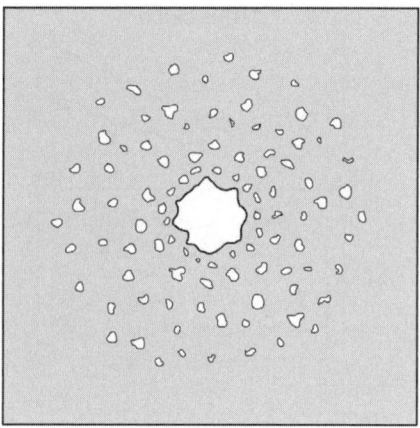

Fig. A2 Typical plate damage from a multiple-particle debris cloud.

fragments from the first two plates. The debris cloud penetration process is repeated at successive plates until none of the particles can perforate the next plate.

Figure A2 illustrates typical damage to a plate impacted by a multiple-particle debris cloud, which includes craters and holes produced by nonperforating and perforating particles, respectively. Depending on the incident fragment number, mass, and velocity distributions, there can also be an enlarged central hole where the plate material is completely removed because of blow-out and/or petaling of the plate. For hypervelocity impacts, there may also be material spalled from the rear of the plate with and without individual debris particle perforations, depending on impact speed. The plate will also be plastically deformed (dished) to an extent dependent on the magnitude and distribution of the impulse transmitted to the plate by the impacting particles.

III. FATEPEN Penetration Model Overview

Figure A3 contains a flow chart mapping the main penetration computational loop in FATEPEN. Figure A4 illustrates debris cloud and plate damage characteristics as predicted by FATEPEN. The example shown in Fig. A4 presumes a single intact fragment (called the primary penetrator particle) impacting the first plate. The impact velocity was sufficient to fracture both the fragment and the plate plug at the first plate. The residual primary particle and the higher velocity secondary particles behind the first plate were capable of penetrating the second plate. The primary particle lost mass because of erosion and extrusion shear mechanisms at the second plate, but it did not fracture.

As shown in Fig. A3, a typical run begins by specifying the initial primary fragment characteristics, the plate array characteristics, and the encounter conditions. The PC version of FATEPEN is an interactive program, and the user may select preprogrammed penetrator characteristics from a default catalog or define new fragments by editing the catalog entries or reading in previously saved penetrator files. Likewise, plate array characteristics can be changed by editing the default

PENETRATION FROM FRAGMENTATION MUNITIONS 261

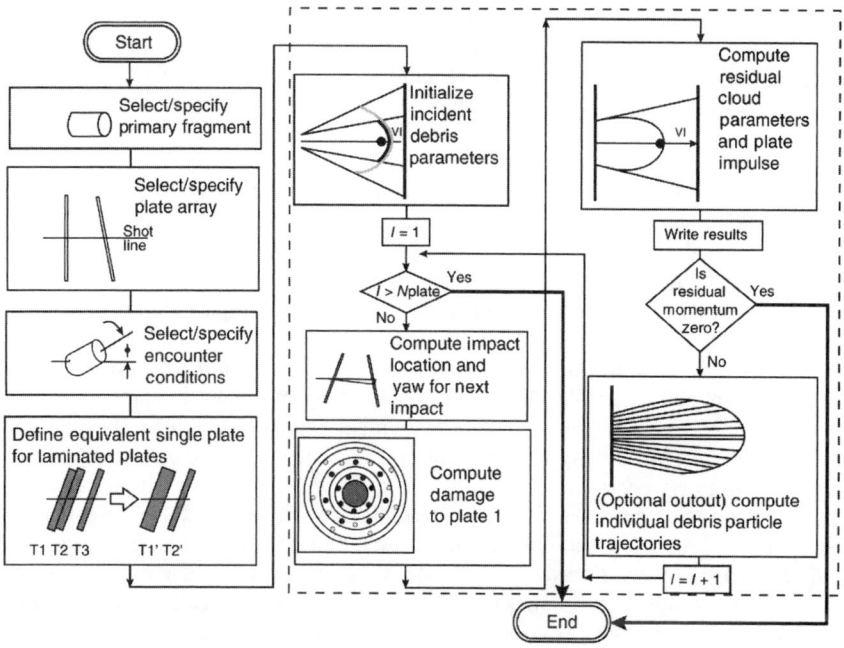

Fig. A3 Main penetrator computation in FATEPEN.

plate array or by reading and editing previously saved FATEPEN plate array descriptions. The default penetrator catalog and plate array are both easily changed using the file options within FATEPEN. Possible penetrator materials include various alloys of steel and tungsten, depleted uranium, tantalum, titanium, aluminum,

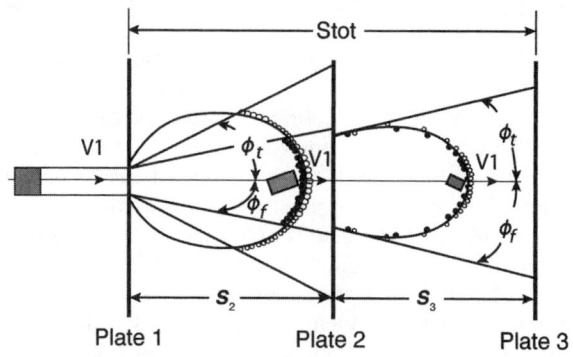

■ Primary penetrator particle
• Larger secondary penetrator particles
• Smaller secondary penetrator particles
° Plate particles

Fig. A4 FATEPEN debris cloud characteristics.

and aluminum/Teflon. Allowable fragment shapes are parallelepipeds, cylinders, tapered cylinders, and spheres. Possible plate materials include steel, aluminum, titanium, magnesium, doron, phenolic, pine, oak, cast iron, copper, lead, tuballoy, unbonded nylon, bonded nylon, lexan, cast Plexiglas, stretched Plexiglas, bullet-resistant glass, face-hardened steel, and fluid and graphite epoxy. The user is prompted for the impact velocity (or range of impact velocities). The user can also specify a range of penetrator masses and orientations. The option is also given to specify either the initial fragment impact yaw angle (for cylinders, tapered cylinders, and parallelepipeds) or the initial impact presented area (for cylinders). If presented area is selected, FATEPEN computes the associated impact yaw angle. Finally, the user is given the option to override computed impact angles by specifying the impact angle at the first plate and one impact angle for all subsequent plates. The impact angle is the angle between the impacting face of the fragment and the plate surface, and it partially determines the severity of impact fracture.

After the primary fragment, plate array, and encounter conditions are specified, FATEPEN searches the input plate array for laminated plates (i.e., occurrences of zero spacing between plates) and replaces them with equivalent single plates for the penetration calculations (the original plate array is returned after the penetration calculations for input/output and further editing). Then FATEPEN assigns values to parameters describing characteristics of the secondary particles in the debris cloud incident on the first plate. The parameters are currently all set to zero so that only the primary fragment impacts the first plate, as shown in Fig. A4. However, any set of initial debris cloud parameters, which conform to the general FATEPEN debris cloud model (as shown behind the first plate in Fig. A4), could be substituted for the current null specification. In general, the debris cloud may include the primary penetrator particle, two sizes of secondary penetrator particles, and one average size for plate particles. The primary particle is presumed to be located at the debris cloud's leading edge. The secondary penetrator particle trajectories are bounded by the cone half-angle ϕ_f and the plate particle trajectories are bounded by the cone half-angle ϕ_t. The secondary particles are presumed to emanate from the plate where impact fracture first occurred (the first plate in Fig. A4). The spatial distribution of the secondary particle trajectories is governed by a distribution function, which presumes the particle trajectories' areal density is inversely proportional to the radius from the shotline.

Following definition of the initial debris characteristics, the main computational loop is entered. The primary fragment velocity and angular momentum vectors are first used to compute the impact location, obliquity, and yaw at the next plate. Plate damage caused by the incident debris cloud is computed next, as shown in Fig. A5. Possible plate damage includes holes and/or craters made by individual particles and a central hole-out region caused by the particles acting in unison.

Residual debris characteristics are determined after the plate damage calculations. Generally, some of the incident debris particles will penetrate, and some will be stopped, as shown at the second plate in Fig. A4. Those that penetrate will generally lose mass and velocity and drive additional new plate particles into the residual debris cloud. Secondary penetrator particles are allowed to lose mass to shock-erosion, extrusion-shear, and lateral erosion mechanisms, but they are presumed not to fracture. The secondary penetrator particles may drive intact or fractured plate plugs into the residual debris cloud, depending on the impact

PENETRATION FROM FRAGMENTATION MUNITIONS

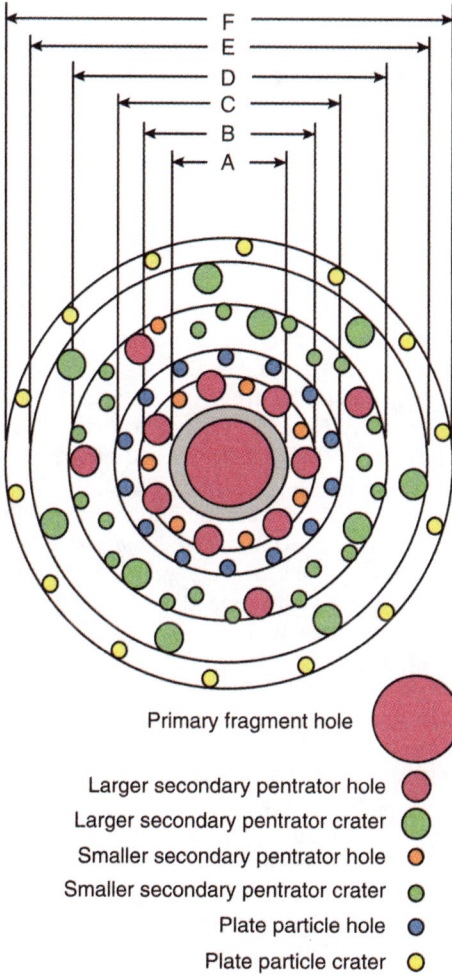

Fig. A5 FATEPEN plate damage characteristics.

conditions. Secondary plate particles are allowed to penetrate a plate (i.e., produce holes) but are presumed to give up all their momentum. The primary penetrator particle, monitored separately, may fracture on any impact in addition to losing mass to the aforementioned mechanisms. It may also generate single or multiple plate particles. When the primary penetrator particle and/or its plate plug fractures, it produces a new debris cloud, as shown behind the first plate in Fig. A4. The primary fragment residual velocity and angular momentum vectors are computed to determine the encounter conditions for the next impact. Thus, behind any particular plate, it is possible to find the following:
  1) A residual primary penetrator fragment or particle.
  2) New secondary penetrator particles resulting from impact.

3) Fracture of the primary particle at the current plate.
4) Secondary plate particles driven from the current plate by the primary particle.
5) Residual secondary penetrator particles that were generated at a previous impact.
6) Secondary plate particles produced by these residual penetrator particles.

The sizes, velocities, and dispersion angles of each type of secondary particle are compared, and an averaging process is used to recast the residual debris particles into the standard debris cloud, comprising the primary particle, two sizes of secondary penetrator particles, and one size of plate particles. After the standard residual debris cloud particle sizes and velocities have been selected, the associated numbers of each particle are adjusted so that the individual particle momenta and the total residual cloud momentum are consistent with the values determined by the individual particle penetration calculations. The impulse delivered to the current plate is then computed as the difference between debris cloud momenta in front of and behind the plate. Now, an optional debris trajectory routine can be called to generate and store individual debris particle trajectory descriptors that may be used outside FATEPEN to produce graphical displays of impact patterns and debris cloud profiles. Finally, the computation returns to the beginning of the main loop to compute the impact location and damage to the next plate. As shown in Fig. A3, the main computational loop is repeated until the plate array is completely penetrated or all particles are stopped.

References

Brown, C. J. (Oct. 1977), "Estimates of Residual Mass and Velocity for Tungsten and Steel Fragments," U.S. Army Ballistic Research Lab., BRLM Rept. No. 2794, Aberdeen Proving Ground, MD.

Dickinson, D. L. (March 1979), "Investigations of High Velocity Fragments Impacting Plate Arrays," Naval Surface Weapons Center, Rept. NWSC TR-79-66, Dahlgren, VA.

Dickinson, D. L., Yatteau, J. D., and Recht, R. F. (1987), "Fragment Breakup," *International Journal of Impact Engineering*, Vol. 5, Nos. 1–4, pp. 249–260.

Gurney, R. W. (Sept. 1954), "The Initial Velocities from Bombs, Shells and Grenades," U.S. Army Ballistic Research Lab., BRL Rept. No. 405, Aberdeen Proving Ground, MD.

Joint Technical Coordinating Group for Munitions Effectiveness (Nov. 1977), "Penetration Equations Handbook for Kinetic Energy Penetrators," 61JTCG/ME-77-16, Aerial Target Vulnerability Subgroup.

Piekutowski, A. J. (1990), "A Simple Dynamic Model for the Formation of Debris Clouds," *International Journal of Impact Engineering*, Vol. 10, pp. 453–571.

Recht, R. F. (1978), "Taylor Ballistic Impact Modeling Applied to Deformation and Mass Loss Determinations," *International Journal of Engineering Science*, Vol. 16, No. 11, pp. 809–827.

Recht, R. F. (March 1988), "High Obliquity Ballistic Perforation Models (Program FPIIM)," Denver Research Inst., Final Rept. Contract N60921-86-D-A070, Task B009, Naval Surface Warfare Center, Dahlgren, VA.

Recht, R. F. (1990), "Analytical Modeling and Plate Penetration Dynamics," *High Velocity Impact Dynamics*, ed. J. A. Zukas, Wiley, New York, pp. 443–514.

Recht, R. F., and Finnegan, S. A. (May 1986), "Penetration Equations for Tungsten Fragments," Denver Research Institute for the Naval Weapons Center, Rept. NWCTP6768, China Lake, CA.

Recht, R. F., Ipson, T. W., and Wittrock, E. P. (Dec. 1969) "Transformation of Terminal Ballistic Threat Definitions into Vital Component Malfunction Predictions," Denver Research Institute for the Naval Weapons Center, Rept. NWC-TP-4871, China Lake, CA.

Weiss, H. K. (Feb. 1949), "Justification of An Exponential Fall-Off Law for Number of effective Fragments," U.S. Army Ballistic Research Lab., BRL Rept. No. 697, Aberdeen Proving Ground, MD.

Yatteau, J. D. (Feb. 1982), "High Velocity Multiple Plate Penetration Model," Denver Research Institute for the Naval Surface Weapons Center, Rept. NSWC-TR-82-123, Dahlgren, VA.

Yatteau, J. D. (April 1991), "Effects of Fragment Shape on Impact Fracture," Applied Research Associates, Inc. for the Naval Surface Warfare Center, ARA Letter Rept. 559152, Dahlgren, VA.

Yatteau, J. D., and Dickinson, D. L. (1993), "An Engineering Model to Predict Damage to Plates Impacted by High Velocity Debris Clouds," Proceedings of the 1992 Hypervelocity Impact Symposium, *International Journal of Impact Engineering*, Vol. 14, Nos. 1–4, pp. 831–842.

Yatteau, J. D., and Dunn, J. A. (Feb. 1991), "Penetration Methodology for Bar Fragments," Applied Research Associates, Inc. for the Air Force Armament Lab., Rept. AFATL-TR-90-100, Eglin Air Force Base, FL.

Yatteau, J. D., Dunn, J. A., and Dickinson, D. L. (March 1994a), "Terminal Ballistic Impact Fracture of Steel Cubes," Applied Research Associates for the Naval Surface Warfare Center, Rept. NSWC-TR-91-397, Dahlgren, VA.

Yatteau, J. D., Zernow, R. H., Dzwilewski, P. T., Patel, G. A., and Recht, R. F. (April 1994b), "Analytical Support for the Patriot Program," Applied Research Associates, Inc., Final Rept. Raytheon/ARA Subcontract 71-613-CD-0072, Littleton, CO.

Yatteau, J. D., Zernow, R. H., Ford, S. R., and Dickinson, D. L. (May 1995a), "Debris Cloud Constituents and Plate Damage Throughout a Spaced Plate Array Impacted at Speeds Between 1.5–4.0 km/sec," *15th International Symposium on Ballistics*, Jerusalem, Israel.

Yatteau, J. D., Zernow, R. H., and Frew, K. C. (Aug. 1995b) "MUVES/FATEPEN Analyst's Guide," Applied Research Associates, Littleton, CO.

Yatteau, J. D., Zernow, R. H., and Recht, R. F. (Jan. 1991), "Compact Fragment Multiple Plate Penetration Model (FATEPEN 2), Volume I – Model Description; Volume II – User's Manual," Naval Surface Warfare Center, Rept. NAVSWC TR 91-399, Dahlgren, VA.

Yatteau, J. D., Zernow, R. H., Recht, G. W., and Edquist, K. T. (Feb. 2005), "FATEPEN (Version 3.0.0), Terminal Ballistic Penetration Model, Volume I – Analyst's Manual; Volume II – User's Guide; Volume III – Validation Document," Applied Research Associates, Dahlgren, VA.

Appendix B

Behind-Armor Debris Characterization

AS DESCRIBED in Chapter 2, when a munition perforates armor (see Fig. B1), a spray of potentially lethal fragments is formed behind the armor. These fragments, referred to as BAD, consist of broken pieces of penetrator and armor material ejected by the penetration process and/or shock wave release. The importance of this debris as a damage mechanism is increased by the fact that it spreads away from the residual penetrator shotline, thereby potentially interacting with a large percentage of the vehicle's vulnerable contents, including the crew.

To characterize this threat and assess vulnerabilities, one would, of course, like to know the mass, presented area, and velocity (i.e., both speed and direction) of all the fragments that constitute the debris cloud. Unfortunately, these characteristics are often difficult to measure without destroying expensive instruments. So, thin metallic "witness plates" have instead been used rather extensively to capture the effect of spall fragments by recording holes in one or more plates. In addition, but less frequently, "softcatch" experiments that capture the debris fragments in soft materials such as Celotex are performed. Each fragment is subsequently weighed, and its presented area is measured from 16 different viewpoints by an Icosahedron gauge.

I. Witness Pack

A typical arrangement for a witness pack is shown in Fig. B2. This is a stack of five thin metallic plates measuring 4 ft by 4 ft in breadth and with thicknesses of 1/32 in., 1/32 in., 1/16 in., 1/16 in., and 1/8 in., separated by 1 in. Styrofoam spacers.

For oblique shots, the witness plates are typically 4 ft by 8 ft. If the target is at an obliquity angle of θ, then the witness pack will usually be at an obliquity angle of $\theta/2$. An example of the holes left on the first plate of a witness pack is shown in Fig. B3.

It is important to recognize that the hole size left in a witness plate does not uniquely characterize the fragment that produced it. The same size hole could be produced by a relatively large fragment at a low velocity or by a smaller fragment at a higher velocity; there is simply no way to tell from the hole alone. Although it is true that multiple plates provide more information for bounding the velocity, it is not often the case that holes can be traced between successive plates. A fragment that impacts and perforates one plate can also push out a plate plug mass, which

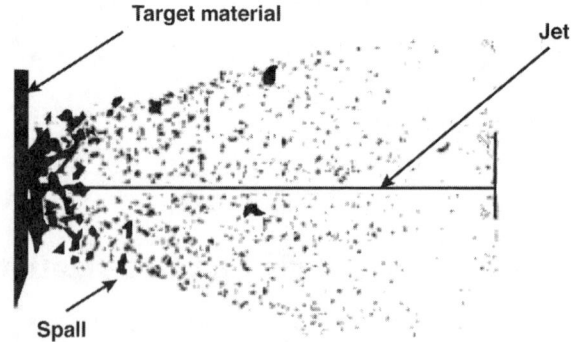

Fig. B1 Radiograph of BAD from an SC perforation of RHA.

then acts much like (and is easily disguised as) another debris fragment. Another complicating factor is that what looks like a single hole on the witness plate may in fact be two or more merged holes. It is also entirely possible to have debris fragments that may leave dents but that do not perforate the first witness plate. For these and other reasons, witness plates are not the ideal medium for capturing information about BAD. Nevertheless, they have been used extensively over the years, and our task is to glean as much information from their use as we can.

Past efforts attempted to derive the mass and velocity of each fragment from the hole it produced in one or more plates. This procedure requires certain assumptions concerning the presented area of the fragment when it impacts. Unfortunately, the

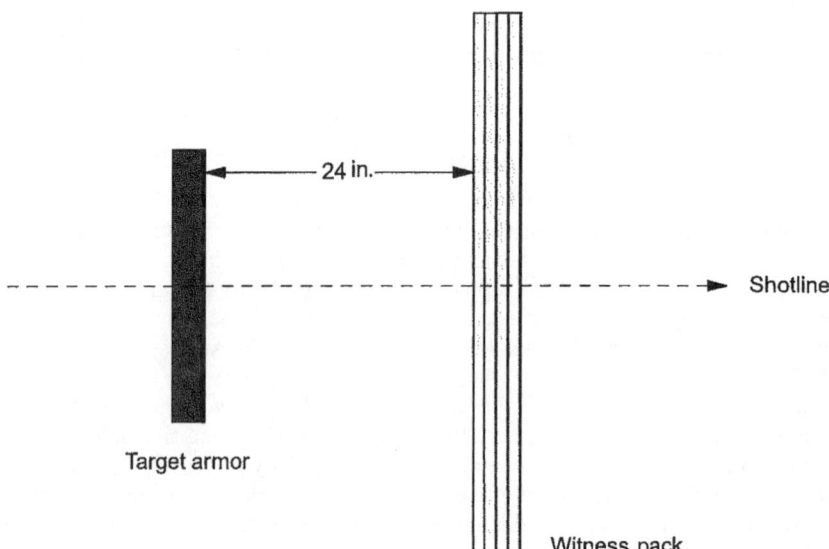

Fig. B2 Typical arrangement of target armor and witness pack for normal obliquity.

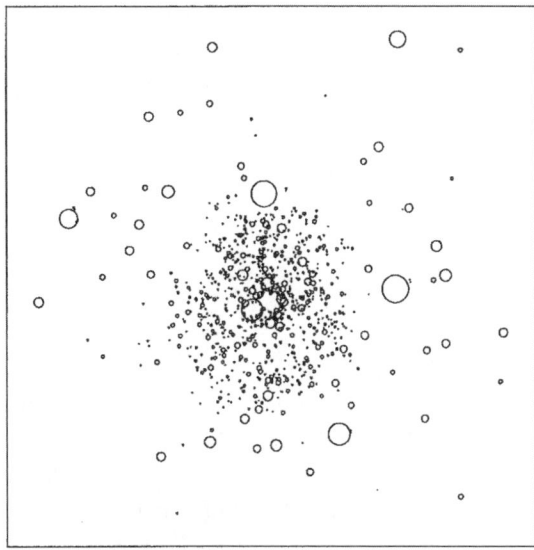

Fig. B3 Witness plate holes captured from BAD because of a KE long rod perforating RHA.

mass, so derived, is highly sensitive to this assumption. Here we describe a different approach that has been taken with these data. Instead of attempting to derive the mass and velocities of individual fragments, a model is first constructed to generate what would be produced on a simulated witness pack. We then compare the two patterns, the simulated witness pack with the actual witness pack.

II. Analysis of Witness Plates

In spite of the limitations of witness plates, these plates do provide some positive insights, as discussed in the following subsections.

A. Spatial Distribution of Holes

For a normal obliquity impact, we would expect the pattern of holes to be circularly symmetric about the shotline, and that is basically what we see. There is not perfect symmetry, however, and we expect some irregularities, which we attribute to the stochastic nature of the fracturing progress. Now if we sort the holes (without regard to size) in terms of their radial distance from the centerline, then we can count the accumulated number of holes as a function of radial distance. What we find is that this distribution can be fit well with a Weibull function. More precisely, let P be the accumulated number of holes divided by the total number of holes on the plate and let θ be the cone angle measured from the centerline to the center of the hole. From geometry, we have that $\theta = \tan^{-1}(r/z)$, where r is the radial distance from the centerline, as measured on the witness plate, and z is the distance

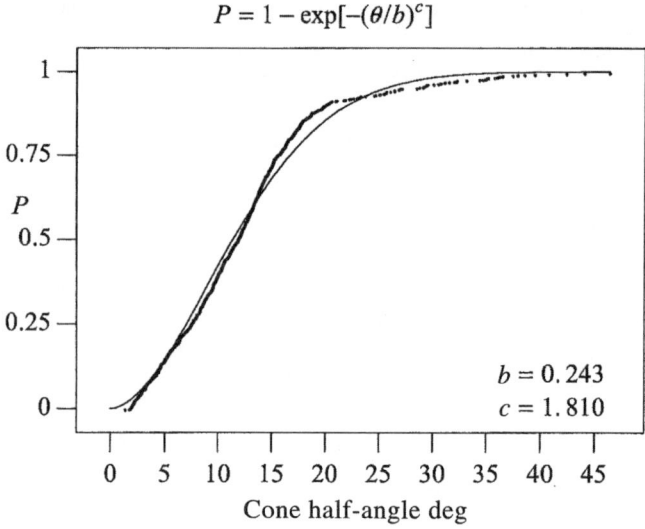

Fig. B4 Cone angle distribution of holes fit with Weibull function.

between the back plane of the target and the front of the witness plate (typically 24 in. to the first plate, as shown in Fig. B3). The plot shown in Fig. B4 is typical of the cone angle fits, where the individual points represent the data and where the curve is the Weibull function.

B. Hole-Size Distribution

Another aspect of witness plates that we can analyze is the distribution of hole sizes. Here we group holes according to size from smallest to largest, regardless of location on the plate, and we find that this can be fit with a lognormal distribution. Figure B5 shows the fit to the witness plate data, depicted in Fig. B3.

C. Correlation Between Hole Size and Hole Location

We also need to determine if there is any correlation between the angular position of a hole and its size. For example, we might expect that smaller holes are closer to the centerline and larger holes are predominately at larger angles. However, for the most part, that is not what we find. Usually, the correlation between the cone angle and hole size is small, indicating a lack of correlation. For the witness plate data shown in Fig. B3, the correlation coefficient is 0.227. (Perfect correlation would have a value of 1, no correlation would have a value of 0, and perfect anticorrelation would have a value of −1.)

III. Description of the Mathematical Model

We now describe the main features of the mathematical model, what parameters are calibrated, and how it is used in simulation.

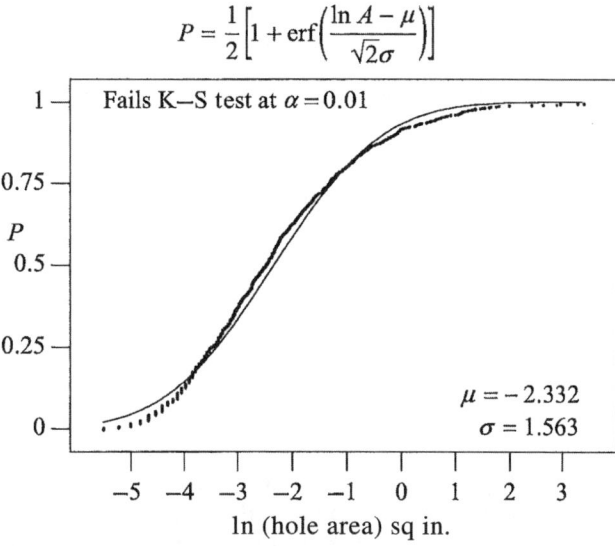

Fig. B5 Hole-size distribution fit with lognormal function.

A. Data-Driven, Stochastic Model

This BAD model is a phenomenological model based upon data collected by witness packs, radiographs, and measurements of spall fragments. It is highly data driven; no attempt of trying to derive the model from first principles is made. The model is also stochastic in nature, consisting of a number of probability distribution functions that are sampled in Monte Carlo fashion to generate the debris fragments.

B. Number of Fragments

The total number of fragments generated by the model, including both spall and broken penetrator pieces, N, is derived from the holes recorded on witness plates, along with penetration equations for perforating witness plates. Data indicate that this number goes roughly as the square of the residual velocity of the penetrator:

$$N = nv_r^2 \tag{B1}$$

where the coefficient n is calibrated from witness plate data.

C. Fragment Type

Each fragment may be either spall (made of target material) or a piece of the penetrator. Let p be the probability that a given fragment is from the penetrator. Then $0 \le p \le 1$, where $p = 0$ for spall only, $0 < p < 1$ for a mixture of spall and penetrator pieces, and $p = 1$ for penetrator pieces only.

The type t of each fragment is a Bernoulli trial:

$$T = \text{Bernoulli}(p) = \begin{cases} 0 & \text{for spall} \\ 1 & \text{for penetrator piece} \end{cases} \quad (B2)$$

and the corresponding weighted fragment density is

$$\rho = t\rho_{\text{pen}} + (1-t)\rho_{\text{spall}} \quad (B3)$$

D. Fragment Shape Factor

Fragment presented area is characterized in terms of a dimensionless shape factor γ, defined by

$$A = \gamma(m/\rho)^{2/3} \quad (B4)$$

where A is the presented area of the fragment, m is its mass, and ρ is the material density. Experimental measurements indicate that fragment shape factors have a lognormal distribution and are uncorrelated to the mass of the fragment. It is convenient to specify the minimum and maximum shape factors, $\gamma_{\min}$ and $\gamma_{\max}$, and then compute the values of the parameters that go into the shape factor distribution function. Thus, the cumulative probability of having a shape factor γ or smaller is given by

$$P(\gamma) = \frac{1}{2}\left[1 + \text{erf}\left(\frac{\ln(\gamma) - \mu_\gamma}{\sqrt{2}\sigma_\gamma}\right)\right] \quad (B5)$$

Here, μ_γ and σ_γ are the mean and standard deviation, respectively, of the natural logarithm of the fragment shape factors. If μ_γ and σ_γ are not known, it is possible to calculate them, provided the minimum and maximum shape factors are known. If we want the cumulative distribution between the minimum and maximum shape factor to represent 95 percent of the total shape factor distribution, then it can be shown that

$$\mu_\gamma = \frac{1}{2}\ln(\gamma_{\min}\gamma_{\max}) \quad \text{and} \quad \sigma_\gamma = \frac{\ln(\gamma_{\max}/\gamma_{\min})}{2\sqrt{2}\,\text{erf}^{-1}(0.95)} \quad (B6)$$

where $\text{erf}^{-1}(0.95)$ is the inverse error function evaluated at 0.95 and has the approximate value of 1.3859.

E. Fragment Presented Area

Fragment presented area, as recorded by witness plates, is also found to be lognormally distributed. The cumulative probability of having an area A or smaller is given by

$$P(A) = \frac{1}{2}\left[1 + \text{erf}\left(\frac{\ln(A) - \mu_A}{\sqrt{2}\sigma_A}\right)\right] \quad (B7)$$

Here, μ_A and σ_A are the mean and standard deviation, respectively, of the natural logarithm of the hole areas. Analogous to the shape factor distribution, these can be calculated from the minimum and maximum hole areas. In practice, if we stipulate that the cumulative distribution between the minimum and maximum area represents 95 percent of the total hole area distribution, then it can be shown that

$$\mu_A = \frac{1}{2} \ln(A_{min} A_{max}) \quad \text{and} \quad \sigma_A = \frac{\ln\left(A_{max}/A_{min}\right)}{2\sqrt{2}\,\text{erf}^{-1}(0.95)} \tag{B8}$$

F. Angular Distribution

The angular distribution of fragments about the center of the target hole usually fits a Weibull distribution. Accordingly, the cumulative probability that a fragment is at an angle θ or less from the shotline is given by

$$F(\theta) = 1 - \exp\left[-\left(\frac{\theta}{b}\right)^c\right] \tag{B9}$$

where b is the scale parameter and c is the shape parameter. These parameters are not intuitive, and a better choice is the cone half-angles that enclose 50 percent and 95 percent of the fragments, θ_{50} and θ_{95}. The relationship between the two sets of parameters is

$$\theta_{50} = b[-\ln(0.5)]^{1/c} \quad \text{and} \quad \theta_{95} = b[-\ln(0.5)]^{1/c} \tag{B10}$$

or

$$c = \frac{\ln\left[\frac{\ln(0.05)}{\ln(0.5)}\right]}{\ln\left(\frac{\theta_{95}}{\theta_{50}}\right)} \quad \text{and} \quad b = \frac{\theta_{95}}{[\ln(0.05)]^{1/c}} \tag{B11}$$

G. Velocity Field

Individual fragment velocities are not directly measured in BAD experiments, so we do not know how the velocities are distributed over the debris cloud. (We have even less information about the distribution of fragment velocities at a particular angle.) One can reasonable expect, however, that the velocity is correlated to angle so that it is a maximum along the shotline and decreases with increasing cone angle. A simple function that has this property, and also gives a value of zero when the angle is 90 deg, is the cosine function. Accordingly, the velocity field at an angle θ is taken to be

$$v(\theta) = v_{max} \cos \theta \tag{B12}$$

where v_{max} is taken to be the residual velocity of the penetrator, v_r. In practice, witness plate holes are not typically seen beyond a maximum come angle θ_{max}, where θ_{max} is typically 45 deg for nonoblique shots. It is possible to estimate θ_{max} as the cone angle that encompasses 99 percent of the witness plate holes from the angular distribution. Using equation (B11), we find

$$\theta_{max} = b[-\ell n(0.01)]^{1/c} \tag{B13}$$

The user can override this estimate by specifying a value for θ_{max}.

H. Model Calibration

A total of 13 parameters have to be calibrated for each threat-target combination:

B	Weibull scale parameter.
C	Weibull shape parameter.
n	Coefficient for total number of fragments in equation (B1).
T	Fraction of total number of fragments that are broken pieces of the penetrator.
ρ_{pen}	Penetrate or material density.
ρ_{spall}	Spall material density.
γ_{min}	Minimum shape factor in a lognormal distribution.
γ_{max}	Maximum shape factor in a lognormal distribution.
A_{min}	Minimum fragment presented area in a lognormal distribution.
A_{max}	Maximum fragment presented area in a lognormal distribution.
θ_{50}	Cone half-angle containing 50 percent of the fragments.
θ_{95}	Cone half-angle containing 95 percent of the fragments.
θ_{max}	Cone half-angle containing 99 percent of the fragments or a cutoff value if user-specified.

IV. Summary

The current BAD characterization is designed to be a stochastic simulation of the debris process. This permits us to calibrate the model to the trends found in existing witness plate data while at the same time permitting variations with each model run. This approach makes optimal use of past data, while providing a framework for enhancements as the debris process is better understood and new measurement technologies become available. For more information on this subject, the reader is referred to Shnidman (1988).

Reference

Shnidman, R. (1988) "Direct Fragment Lethality Inference from Witness Plate Array Data," *Proceedings on the 25th ADPA Survivability and Vulnerability Symposium*, San Diego, CA.

Appendix C

Estimating Component Probability of Damage Given a Hit

KRUSE and Brizzolara (1971) provide the following analytical procedures for estimating probability of component dysfunction given a hit $P_{CD|H}$ for a component struck by single fragments from random directions with the following assumptions: 1) the component is treated as having six basic faces; 2) attacks are considered at 0-deg and 45-deg obliquity; and 3) each attack direction is considered equally likely. The basic faces for any component are obtained by projection of the component onto the six faces of a cube enclosing the component (Fig. C1).

Thus, assumption 1 allows a 3-D component to be represented by six 2-D faces. Assumption 2 produces 30 attack directions consisting of one at 0 deg and four at 45 deg on each of the six faces. The total presented area A_p for the 30 attack directions is derived as follows:

$$A_p = a + 4 (\cos 45)a$$
$$A_p = a (1 + 2.83)$$
$$A_p = 3.83a \tag{C1}$$

where a equals the sum of the presented areas of the six basic faces.

The total critical damage area (previously called kill area) is derived from the critical portions of each basic face that are vulnerable to attack from 0-deg and 45-deg directions. The total critical damage area for an individual face is

$$(A_{cd})_i = \sum_{\alpha=1}^{N} a_{i\alpha} + 2.83 \sum_{\alpha=1}^{N} b_{i\alpha} \tag{C2}$$

where

$a_{i\alpha}$ = an area that is a portion of a basic face i damaged by the 0-deg obliquity impact of a fragment having a specific set of penetration parameters. $i = 1, 2, \ldots, 6$ and $\alpha = 1, 2, 3, \ldots, N$.

$b_{i\alpha}$ = an area that is a portion of a basic face i damaged by the 45-deg obliquity impact of a fragment having a specific set of penetration parameters. $i = 1, 2, \ldots, 6$ and $\alpha = 1, 2, \ldots, N$.

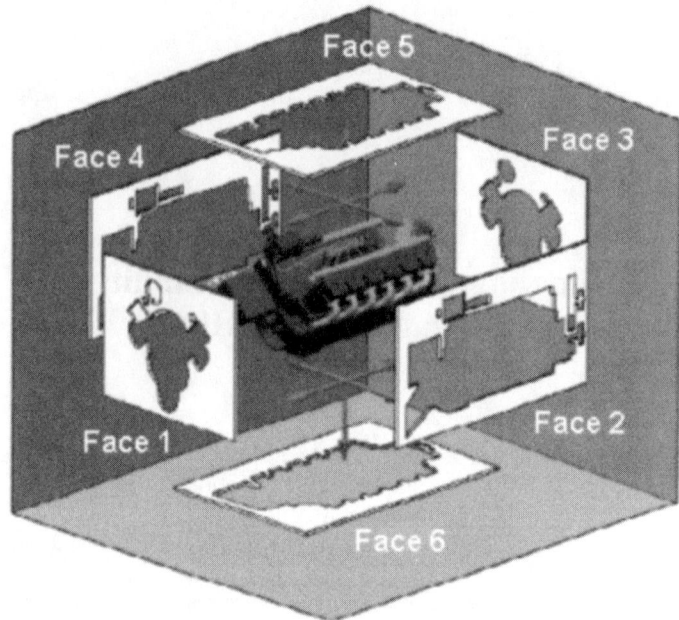

Fig. C1 Projection of a component onto six faces of a cube.

The total critical damage area for a component is

$$A_t = \sum_{i=1}^{6} (A_{cd})_i$$

and

$$P_{\text{CD|H}} = A_t/A_p \qquad (C3)$$

Thus, the vulnerability analyst must perform the following:
1) Determine how the component functions.
2) Designate critical regions and areas. Critical areas are the projections of critical regions onto the basic faces.
3) Establish damage criteria for rendering the component nonfunctional at some prescribed level. Of all the steps in this procedure, this can be the most challenging. A damage criterion can be a simple definition of the type and amount of damage (such as a minimum hole size) required to render a component dysfunctional, or it could be a required performance level calculated by an algorithm or engineering model using physical damage parameters as input.
4) Record barrier thicknesses and materials in the order in which they would be struck by a fragment. The last barrier would represent the critical region of the component.

5) Make decisions regarding the critical damage. This involves determining for what specific penetrator parameters do critical areas become critical damage areas.

$P_{CD|H}$'s determined in the manner described previously, for components found in ground mobile targets, have been in the form of cumulative step functions that correlate single-fragment mass-velocity combinations to a $P_{CD|H}$ value. An example of a two-step function is shown in Fig. C2.

"Two-step" means that for each fragment mass, there are two velocity increments that correlate to different $P_{CD|H}$ values. $P_{CD|H}$ functions have generally been supplied in two- and four-step formats depending on the complexity of the component. A more complex component would have a four-step function. Note that $P_{CD|H}$ functions should actually be continuous. The steps in the functions occur because only a few attack obliquities (0 deg and 45-deg) are analyzed for each basic face of a component.

The 129.6-gram (2000-grain) fragment data in Fig. C2 indicate that the fragment has the necessary penetration capability at 122 m/s (400 fps) to cause dysfunction at the 0-deg obliquity attack condition. The same fragment traveling at 183 m/s (600 fps) would have the same capability at 45-deg obliquity as well, and the $P_{CD|H}$ value is greater. Only two attack obliquities were originally chosen to prevent the component analysis (which was performed entirely by hand) from becoming too complex. Later, when the methodology was placed in COMPKIL, the same conditions were adopted to conserve computational resources. An obvious improvement to the methodology would be to increase the number of attack obliquities. This would provide equation forms of the functions to V/L codes, instead of tabular data that must be interpolated.

In addition to the case where all attack directions are equally likely, Kruse and Brizzolara (1971) also provide the equations necessary to determine A_p and A_t for several directional fragment attack patterns. Note, however, that because V/L codes do not keep track of which basic face of a component a fragment strikes, directionally dependent $P_{CD|H}$'s are not possible. An approach to account for this

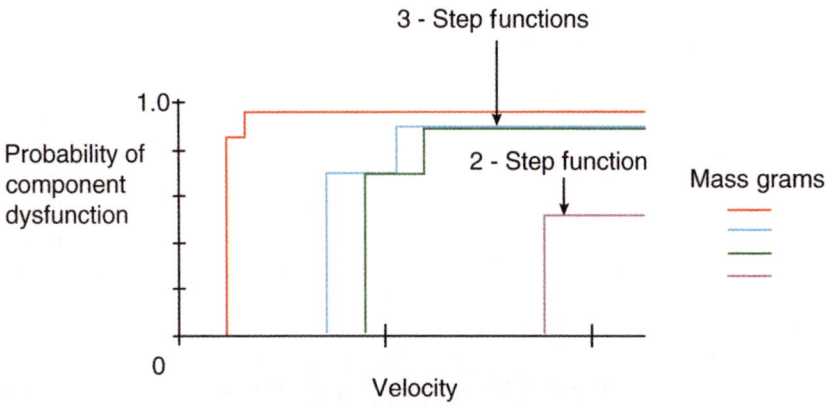

Fig. C2   Mass-velocity $P_{CD|H}$ functions.

deficiency would be to describe directionally dependent components in great detail and derive $P_{\text{CD|H}}$'s for critical subcomponents.

Component dysfunction because of penetrating damage from bullets and EFPs can be determined in the same manner as fragments. In fact, the previously described steps would be identical for those penetrators. The last step would be different because of the fact that different penetration equations would apply. Because bullet penetration equations were not placed in COMPKIL, a fragment shape factor was developed for a particular bullet such that the fragment penetration equations in COMPKIL provided reasonable results for the bullet in question.

Because of the construction of most components in ground-mobile targets, SCJs generally have sufficient penetration capability to cause component dysfunction. What remains is to determine hole size. Again, the first four steps described previously apply, and the hole size is determined using an SC penetration equation. In the past, if it was unclear whether a particular SCJ had sufficient penetration capability, the barriers identified in the fourth step were converted to mild steel equivalency, and the appropriate penetration equation was used to determine if the depth of penetration, along with hole size, was sufficient for component dysfunction.

I. Graphical Method

An ideal situation for a vulnerability analyst exists when experimental data for threat mechanism attacks on components of interest are available. The analyst can attempt to validate a $P_{\text{CD|H}}$ that was developed analytically or, if enough data exist, construct a $P_{\text{CD|H}}$ directly from the data. Flint (1994) calls the latter a graphic vulnerability process and describes it as:

> ... a procedure in which homogeneous regions of the component are outlined on a scale view of the whole component and the region areas are measured and recorded. An estimate of the fragment effect, or $P_{\text{CD|H}}$, is multiplied by the area to provide the vulnerable area contribution for a hit on a homogeneous area. All other regions that are thought to be vulnerable are similarly processed for their vulnerable area contributions and the sum of all contributions is the total estimated vulnerable area (AV) for the fragment mass, velocity, and view considered. The estimated $P_{\text{CD|H}}$ is AV/AP where AP is the total presented area.

To demonstrate this process, we use Flint's data for a 15-grain fragment vs a power supply at 0-deg azimuth. Flint identified 14 homogeneous (critical) areas that were regarded as vulnerable to attack from 15-grain fragments. Next, the $P_{\text{CD|H}}$ was determined for the individual areas if impacted by a 15-grain fragment traveling at each of three velocities. Flint notes that for one of the areas it was determined upon inspection of the component that only about 60 percent of the area was vulnerable to a 15-grain fragment. Thus, the maximum $P_{\text{CD|H}}$ for the particular subarea would be 0.6. This value was multiplied by the homogeneous area to obtain the contributing A_V. After this was performed for each subarea or critical area of the component, the overall component $P_{\text{CD|H}}$ could be calculated and plotted. Figure C3 shows the $P_{\text{CD|H}}$ function for the power supply subjected to a 15-grain fragment attack at a 0-deg attack azimuth.

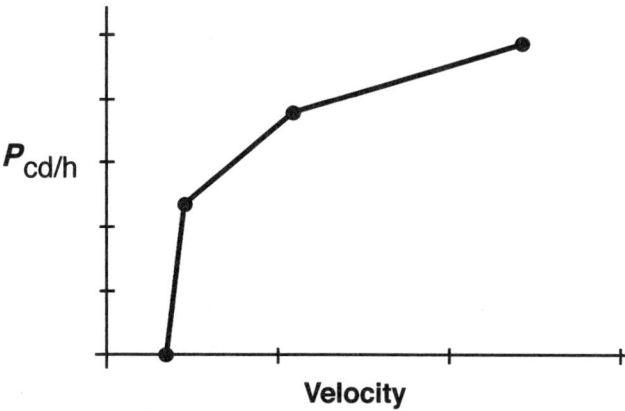

Fig. C3 Power supply $P_{CD|H}$ for 15-grain fragment, 0-deg azimuth.

It is important to note that the critical regions/areas and their associated $P_{CD|H}$ were obtained from scaled drawings and physical inspection of the power supply. This situation was the ideal situation.

II. Multiple Fragment Effects

The $P_{CD|H}$ function described earlier provides the probability that a single fragment is capable of producing sufficient damage to render a particular component nonfunctional at a prescribed level of capability. Historically, V/L codes analyzed the effects of fragment impacts on a target one at a time. If a particular fragment striking a critical component did not meet the penetration requirement, as provided in the $P_{CD|H}$, it was discarded as a nonlethal fragment, and the V/L code moved on to analyze any subsequent fragments impacting the component. It has always been obvious, however, that this practice was flawed for certain types of components, such as fluid containers. If a single fragment, for example, does not create a hole large enough to drain the fluid out of a container in a prescribed time limit, then perhaps two, three, or more fragments could surpass the hole size criterion. In recent years, modules for fluid containers (Hunt 1992) and tires (Grote et al. 1996) have been incorporated directly into V/L codes that can account for, and keep track of, multiple fragments at a time. These component dysfunction models are not $P_{CD|H}$ but rather physics-based models that provide a resulting level of capability or performance of the component. This approach to component dysfunction fits in nicely with the strategy of having V/L codes supply results/predictions in terms of measurable capabilities.

There is reason to believe that the development of component failure criteria that can be implemented within a vulnerability code may be the most difficult piece of the shock problem to solve. The failure criteria could be as simple as defining threshold acceleration levels. The difficulty arises in trying to determine what that level would be for various types of components or even similar components.

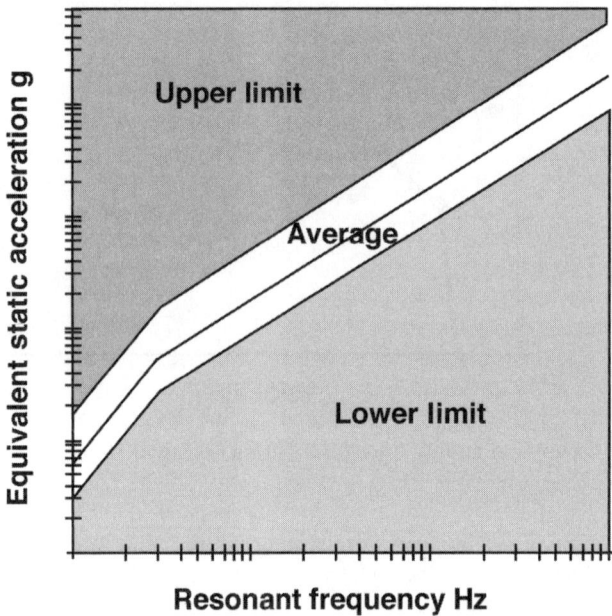

Fig. C4 Shock response spectra of "default" ballistic shock limits.

Walton (1989) has spent considerable time and effort in developing a ballistic shock protection requirement for armored vehicles. This requirement defines a Shock Response Spectrum (SRS) that components must be able to withstand. Figure C4 graphically presents the shock protection requirement (U.S. Department of Defense 2000). Actual components and their associated mounting hardware are subjected to typical shock spectrums, as shown in Fig. C4, and must remain functional afterward. This process is a pass/fail approach to component qualification and has some potential utility to aid component and mounting hardware design and specification.

The SRS protection requirement cannot be implemented as a component failure criterion for use in predictive models. One could predict, in the absence of models, that if components were exposed to an SRS at or below the requirement, the components should survive; although they should have the potential for failure. Some problems associated with implementing SRS in a predictive model are as follows:

1) The SRS requirement is not a failure criterion.

2) The SRS is not unique. Different acceleration-time histories can produce the same SRS.

3) The SRS is used for frequency content up to 10 kHz and it is believed that electronics will be vulnerable at much higher frequencies.

4) Currently, the shock-propagation techniques under consideration do not propagate frequency information that would be required for an SRS.

If shock failure spectra could be developed, and if frequency information could be added to current propagation methods under consideration, shock spectra may be a candidate for failure criteria.

References

Flint, J. B. (April 1994), "The Vulnerability of a High-Voltage Power Supply to Fragment Attack and Modeling Guidelines," Air Force Development Test Center, Rept. AFDTC-TR-94-22, Eglin Air Force Base, FL.

Grote, R. L., Moss, L. C., and Davisson, E. O. (Nov. 1996), "Tire Penetration and Deflation Models for Wheeled Vehicles Subjected to Fragment Attack," U.S. Army Research Lab., Rept. ARL-TR-1233, Aberdeen Proving Ground, MD.

Hunt, J. E. (Oct. 1992), "PKGEN, A Computer-Aided Generator for Component Probability of Kill Functions," *Proceedings of the Second Ballistics Symposium on Classified Topics*, American Defense Preparedness Association, pp. 325–333.

Kruse, L. R., and Brizzolara, P. L. (Dec. 1971), "An Analytical Method for Deriving Conditional Probabilities of Kill for Target Components," U.S. Army Ballistic Research Lab., BRL Rept. No. 1563, Aberdeen Proving Ground, MD.

U.S. Department of Defense (Jan. 2000), "Department of Defense Test Method Standard for Environmental Engineering Considerations and Laboratory—Part 2, Laboratory Test Methods, Ballistic Shock," MIL-STD-810F (Draft).

Walton, W. S. (Nov. 1989), "New Ballistic Shock Protection Requirement for Armored Combat Vehicles," *Proceedings of the 60th Shock and Vibrations Symposium*, Pt. 2, pp. 299–309.

Appendix D

Case Study: MUVES-SQuASH VV&A for the Bradley Fighting Vehicle (BFV)

A NOTABLE V/L VV&A effort was the 1998 comparison of data from the previously discussed MUVES–SQuASH model data with LFT data for the BFV. Although the effort is sometimes regarded as simply a preliminary step in the evaluation and accreditation of the MUVES–SQuASH model (resource limitations prevented the full implementation of its original design), the work is nonetheless notable for its statistical approach and planned execution of the VV&A process.

The following text, much of which has been taken directly from an ARL technical report by Baker et al. (1998), is included here to provide some detail on the process and procedures used (or intended to be used) for V/L VV&A. Note that because the numerical values of the inputs and results are of a sensitive nature (and are insignificant for our purposes), they have been omitted.

I. MUVES Functionality and Code

The levels and operators of the V/L Taxonomy (an earlier form of the MMF) provided a useful structure for addressing the VV&A of the MUVES–SQuASH model. Figures D1 and D2 show the taxonomy levels as applied to the mapping of MUVES–SQuASH methodology.

In the case of direct-fire weapons, MUVES begins execution at level 1, as this is typically the information the analyst provides. (In the case of indirect-fire weapons, execution begins at level 0, the initial conditions of the threat at launch time.) At each threat-component interaction, an appropriate interaction module (IM) is selected to compute the physical interaction with the component.

However, the damage to the component, in terms of whether or not it "survived" the encounter, is not computed at this point in the process. Rather, all of the information to compute the overall damage is placed in a damage packet, which is placed in a queue. Only after all threats (e.g., main penetrator, broken penetrator pieces, BAD spall, and secondary spall) have been "played" is the damage packet pool evaluated. Also, at each encounter, the threat can be degraded (e.g., mass and velocity decreased), deflected (e.g., if it perforates the component), or ricocheted (e.g., if it is at a high enough obliquity angle). The degraded threat, plus any new threats, is then propagated to the next encounter. In general, there is not a single

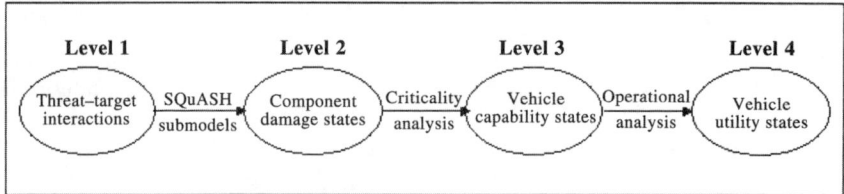

Fig. D1 MUVES–SQuASH methodology in the context of the V/L Taxonomy.

undeviated path that the threat traverses. MUVES handles this by providing 1) dynamic ray-tracing (i.e., each time a new threat is spawned, it creates a new ray to propagate the threat), and 2) recursive processing (i.e., each new threat is put on the same footing as the main penetrator).

Each threat continues encountering components until it either comes to rest or exits the vehicle. After all threats have been processed in this way, the queue of damage packets is sorted by component. The appropriate evaluation module (EM) is selected to compute the component's P_K. These P_K's can be used in a Monte Carlo simulation to represent either fully functional or killed components (i.e., the component damage states). After all the critical components have been processed in this manner, the P_K's are passed to the fault trees representing the various systems of the vehicle. The program then computes the probability of "deactivating"

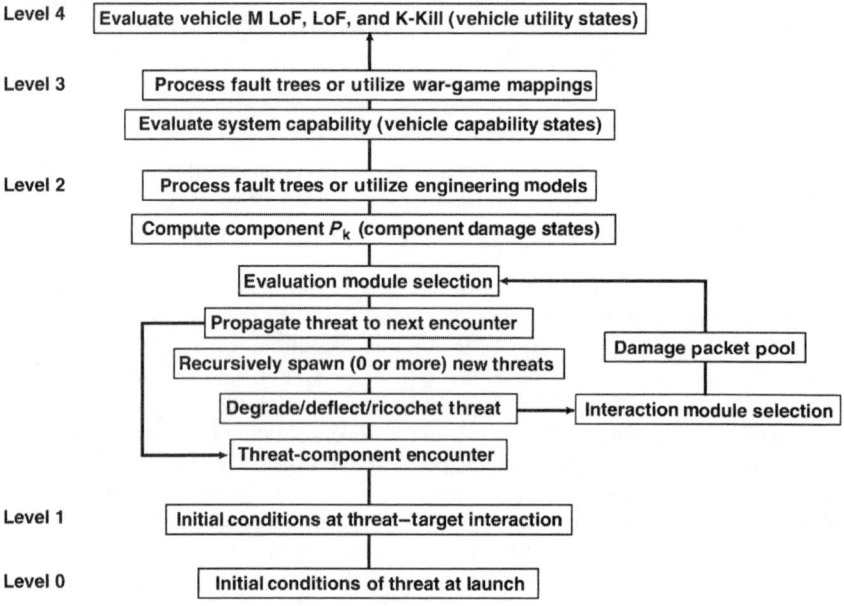

Fig. D2 Flowchart of MUVES methodology.

each fault tree by combining all the individual component P_K's. This assumes statistical independence among the component kills.

Alternatively, one can perform Monte Carlo draws on the P_K's to simulate the component damage states and then evaluate the fault trees to get the system damage states. It is at this state that the DAL enters the computation. By combining the DCU values of the DAL with probability of system damage, the program computes estimated mobility and firepower LoF along with the probability of K-Kill of the vehicle. MUVES also provides many postprocessing tools for analyzing the program output.

At the time of this assessment, MUVES was reported to contain approximately 400,000 lines of source code. Approximately 10 percent of it was general utility and nonspecific vulnerability code (e.g., error handling, memory management, etc.), which was well tested and proven to be reliable. Approximately 75 percent of the code was MUVES-specific vulnerability code that was used across all approximation methods (e.g., user interface, system evaluation package, and physical interactions package). The remaining 15 percent of the code was approximation-method-specific vulnerability code (e.g., damage evaluation methodology used only by the MUVES–SQuASH model).

There were four areas that assessors identified for examination as part of the MUVES code verification. They were 1) implementation, to verify that the code was an accurate translation of the underlying mathematical or logical model, 2) documentation, to ensure that the internal and external documentation was complete and consistent with the code itself, 3) boundary conditions, to verify that the program properly handled input lying outside of its intended range by invoking appropriate warning or error messages, and 4) explicit calculations, to compare selected cases where the answer was known or could be calculated.

II. V&V of MUVES–SQuASH Submodels and Program Inputs

Although time and money constraints ultimately prevented the assessors from being able to examine the MUVES–SQuASH submodels in detail, the original plan was to perform verification and laboratory validation of the following submodels as stand-alone programs: armor penetration (KE rounds, SCJs, and EFPs); penetrator breakup; ammunition explosions; BAD; fragment penetration, deflection, and ricochet; component P_K; crew incapacitation; hydraulic, fuel, and water lines; and wire bundles.

In addition, although program input is normally considered a responsibility of the user and is not a topic for program V&V, input such as the target description (which usually consists of thousands of solids and is highly complex) is such a major part of the successful performance of MUVES that V&V evaluators decided to include it in their scope of interest. The following input categories were identified for testing: target description; threat characteristics; component P_{KIH} values; degraded states, fault trees, and criticality analysis; and performance metrics.

In addition, the results of component P_K's, which are generated offline with a separate program, are used to produce tables or functions that are then used as MUVES–SQuASH input. Therefore, the stand-alone procedure used for this purpose was identified as needing to be included in the V&V process.

III. Overall MUVES–SQuASH Model Validation

After all the submodels have been developed, tested, and fully assembled in the MUVES–SQuASH infrastructure and the input has been examined, the overall model can be tested. The idea is to test the data flow between submodels, check that each submodel has the necessary input, check for possible side effects, and check for methodology gaps in the overall structure.

In the original plans for this effort, assessors wanted to use a three-pronged model validation approach, consisting of both bottom-up (submodel V&V), top-down (comparison with LFT) interrogation, and a separate procedure for the examination of model inputs. The bottom-up approach consists of a validation of each submodel as a stand-alone model. After this is completed, the evaluator is up to level 2 in the taxonomy. The mappings to level 3 (criticality analysis) and level 4 (operational analysis) are then merely MUVES inputs and are examined using the top-down approach, which compares the overall model output to LFT results.

Unfortunately, resource limitations dictated that VV&A participants exercise only the top-down approach, and model outputs were compared with LFT results at levels 2, 3, and 4.

IV. Statistical Comparison to LFT

The statistical methods that were used in this portion of the V&V effort are best described as exploratory data analysis. They included graphical analysis and the comparison of common properties (e.g., the mean and variance of the distributions of the empirical data and model output). Although these methods were easy to use and were quite practical, the impact of errors in judgment was difficult to assess. Accordingly, in some of the comparisons, a determination was made as to where the LFT result would occur within the distribution of model outcomes [hopefully, it would be somewhere near the peak (mode)]. In so doing, assessors were able to establish some degree of confidence that the model outputs were consistent with LFT results.

At the time of this effort, there had been two series of Bradley live-fire tests involving several variants of the vehicle. [The variants were M2A0 and M3A0 (initial production versions), M2HS and M3HS (high survivability), and M2ASTB and M3ASTB (advanced survivability test bed). Following the LFT, design changes identified through the LFT were incorporated into the HS designs, and they were redesignated M2A2 and M3A3, the third production variant. (The second production variants, M2A1 and M3A1, are identical to the M2A0 and M3A0 with regard to survivability. No A1 variants were live-fire tested.)] Tests against the M3HS variant were used because a large number of shots incorporating various threats and impact locations were available for comparison. Also, the M3HS most closely resembled the A3 version, which was scheduled to undergo testing, and preshot model predictions by MUVES–SQuASH, shortly after this work was performed.

The shots to be modeled were chosen as the "most interesting" by the same vulnerability assessors who conducted the original damage assessments; results with respect to perforation were unknown to the modelers. Originally, 16 shots were selected. Although live-fire events, as pointed out previously, are themselves stochastic processes, evaluators generally get no replicated shots. In this instance, however, the assessors were able to choose an additional three shots against other

Bradley variants (M3A0 and M2HS), which, because the configuration of the vehicle was similar in the vicinity of impact, served as replicates to shots against the Bradley M3HS variant.

Thus, 16 model outputs were compared with 19 live-fire shots (including the three replicated shots). The comparisons involved six different threats and looked at component capability (level 2), vehicle capability (Level 3), and vehicle utility (level 4).

V. Model Input

MUVES contains many different penetration algorithms, allowing the user to choose those that would be most appropriate. Obviously, this complicated the V&V effort for MUVES–SQuASH because only a subset of these algorithms would be exercised in any particular set of comparisons. Penetration algorithms used in this study included the LOS model for KE rounds and the Fireman–Pugh and Barb models for SC threats. The LOS model weights the LOS by an adjusted density to provide the same penetration as experimental data. The Fireman–Pugh model fits a curve to experimental data.

A (then) newly developed stochastic model for BAD was also implemented, including recent modifications for SC threats. Using shape factors for steel, aluminum, and tungsten, the COMPKIL model generated component kill probabilities, which were formatted into tables for use in MUVES as input files. Probabilities of crew incapacitation were obtained using the ballistic dose model, and the criticality analysis and DAL allowed the assessors to evaluate level 3 and 4 results.

MUVES–SQuASH was run using 1000 replications for each of the 16 different shots, with the shotlines being chosen from a circular normal distribution with a mean μ equal to the hit location of the appropriate live-fire shot and a standard deviation σ equal to the penetration radius (for KE threats) or the radius of the jet (for SC threats). The value for σ was chosen based upon a brief sensitivity analysis. Smaller values produced a vulnerable area in which some critical components that were seen were not hit. The values chosen optimized (in some sense) the results and were used for providing live-fire preshot predictions. The choice of a circular normal distribution meant that approximately 40 percent of the shotlines were within the 1σ of the mean or that approximately 400 of the model replications generated a shotline that was within the area of the main penetrator for the live-fire shot. The remaining shots, outside that area, provided the stochastic variation necessary to consider slight differences between the geometric description of the Bradley and the individual vehicle chosen for live fire. This was especially important because the shotlines had no cross-sectional area; thus, small changes in shot locations could have had large consequences. The choice of distribution (circular normal, bivariate uniform, etc.), as well as the spread of the distribution, was left to the discretion of the user.

VI. Grading Criteria

For some comparisons of the model output and LFT results, the distribution of model outcomes was overlaid onto the LFT result to determine if the result fell within the distribution. For others, outcomes were merely counted. In all comparisons, however, the followed grading criteria were given: green box (the live-fire event is a *likely* event within the distribution of model outcomes); yellow diamond (the live-fire event is a *reasonable* event within the distribution of model outcomes), red circle (the live-fire event is an *unlikely* event within the distribution of model outcomes), and red X (the live-fire event is an *unrealized* event within the distribution of model outcomes).

Note that the threshold values for likely, reasonable, unlikely, and unrealized events are defined in succeeding text.

Although these criteria were somewhat subjective, the assessors' expectation was that, for a model to be valid, it would be able to achieve the highest grade (the box) most of the time, with the triangle also expected to be a frequent occurrence. On the other hand, more than an occasional circle or X (a grade that indicates the model completely missed the prediction) would be an indication that the model should be examined more closely to see if there was a potential problem with it. Assessors did note, however, that because of the stochastic nature of live-fire events (most of which are without replication), the realization of individual circles or X's was not cause for alarm. It was only if these trends developed over a number of shots that evaluators would have reason to lose confidence in a model.

VII. Box Plot

The box plot was the graphical method of data analysis used to evaluate MUVES–SQuASH as a predictor of crew vulnerability. Crew incapacitations were compared on a shot-by-shot basis. A crew member was deemed incapacitated if he/she were unable to perform at least 75 percent of the assigned duties for the assault role of an infantry soldier. Crew member survivability (as opposed to incapacitation) was not considered in the study.

Thus, for every replication of the model, MUVES–SQuASH output the probability of incapacitation for each crew member. The box plot can provide a quick, compact display of the prominent features of a distribution. The median shows the location of the central tendency, and the length of the box gives an indication of the spread of the bulk of the data. Outside points allow for the consideration of the outliers. The overall plot indicates the degree of symmetry within the distribution. Box plots are useful in situations such as MUVES–SQuASH output where it is impractical to display all the details of the distribution (kill probabilities of each component for every replication).

VIII. Comparisons of Component Capability (MMF Level 2)

A. Vehicle Perforation

The first comparison made in this analysis of component capability dealt with perforation. The assessors wanted to determine whether or not the model correctly predicted the live-fire target-perforation result. They purposely chose five shots that did not perforate into the target interior, and MUVES–SQuASH predicted all five correctly.

Although the live-fire event produces either a perforation or a nonperforation; the model produces a probability of perforation. It does this simply by counting the number of perforations in the 1000 replications. The assessors established the grading criteria so that if the model predicted the live-fire event with a probability of greater than or equal to 0.5, then the live-fire result was the more likely of the two possible events according to the model, and MUVES–SQuASH would get the green box as a grade. Had the probability of the live-fire event been less than or equal to 0.05, then the grade would have been a red circle; had the probability been equal to 0.0, the grade would have been an X. Several of the shots in this evaluation represented an excellent example of the statistical variation present not only in the model but also in the live-fire process, and it serves as a warning that assessors must be careful when compelled to evaluate a model based on a small amount of data.

B. Critical Component Hits

As mentioned previously, critical components are those components that, when killed, contribute to mobility loss, firepower loss, or catastrophic damage. In other words, those components that appear in one or more fault trees used to assess damage to the vehicle. The assessors were interested in what percentage of the critical components killed in the LFT were hit in at least one replication of the model, figuring that if MUVES–SQuASH could not even hit such components, it would be unnecessary to make further comparisons.

The assessors decided that if the model hit 90 percent of the critical components killed in the live-fire test, it would be graded as a green box. If it were unable to hit at least 75 percent of such components, it would receive a red circle grade. Although MUVES–SQuASH hit many critical components in this test, assessors considered only that set of critical components killed in the live-fire shots. About half of the components that were missed in the model fell into one of three categories: 1) hydraulic lines, 2) wires, or 3) specific crew body parts.

Hydraulic lines and wires generally have a small diameter, and their locations often vary slightly from vehicle to vehicle, making it difficult to produce an accurate CAD geometric target description. Crew members are modeled using six body parts: 1) head, 2) upper torso, 3) lower torso, 4) pelvis, 5) arms, and 6) legs. Unfortunately, the description and placement of the wooden mannequins for a live-fire shot often differ significantly from their CAD descriptions and placement within the vehicle. Accordingly, MUVES–SQuASH occasionally predicted hits and kills on crew members while missing the precise body parts "killed" in the LFT.

Obviously, assessors could have relocated the crew and rerun the model, producing results that would have been closer in agreement; however, they were attempting to establish MUVES–SQuASH as a valid predictive tool and were thus interested in why the results differed (e.g., the height of adjustable seats and the positioning of crew members' legs).

C. Critical Component Kills

The third comparison performed involved critical component kills. Assessors calculated the percentage of the number of critical components killed in the LFT for which the model produced some nonzero P_K. This means that the component had a nonzero P_K in at least one replication of the model. This was a rather liberal comparison because it did not require a high probability (i.e., anything larger than zero would do). However, it should be remembered that, as long as the P_K was nonzero, there was the possibility that it would be in a nonfunctional state in the Monte Carlo simulation.

For this comparison, assessors reduced the requirement for the green box grade to 80 percent of the critical components killed in the LFT, as this corresponded to approximately the 90 percent required for agreement on probability of hit P_H times a 90 percent requirement for agreement on $P_{K|H}$. Similarly, a comparison received a grade of a red circle if the percentage of kills was less than or equal to 50 percent, approximately the 75 percent required for agreement on P_H times a 75 percent requirement for agreement on $P_{K|H}$.

The model kills consisted of not only the 15 critical components seen in the LFT but also an additional 40 critical components (referred to as "unobserved model kills") that were not killed in the LFT but that received some nonzero P_K in MUVES–SQuASH. Of course, the kill probability for some of these components was on the order of 0.001, but such a probability indicated that the component had some nonzero P_K in at least one replication of the model. Furthermore, the examination of reasons for these small numbers could point to areas in the model that required further investigation.

Again, the comparisons here were relatively predictable except for the aforementioned difficulty in accurately predicting the exact positioning of the crew. For example, one shotline went behind both the commander and gunner, while another went under the commander's adjustable seat but hit the gunner's legs.

D. Crew Incapacitations

For the fourth comparison, assessors compared, for each shot, the expected number of crew incapacitations, which was obtained by adding the mean probability of incapacitation for each crew member. (Note that the version of the BFV that had been tested in the LFT had a crew of five: commander, driver, gunner, left scout, and right scout.) It was felt that if the expected numbers of crew incapacitation were within 0.5, the comparison would get a green box grade; if they weren't, they would get a red circle grade.

In addition to the previously mentioned difficulty for MUVES–SQuASH to predict incapacitation for movable crew members, there were two other problems associated with this evaluation. First, crew members had been equipped with body armor [combat vehicle crewman (CVC) helmets and vests] in the LFT; however, they were not modeled as such in MUVES–SQuASH. Second, the LFT assessments were computed using the previously discussed Sperrazza–Kokinakis correlations, which are actually average incapacitations and not true probabilities of incapacitation, but MUVES–SQuASH made use of the ballistic dose correlations, which are true probabilities.

Although assessors were unsure as to what degree these problems affected their comparisons, they did recognize that future comparisons must be more rigorous in comparing "apples to apples."

IX. Comparisons of Vehicle Capability (MMF Level 3)

The fifth statistical comparison to LFT dealt with degraded states (DSs), an MMF level 3 metric. To illustrate the way the DS probabilities were computed, the

Table D1 Degraded states of the BFV communication system

DS	Description of dysfunction
X_0	No communication damage
X_1	No internal communication
X_2	No external communication > 300 ft
X_3	No external communication
X_4	X_1 and X_2
X_5	X_1 and X_3

example of the vehicle's communication system is used (see Table D1). First note that there are only three primary states for the communication system: X_1, X_2, and X_3 (where X_2 is a subset of X_3). All other states are a combination of these three states. The fault trees for these states are shown in Fig. D3.

The critical components are simply labeled 1 through 8 in these diagrams. The actual components had both a MUVES–SQuASH name and a component identification number. Note that component 1 is used in the first and the third fault trees. It is important to recognize that these are not two different copies of the same component, but rather they represent the same component playing two different roles. (In this particular case, it happens to be the electrical power that serves both internal and external communication.)

Consequently, assessors have to be careful in evaluating these fault trees. If component 1 is killed in the first fault tree, then logically it must also be killed in the third fault tree. If a P_K is merely assigned to component 1 and assessors use probability algebra to evaluate and combine the fault trees, the wrong answer will

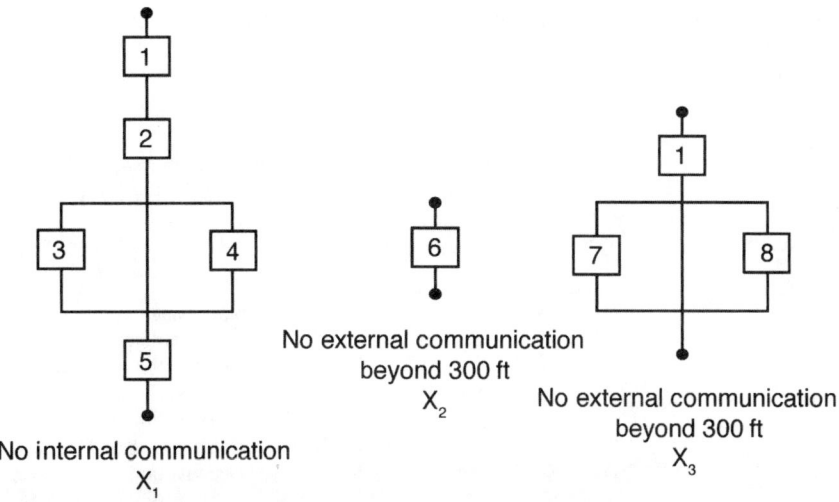

Fig. D3 Primary fault trees of the communication system.

be derived. This is an example of what may be called functional dependence of components.

Another type of dependency, one that is not apparent from the fault trees, is what may be called spatial dependence. This corresponds to having two critical components in close proximity to one another in the vehicle so that hitting one is statistically correlated to hitting the other. For example, spall fragments are usually spread out over a cone. If two components lie close to one another, then perhaps they both will get hit. Or, depending on the direction of the fragments, one component may shield the other. These types of correlations show up as statistical dependencies. Using probability algebra with the individual component P_K's will not account for these dependencies either. Whenever there are dependencies among the components (functional or spatial), the probability of killing both components simultaneously is not equal to the product of the separate probabilities, therefore, probability algebra is not a viable option. Instead Monte Carlo simulations should be used.

The Monte Carlo method is capable of dealing with dependencies; it simply takes a little longer to evaluate. The method consists of assigning binary kill/no-kill states to each critical component in the fault trees by drawing on their respective P_K's and only then evaluating the fault trees. Fault-tree evaluation is simply a determination as to whether there is any continuous path that can be traversed between the input node (at the top) and the output node (at the bottom). If there is at least one such path, then the tree is not deactivated and the represented functionality is not lost. All trees are evaluated, and thus combinations are revealed. There is no need for probability algebra because all evaluations are done with specific realizations.

Because there are three primary fault trees in the communication system (shown in Fig. D3), and each fault tree has only two outcome states [either killed (deactivated) or not killed (activated)] there are eight (i.e., 2^3) possible outcome states for the communication system (see Table D2).

In this way, each and every deactivation state is accounted for. This ensures that the set of degraded states is mutually exclusive and exhaustive so that, no matter what the damage to the vehicle, the communication system will be in one and only one DS.

Table D2 Partitioning the outcome space into a mutually exclusive set of states

Deactivation state	Fault tree state			Description of primary fault trees
	X_1	X_2	X_3	
S_{000}	0	0	0	No damage
S_{100}	1	0	0	Deactivation of X_1 only
S_{010}	0	1	0	Deactivation of X_2 only
S_{001}	0	0	1	Deactivation of X_3 only
S_{110}	1	1	0	Deactivation of X_1 and X_2
S_{101}	1	0	1	Deactivation of X_1 and X_3
S_{011}	0	1	1	Deactivation of X_2 and X_3
S_{111}	1	1	1	Deactivation of all three

Table D3 Mapping from deactivation state to degraded state

Deactivation state	DS	Description of DS
S_{000}	X_0	No communication damage
S_{100}	X_1	No internal communication
S_{010}	X_2	No external communication > 300 ft
S_{001} or S_{011}	X_3	No external communication
S_{110}	X_4	X_1 and X_2
S_{101} or S_{111}	X_5	X_1 and X_3

These eight deactivation states are then mapped into the corresponding DS (see Table D3).

The procedure, then, is to get a vector of component P_K's from MUVES–SQuASH (where n is the number of critical components),

$$P = (p_1, p_2, p_3, \ldots, p_n)$$

perform a Bernoulli trial (an experiment having two possible outcomes) on each component to get the vector of component kills,

$$k = (k_1, k_2, k_3, \ldots, k_n)$$

substitute these into the three primary fault trees, evaluate the deactivation states, and, from the mapping, evaluate the DS. It has been seen that any vector of component kills will map into a unique DS.

Of course, this exercise can be applied to systems other than the communication system (see Table D4).

Thus, for each MUVES–SQuASH replication, the assessors were able to find the unique DS outcome, labeled $M_m F_f A_a X_x C_c S_s$. By repeating the process a number of times, they were able to generate a probability distribution of outcomes, which could be compared with the result from the LFT assessment. MUVES–SQuASH provided a distribution of vehicle DSs, ranked from most likely to least likely. Assessors established grading criteria such that if the LFT result fell in the tail of the distribution where the cumulative probability (CP) was greater than or

Table D4 Number of degraded states – all systems

System	Number of primary fault trees	Number of deactivation states	Number of DSs
Mobility (M_m)	4	16	7
Firepower (F_f)	6	64	21
Acquisition (A_a)	2	4	4
Communication (X_x)	3	8	6
Crew (C_c)	3	8	4
Scout (S_s)	2	4	3

equal to 95 percent (the least likely 5 percent), then the comparison (and, hence, the model) received a red circle grade. Conversely, if it fell in the upper half of the distribution (the most likely 50 percent), then the comparison received a green box grade.

If the assessed value was an unrealized result (i.e., does not appear in the distribution), then the comparison received a grade of X. Unfortunately, this distribution is not continuous, and it would be merely coincidental if these thresholds appeared in the cumulative distribution function. Where two live-fire assessments were possible, average grades were assigned.

X. Comparisons of Vehicle Utility (MMF Level 4)

A. Mobility/Firepower LoF

The sixth, seventh, and eighth statistical comparisons dealt with M-LoF, F-LoF, and probability of K-Kill. MUVES–SQuASH calculated M-LoF and F-LoF by mapping the set of damaged components into the BFV DAL. The corresponding M-LoF and F-LoF values from the LFT were obtained in an analogous manner using this same DAL.

As a reminder, the DAL process consists of three parts:

1) Identifying the critical components of the vehicle and the way that they are logically connected using fault trees that support the various mobility and firepower subfunctions.

2) Identifying the dysfunctional states and assign corresponding DCU values.

3) Combining the individual DCU values into a single number representative of the vehicle as a whole.

The first part represents input to the MUVES–SQuASH model, while the second part is performed within the various submodels using the DAL. The third part is dealt with through the application of the survivor rule. This rule is stated as follows:

$$\langle \text{LoF} \rangle = 1 - \prod_{\text{System} \in \text{DAL}} (1 - \langle \text{LoF} \rangle_{\text{System}})$$

where the quantity $\langle \text{LoF} \rangle_{\text{System}}$ is defined by

$$\langle \text{LoF} \rangle_{\text{System}} = \sum_{\text{State} \in \text{System}} P_{\text{state}} \cdot \text{DCU}$$

where P_{state} is the probability of occurrence for an individual state within a system. Because the states within a system are complete (i.e., mutually exclusive and exhaustive), assessors are guaranteed that the quantity $\langle \text{LoF} \rangle_{\text{System}}$, as defined, cannot exceed the value of 1. The survivor sum will then ensure that $0 \le \langle \text{LoF} \rangle \le 1$.

Mobility LoF represents a percentage loss of mobility for the BFV; an M-LoF of 1.0 (i.e., complete loss of mobility) defines the vehicle as incapable of executing controlled movement within 10 min of being hit, not repairable by the crew on the battlefield. Firepower LoF represents the percentage loss of firepower for the BFV; an F-LoF of 1.0 (i.e., complete loss of firepower) defines the vehicle as incapable of delivering controlled fire within 10 min of being hit, not repairable by the crew on the battlefield.

B. K-Kill

The ninth comparison considered the probability of a K-Kill to the BFV. As mentioned previously, a K-Kill is defined as a complete loss of both mobility and firepower functions to the vehicle, not economically repairable. Because every individual replication and shot either does or does not produce a K-Kill, the set of grading criteria used here was similar to that used in the perforation comparison. That is, if the model predicted the live-fire event with a probability greater than or equal to 0.5, then the live-fire result would be the more likely of the two possible events according to the model and would get a green box grade. Had the probability of the live-fire event been less than or equal to 0.05, then the grade would have been an X.

One example of a lesson learned during this part of the V&V effort had to do with penetration of the ammunition in the BFV. Assessors were unable to predict the LFT outcome for one of the shots with a large overmatch. The live-fire event did not produce a K-Kill, but MUVES–SQuASH predicted such an outcome for all replications. Previous testing had been performed against stowed ammunition boxes and some surrogate ready ammunition boxes and demonstrated that, if this particular main threat penetrated the stowed ammunition with a residual greater than 4 in., a K-Kill would occur. This criterion (a step function based on residual penetration) was used in the model for the ready ammunition as well. However, the ammunition in such earlier testing had been tightly packed, always resulting in fratricide. In the LFT, it was hanging in the ready ammunition compartment. Although the main threat did indeed penetrate the ready ammunition with a sufficient residual, the impacted rounds merely dropped to the floor, and a K-Kill was averted. The model was modified to reflect this fact. In particular, the step function was changed for the ready ammunition compartment.

XI. IDA and VAST Comparisons

The assessors also performed two other tests on the MUVES–SQuASH results. The first test was the Institute for Defense Analysis (IDA) test, which examined the magnitude of the component P_K's to give assessors a better understanding of MUVES–SQuASH's effectiveness in predicting level 2 metrics (i.e., component damage). Based on the model's mean probabilities of the component kills, the IDA test categorized each critical component into one of six bins and then summed the P_K's over all components within each bin to obtain an expected number of kills. This was then compared with a corresponding bin containing the actual number of critical components from the subset of components represented in that bin killed in the LFT. Thus, if the P_K for component C, as determined by the model, fell into the fourth bin ($0.4 < $ MUVES–SQuASH $P_K \leq 0.6$), then the live-fire result for component C [either kill ($P_K = 1$) or no kill ($P_K = 0$)] would fall into the corresponding fourth bin. Live-fire outcomes were also summed within the bins.

The IDA test was not included in the assessors' overall comparison matrix because of difficulty in establishing adequate grading criteria; however, assessors did recognize its significance in the overall VV&A process. Specifically, the comparison generally showed the model to be underpredicting some component kill probabilities. Possible reasons for this included low component P_K's (input to the model), undersampling (resources for the model), and shotlines with no cross-sectional area (design within the model). For further information on the IDA test, readers are encouraged to consult the ARL report by Baker et al. (1998).

The final test of the MUVES–SQuASH model involved comparing its results to the previously discussed VAST model, an older deterministic model designed to evaluate vulnerability of AFVs to KE penetrators and SC warheads. Performing this comparison was reasonable because the same data BFV live-fire data used for the MUVES–SQuASH comparison had been used for VAST. The specific actions and results of the VAST SQuASH comparison are detailed in the ARL report by Baker et al. (1998). However, in general, assessors concluded that the deterministic VAST model produced higher predictions than both the MUVES–SQuASH and the LFT results, and, therefore, the stochastic MUVES–SQuASH model represented a significant improvement over VAST.

XII. Conclusions of the MUVES–SQuASH VV&A for the BFV

The comparison of the MUVES–SQuASH model results to the BFV LFT results (and, to a lesser degree, the VAST results) was beneficial in several regards. In general, the exercise helped reaffirm the idea that a strategic mix of modeling and testing is the most cost-effective way to conduct Army analysis. In addition, even though assessors were unable to fully implement the scope of their VV&A plans, the exercise reiterated the fact that a comprehensive and properly executed VV&A plan is the key to establishing the credibility of a model.

The exercise also helped highlight several specific issues related to V/L, including the following:

1) Comparisons involving crew members can be difficult because of variable positioning (e.g., adjustable seats), differences in dimensions of crew simulators (e.g., plywood mannequins) and modeled crew members, differences in body armor tested and modeled, and differences in methodology to assess crew incapacitation.

2) Shotlines with no cross-sectional area make model outcomes highly dependent on geometric detail (i.e., small displacements in shot location can result in major differences in metrics).

3) Significant variability can exist within LFT.

Reference

Baker, W. E., Saucier, R., Muehl, T. M., and Grote, R. L. (1998), "Comparison of MUVES-SQuASH With Bradley Fighting Vehicle Live-Fire Test Results" U.S. Army Research Lab., Rept. ARL-TR-1846, Aberdeen Proving Ground, MD.

Appendix E

Details and Developments in the Missions and Means Framework

AS NOTED in the body of this book, the onset of the LF program provided an important stimulus to focus V/L analysis efforts. Probably the first critical issue was to separate vulnerability calculations into three related, but distinctly different classes of, metrics: damage, performance, and utility. The aforementioned V/L Taxonomy became the structure used to frame these relationships. With that clarification, the supporting mapping functions could begin to be developed according to the relevant physics, engineering, and utility considerations.

Starting in 1985, during the first decade of these developments, efforts were concentrated mainly on ballistic mechanisms for single-event interactions to represent individual live-fire shots. As further progress was made, two further issues emerged. The first was that a great deal of contextually important information was simply not contained in the V/L Taxonomy with its limited "view" of the world. Another was that there was no clear basis for establishing mission utility, the last level of the taxonomy. It was clear to many vulnerability workers that, whatever the proper basis for mission success, it was not the purview of the vulnerability worker to define effectiveness parameters.

Moreover, when the U.S. Army began pursuing SoS mission solutions in the late 1990s, it became clear that a much fuller metaphor was needed to think about battlefield modeling, one that included an explicit layout of the mission, that included the (cooperative) capabilities of multiple platforms, that represented the opposing forces, and that, optionally, could reflect all levels of war, from the strategic national level to the lowest tactical echelon.

Accordingly, the MMF was developed to provide a more complete metaphor to identify the many key pieces, their requisite order of instantiation, the various interactions, and the dynamics to be expected in time- or event-stepped simulations.

Although the focus of this book is mobile ground platform vulnerability, the dominant use of V/L metrics today (as well as a half century ago) is supporting war-gaming by providing war-game participants with an appreciation of the effects of gamed interactions on forces and platforms. The MMF provides an abstraction that can help participants in various kinds of war-games to think through the many parameters that must be considered, the levels of approximations that may be appropriate, and the ways in which to categorize and structure the many interactions that characterize modern warfare simulation.

Contained in this appendix are three papers. The first is the foundation MMF paper, which was presented at the 2003 Interservice/Industry Training, Simulation, and Education Conference (IITSEC), by Sheehan et al. (2003). This work brought together for the first time the MMF parameter space. This space has changed little since it evolved to the current form.

The second paper, which is by Minchew and Bray and unpublished, describes how the MMF is instantiated in a work-flow structure. By referring to the MMF diagram shown in the first paper, the work-flow begins with specifying the mission and then allocating the means to accomplish it. With these two pieces complete, the mission is fully planned. The final stage, then, is to assess mission accomplishment. Anyone trained in military mission planning will recognize that the MMF simply provides an explicit and logical structure for executing what military operators call the military decision-making process (MDMP). In fact, other than the visible logic structure, there's nothing really new in MMF.

Finally, the third paper by M.L. Minchew, which was submitted for the 74th Barchi Prize at the 2006 *Military Operations Research Society* (MORS) *Symposium*, is a more recent exposition on some of the ways the MMF can be used to understand the relationship between functions and capabilities, the tasks that require them, and the resources that provide them within the context of different mission types.

Paper 1: The Military Missions and Means Framework

J. H. Sheehan
PM Knowledge Integration, DOD DOT&E/C3I & Strategic Systems
4850 Mark Center Drive, Alexandria, VA 22311
Jack.Sheehan@osd.mil

P. H. Deitz
U.S. Army Materiel Systems Analysis Activity, ATTN: AMXSY-TD
Aberdeen Proving Ground, MD 21005-5071
phd@amsaa.army.mil

B. E. Bray
Dynamics Research Corporation
213 C4 Delaware Street, Leavenworth, KS 66048
bbray@drc.com

B. A. Harris
Dynamics Research Corporation
60 Frontage Road, Andover, MA 01810
bharris@drc.com

A. B. H. Wong
U.S. Army Materiel Systems Analysis Activity
Aberdeen Proving Ground, MD 21005-5071
awong@amsaa.army.mil

Abstract

As the Department of Defense (DOD) transforms itself from a forces-based, materiel-centric Cold War posture to a capabilities-based, mission-centric asymmetric-warfare posture, it is increasingly vital that military planners, operators, and analysts concern themselves not only with "doing things right" (i.e., the technical architecture) but also with "doing the right things" (i.e., the operational architecture). Moreover, the historic "right thing" of winning the large-scale conventional engagements in Europe has given way to multiple and diverse "right things" of unconventional combat, homeland defense, peacekeeping missions, and various kinds of military operations other than war (MOOTW). To address these complex new objectives, a framework is needed to comprehensively organize and rigorously specify operational purposes and goals and then explicitly relate, map, and allocate them to the proposed technical means for accomplishment. This paper describes the fundamental elements and usage of the military Missions and Means Framework (MMF), which is increasingly being used to represent the synthesis of military operations and the employment of materiel/forces to accomplish these operations. The MMF provides a disciplined procedure for implementing the transformation guidance in Rumsfeld (2003) and Chu (2003) and the acquisition reform promulgated in Wolfowitz (2003a, 2003b) and Myers (2003).

About the Authors

Mr. Jack H. Sheehan is the Program Manager for Knowledge Integration in the C3I and Strategic Systems Directorate of the office of the Director, DOD Operational Test and Evaluation (DOT&E), under an Interagency Personnel Act (IPA) agreement with the University of Texas at Austin. He has concurrent, collateral assignments within Headquarters, U.S. Army G8 and the USD/AT&L Defense Modeling and Simulation Office (DMSO).

Dr. Paul H. Deitz is the Technical Director for the U.S. Army Materiel Systems Analysis Activity (AMSAA). He is a member of the Senior Executive Service with approximately 40 years of experience in laser effects, ballistic vulnerability, and live fire test and evaluation of U.S. Army combat systems.

Mr. Britt E. Bray, LTC(R) USA is a senior military analyst for the Dynamics Research Corporation after retiring from the Army with 23 years of experience in operations, training, plans, and combat development. A graduate of the Army's Command and General Staff College, he served at company through corps levels and also gained Joint experience in the J3 of a Joint Task Force.

Mr. Bruce A. Harris, COL(R) USA is the Director of the Training and Performance Analysis Division of the Dynamics Research Corporation. He retired from the Army with 29 years of service and is a former Headquarters, Army Staff planner and department chair at the National War College.

Mr. Alexander B. H. Wong is a senior analyst assigned to AMSAA's Office of the Technical Director.

Purpose

This paper presents the MMF for warfare representation. MMF is not a theory of war. Rather, it is a framework for explicitly specifying the military mission and quantitatively evaluating the mission utility of alternative warfighting Doctrine, Organization, Training, Materiel, Leadership, Personnel, and Facilities (DOTMLPF) services and products.

Need

In the post-9/11 world, the United States needs the capability to counter new, emerging concepts and threats while retaining the overmatch to deter existing, conventional capabilities. Relevant to extant U.S. warfighting capabilities, this requires strategic mobility in hours/days, not weeks/months; joint netted lethality in close combat; and combined and noncombatant survivability in complex terrain. In addition, although the nation invests more in military capability than most of the world combined, to achieve national security objectives over the long haul, the suitability of solutions (i.e., total DOTMLPF life-cycle ownership cost) and the time to field capabilities must improve.

Effectiveness and survivability will require unprecedented integration and interoperability across the Services and down to the lowest echelons. Sustainment footprint and personnel efficiency drive suitability. Decision cycles drive fielding time.

Information technology is a key enabler in each case, for both (new) materiel and non-materiel solutions. Unfortunately, the underlying computing technology is changing faster than operational capabilities can be conceived and implemented. Technology "churning" can impede the fielding of effective systems. Non-information technology components of military-grade materiel solutions (the fundamental platforms, vehicles, air frames, and hulls) are expensive, slowly evolving, long-lead elements both in development and manufacture. The pace and disruption of spiral development can make waste here.

Objective

Einstein once said that the "perfection of means and confusion of ends seem to characterize our age." Unfortunately, these words characterize certain DOD transformation initiatives today, where efforts focus largely on the *materiel*—the physical means needed for successful military prosecution—without adequate consideration for (or linkage to) the *missions*—the end actions that must be accomplished to meet objectives. To use the terminology of the engineer's maxim, form (the technical and systems architecture) is often not following function (the operational architecture).

To enable DOD transformations, from concept through actual combat, a framework is needed to help the warfighter, engineer, and comptroller specify a common understanding of military operations, systems, and information and provide quantitative mission assessment of alternative solutions.

A disciplined procedure is required to explicitly specify the mission, allocate means, and assess mission accomplishment. Procedure objectives are to 1) unify the warfighter, engineer, and comptroller understanding of missions and means; 2) account for the tangible, physical, objectively measurable factors (traditional testing and evaluation) and the intangible, cognitive, ultimately subjective factors (traditional warfighter expertise) that constitute mission success; 3) be sufficiently credible, timely, and affordable to make hard decisions—and have those decisions stay made; 4) be sufficiently consistent, concise, repeatable, and scalable to compete effectively with alternative methodologies; and 5) provide a disciplined process to implement the transformation guidance in Rumsfeld (2003) and Chu (2003) and associated acquisition reform in Wolfowitz (2003a, 2003b) and Myers (2003).

Related Efforts

As reported in Deitz (2002), tailored adaptations of the MMF have been successfully applied in these areas:

- Future Combat Systems (FCS) operational requirements definition (Purdy, PM-FCS)

- Joint Training System (JTS) life-cycle (Rothmann, OSD/P&R)
- Naval aviation training (Duke, NAWC-TSD)
- Comanche analysis of alternatives and Army Battle Control System test and evaluation (Krondak, TRAC-FLVN)
- Air Operations Center design (Andrew, ESC-Hanscom)

More recently, Deitz (2003) and Hughes (2003) report that the MMF has been employed to help define the FCS test and evaluation concept.

The Framework

Fundamental Elements

The MMF uses 11 fundamental elements to organize and specify military operations. As shown in Fig. 1, mission content is organized into seven groups (hereafter called **Levels**):

Level-7. Purpose, Mission
Level-6. Context, Environment
Level-5. Index, Location/Time
Level-4. Tasks, Operations
Level-3. Functions, Capabilities
Level-2. Components, Forces
Level-1. Interactions, Effects

In addition, the following four transformations (hereafter called **Operators**) are included:

$O_{1,2}x$: transforms **Level-1** interaction specifications into **Level-2** component states.

$O_{2,3}x$: transforms **Level-2** component states into **Level-3** functional performance.

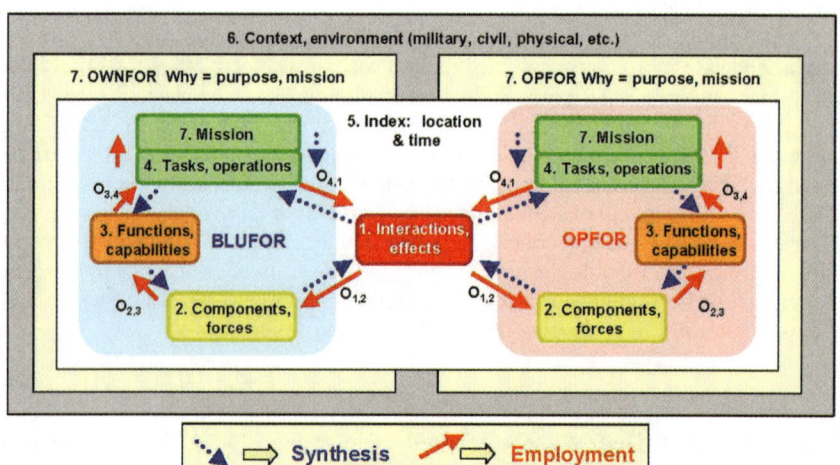

Fig. 1 A two-sided missions and means framework.

$O_{3,4}x$: transforms **Level-3** functional performance into **Level-4** task effectiveness.

$O_{4,1}x$: transforms **Level-4** task sequences into **Level-1** interaction conditions.

Also note in Fig. 1 the MMF's multi-sided nature. The OPFOR coalition influences the outcome of OWNFOR's mission prosecution. The MMF is a symmetric representation of an asymmetric (perhaps decidedly asymmetric) conflict.

A **Level-7** Mission specification package with references to associated **Level-6** Environment and **Level-5** Location/Time specification packages collectively represents the "Missions" part of the MMF; **Level-1** through **Level-4** and the four **Operators** are collectively the "Means" by which Missions are accomplished (hence, the name Missions and Means Framework). The MMF is intended to provide a compatible extension to the Vulnerability/Lethality Taxonomy in Deitz (1989, 1999) and Klopcic (1992), the DOD Architecture Framework in Wolfowitz (2003c), and the Functional Descriptions of the Mission Space (FDMS) in Haddix (2003).

Stocking and Assembly Perspectives

The framework has two names for the mission content specified by each **Level**. Consider a building construction metaphor. A buyer employs a builder to construct a home suitable for a resident. The builder constructs the home by tailoring and assembling standard building materials from a supplier.

Using this metaphor, when mission content in a **Level** is viewed like standard parts in a reference library (by analogy, standard materials stocked in a building supply), the MMF uses the first name for the **Level** content: Purpose, Context, Index, Task, Function, Component, and Interaction. Just as in a building supply (where similar materials are stocked together), when the MMF names, records, and references content from this perspective (hereafter called the Stocking Perspective), the organization within the **Level** is an orthogonal decomposition into homogeneous collections of similar content. For example, in the Stocking Perspective, types of armored ground vehicles would be specified in one **Level-2** branch of the decomposition, types of fighter interceptors would be described in a separate **Level-2** branch, and both would be described as Components.

When mission content in a **Level** is viewed as an assembled package that satisfies a mission requirement (by analogy, a completed room or feature in the home under construction), the framework uses the second name for the **Level** content: Mission, Environment, Location/Time, Operation, Capability, Force, and Effect. Just as in building construction (where diverse materials are assembled to provide a useful kitchen), when the MMF names, records, and references content from this perspective (hereafter called the Assembly Perspective), the organization within the **Level** is a decomposition into heterogeneous packages of diverse content. In the Assembly Perspective, a combined arms ground combat team would be specified in one **Level-2** branch, an aviation strike warfare package would be described in a distinct **Level-2** branch, and both would be described as Forces.

The MMF design employs the DOD Architecture Framework views to define the concepts, rules, constraints, and interfaces needed to assemble parts selected from the Stocking Perspective into packages in the Assembly Perspective. Using architectures, Purpose parts are assembled into Mission packages, Context parts into Environment packages, Task parts into Operation packages, Function parts into Capability packages, Component parts into Force packages, and Interaction parts into Effect packages.

The MMF uses the semantics and syntax of the FDMS to organize Stocking Perspective part specifications into Assembly Perspective package specifications (Haddix, 2003).

Synthesis and Employment Operators

The framework has two distinct versions of each **Operator**: Synthesis (the blue [darker] arrows in Fig. 1) and Employment (the red [lighter] arrows in Fig. 1). An example of the nomenclature for Synthesis is $O_{1,2}S$ and for Employment is $O_{1,2}E$.

Synthesis is the top-down planning and decision-making process warfighters and analysts use to create, define, and design a military evolution to meet mission requirements. Employment is the bottom-up execution and adjudication of actual outcomes when own and opposing missions/means collide in the battlespace.

Synthesis and Employment **Operators** are **not** mathematical inverses (e.g., the $O_{3,4}S$ Synthesis **Operator** is not the mathematical inverse of the $O_{3,4}E$ Employment **Operator**). In the construction metaphor, the algorithms and procedures an architect uses to design a home are not inverses of those used by carpenters and electricians to build the home.

Layered Decomposition

The MMF uses a layered decomposition. Recommended practices are as follows: **Level-4** Tasks, Operations should be layered by the Universal Joint Task List (UJTL) level-of-war (Fig. 2). **Level-2** Components, Forces should be layered by echelons (Fig. 3). **Level-3** Functions, Capabilities layers are designed to provide efficient interfaces for the $O_{2,3}E$ and $O_{3,4}E$ execution. **Level-1**

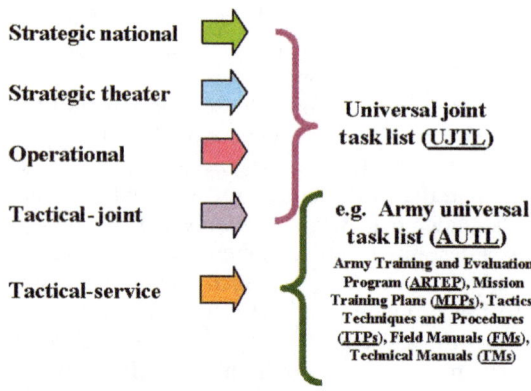

Fig. 2 Task semantics by level of war.

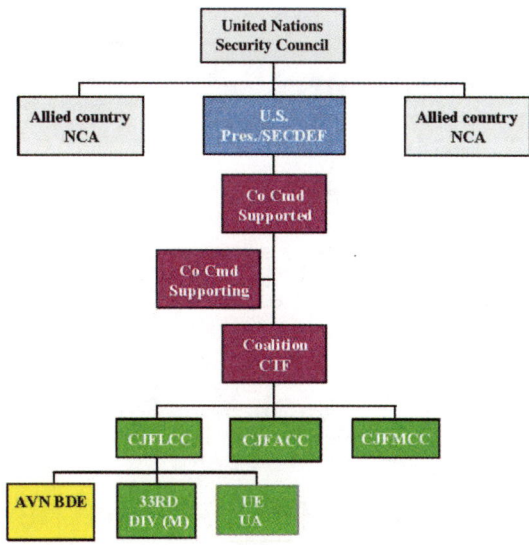

Fig. 3 Components, forces by echelon.

Interactions, Effects layers are designed to provide efficient interfaces for $O_{4,1}E$ and $O_{1,2}E$ execution.

Element Definitions

Level-7: Purpose. Mission defines the "why" and "wherefore" of the military evolution. Definition 3 for mission in JP-1-02 captures the MMF Stocking Perspective intent for Purpose: "An assignment with a purpose that clearly indicates the action to be taken and the reason therefore." The corresponding Army definition for mission captures the MMF Assembly Perspective intent for Mission: "The essential activities assigned to a unit, individual or force. It contains the elements of who, what, when, where and the why (reasons therefore), but seldom specifies how." The focus of Mission within MMF is the "what," "why," and "wherefore" of the required outcomes, not the "who" and "how" to accomplish those outcomes.

Level-6: Context Environment defines "under what circumstances" a Mission is to be accomplished. MMF employs the UJTL-defined and enumerated taxonomy for military, civil, and physical conditions. Individual conditions are captured as (Stocking Perspective) Context. Context becomes an Environment package specification when a collection of Context conditions is assembled into a consistent whole.

Level-5: Index, Location/Time defines "where" in terms of geo-spatial/materiel geometry and "when" in time. Index is a list of individual items, such as the Global Command and Control System (GCCS) GEO-file of 50,000+ key locations in the world. When assembled into a Time-Phased Force Deployment Data (TPFDD) execution matrix, these elements become a Location/Time package specification.

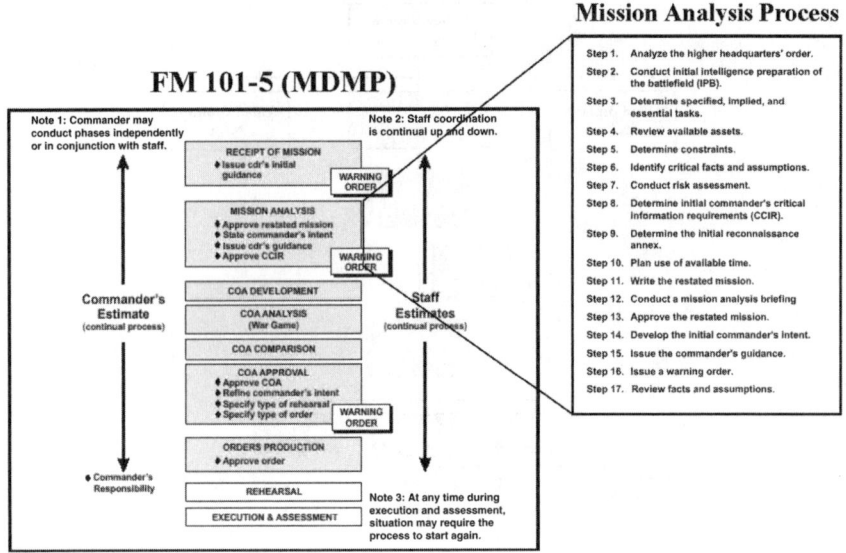

Fig. 4 Specified, implied, essential tasks.

Level-4: <u>Tasks, Operations</u> are the Task-based, outcome-centric specification of the Operations that provide the Means to accomplish the Mission. Tasks and Operations are the "do what" named-with-a-verb "playbook" of military evolutions. Mission analysis, depicted in Fig. 4, is applied to the **Level-7** Mission package (and referenced **Level-6** and **Level-5** package) specifications to identify specified and implied Tasks and to assign conditions (Environment), measures (of Mission accomplishment), and standards (measure thresholds).

The fundamental purpose of **Level-4** is to organize Task outcomes, then evaluate Mission effectiveness. **Level-4** is designed to be a compatible extension of the Mission Essential Task List (METL) methodology defined in the JTS. The Joint Capability Integration and Development System (JCIDS) promulgated in Myers (2003) terms this procedure a Functional Area Analysis (FAA), documented as mission essential tasks, conditions, and standards.

$O_{3,4}S$ is the inverse (i.e., time-backward) planning process the warfighter uses to determine the Functions, Capabilities required to complete Tasks, Operations. The warfighter iterates recursively between mission analysis and course of action development and analysis (Fig. 4). Mission analysis organizes Tasks into Operations packages to achieve measures of Mission outcome. Course of action development uses measures of performance (MOPs) to assign Capability packages to Operations. Course of action analysis uses measures of effectiveness (MOEs) to determine if the assigned collection of Capabilities enables Task execution to meet Mission requirements. JCIDS terms this procedure Capability Assessment.

The output is a collection of required **Level-3** Capability package specifications including (especially) required performance. Recommended practice is to focus on the performance required to enable Tasks to accomplish Missions and to state these required Capabilities as agnostically as possible ("what" is needed without reference to the "how" of specific Service, unit, or weapon implementations).

$O_{3,4}E$ is the forward execution and adjudication process that takes the actual performance that the specified **Level-3** Capability delivered and then determines **Level-4** Task outcomes. Outputs include Capability MOEs for use in a subsequent invocation of the $O_{3,4}S$ inverse planning process. Recommended practice is to use Task-based fault trees where the delivered Capability is compared to the standard established in the Mission analysis for that measure.

Level-3: Functions, Capabilities are the Function-based, performance-centric "how well" specification of the Capabilities that enable Forces to conduct Operations. Move, sense, communicate, engage, and restore are physical Capabilities. Observe, orient, decide, and act are cognitive Capabilities. At the Strategic National level-of-war, move is the Capability to deploy a Marine Expeditionary Force (MEF). At the lowest Tactical levels-of-war, move is the Capability to crawl out of the trench. Capabilities describe the external value provided; Functions specify the internal necessities required to deliver Capabilities.

The fundamental purpose of **Level-3** is to organize Function performance for $O_{3,4}E$ evaluation of Task outcome. JCIDS terms this procedure Functional Needs Analysis (FNA).

$O_{2,3}S$ is the inverse (i.e., time-backward) planning process used to select Components and define Forces to implement Functions and deliver Capabilities required in the **Level-3** specifications. Warfighters call this task-organization, strike package development, or task force design. Engineers call this functional allocation of required operational capabilities. Components are selected based on MOPs (e.g., a C-130J can transport this cube/weight this many miles in this many hours) and assembled into Force package specifications based on MOEs (this deployment task force can deliver the specified brigade cube/weight to the theater in this many lifts over this many days).

The output is a selection of organization, equipment, and network connectivity stated as a collection of required **Level-2** Force package specifications (referencing the associated **Level-6** Environment and **Level-5** Location/Time packages), including (especially) measures and standards for personnel and materiel readiness.

$O_{2,3}E$ is the forward execution/adjudication process that takes the actual readiness the specified **Level-2** Force packages deliver and determines **Level-3** performance Capability. Outputs include Force MOEs for use in a subsequent invocation of the $O_{2,3}S$ inverse planning process. Recommended practice is to use Component-based fault trees where the delivered states are compared to the readiness standards set in the Capability analysis for that measure. Traditional readiness measures focus on materiel repair state and number of available personnel by grade and military occupation skill. As directed by Wolfowitz (2003b), recommended practice is to use METL-based readiness measures/ standards explicitly tied to the Mission, Task, Environment, and Location/Time specifications.

Level-2: Components, Forces are the Component-based, state-centric specifications of the Forces that provide the Means to accomplish a Mission. Components, Forces are the "by whom," named-with-a-noun network of physically and virtually integrated units, personnel, and equipment that are the "players" in military evolutions. **Level-2** defines physical networking (mechanical attachment, communications link) as well as logical networking (command and supporting relationships).

At the National Command Authority layer, individual Services are Components. At the aircraft flight layer, lead and wing are Force packages. A warfighter is a human Component just as a circuit board is a materiel Component.

The fundamental purpose of **Level-2** is to organize Component states for $O_{2,3}E$ evaluation of Function performance. JCIDS terms this procedure Functional Solution Analysis (FSA).

$O_{1,2}S$ is the inverse (i.e., time-backward) planning process used to select Interactions to achieve state changes in **Level-2** Components that will have the intended Effects on **Level-2** Force packages. Warfighters call this target-weapon pairing and Effects-Based Operations. Engineers call this technology selection. Interactions are selected based on MOPs (target X has Y vulnerability to blast) and assembled into Effects package specifications based on MOEs (weapon A can deliver Z lethality > Y vulnerability to target X).

The output is a prioritized list of targets each with associated (hard or soft) engagement/resolution methods. These are stated as a collection of required **Level-1** Effects package specifications (referencing the associated **Level-6** Environment and **Level-5** Location/Time packages) including (especially) measures and standards for delivery and for restoration or damage.

$O_{1,2}E$ is the forward execution and adjudication process that takes the actual Interaction conditions that the specified **Level-1** Effects packages deliver and then determines **Level-2** Component state outcomes. In acquisition programs, this amounts to the actual execution of developmental, operational, and life-fire testing. Outputs include Effects package MOEs for use in a subsequent invocation of the $O_{1,2}S$ inverse planning process.

Level-1: Interactions, Effects are the Interaction-based, phenomena-centric specification of the Effects that Operations have on Forces. Interactions specify how Task execution changes the state of Forces. Interactions are organized by the phenomena (physics, chemistry, biology, psychology, sociology) that generate them. Effects packages organize Interactions based on outcomes they impose on own or opposing forces (sustain, protect, motivate, demoralize, destroy, suppress, neutralize, disrupt, and deceive).

The fundamental purpose of **Level-1** is to organize Interaction phenomena for $O_{1,2}E$ evaluation of Component state changes.

$O_{4,1}S$ is the inverse planning process used to identify Tasks, which, if executed to standard, will generate **Level-1** Effects that will lead to Mission accomplishment as supplied by **Level-4** to **Level-7**.

$O_{4,1}S$ is the "what if, part of the Mission analysis that generates alternative concepts of operation for use in course of action development and analysis.

The output is a **Level-4** specification of Tasks and Operations implied if a particular concept of operation is selected to accomplish the Mission by imposing the stated Effects. Measures and standards are assigned with reference to the associated **Level-6** Environment and **Level-5** Location/Time package specifications.

$O_{4,1}E$ is the forward execution and adjudication that takes the actual **Level-4** Operations package specifications, both OWNFOR and OPFOR, and determines which Interactions will actually occur at which Location/Time and under what

Environment. The output is a **Level-1** Effects package specification. Operation MOEs for use in a subsequent invocation of the $O_{1,2}S$ inverse planning processes are also produced.

Traditional MOE practice has focused on forces-based, materiel-centric measures such as loss-exchange-ratios, force-ratios required to achieve an objective, or time required to complete an operation. The MMF recommended practice is to focus on Mission-centric, Task-based MOEs. Here, MOE measures and standards are the codification of how planned/delivered Task outcome affects Mission success. In many cases, the required Task outcome often involves setting a desired condition that enables a key subsequent Task.

For example, suppose an Operations package calls for a maneuver unit to block a withdrawing enemy force by occupying a choke point along the enemy line of retreat. Within this Operations package, the required outcome for an Intelligence-Surveillance-Reconnaissance (ISR) Task may be to set the condition that the enveloping OWNFOR unit can complete its maneuver in road march formation rather than assault formation. In that case, the recommended MOE measure and standard should be based on whether or not the resulting Common Operating Picture (from the ISR Task report) was sufficiently credible, timely, and affordable for the unit commander to decide to use the faster road march. Setting this condition provides the additional time necessary for the unit commander to set the subsequent conditions for and execute the desired branch Task "defend a prepared position" rather than the less desirable conditions (imposed by later arrival) where the unit commander must execute the branch Task "hasty attack on a maneuvering enemy."

Application of the MMF

As depicted by Fig. 5, the MMF first Synthesizes top-down and then Employs bottom-up. This section provides an example of the processes.

As noted previously, portions of the MMF are already being used in the development and testing of requirements for the Army's planned FCS-equipped Unit of Action (UA). To illustrate, an actual vignette from the scenario used to demonstrate the applicability of the methodology has been selected to describe a strategic

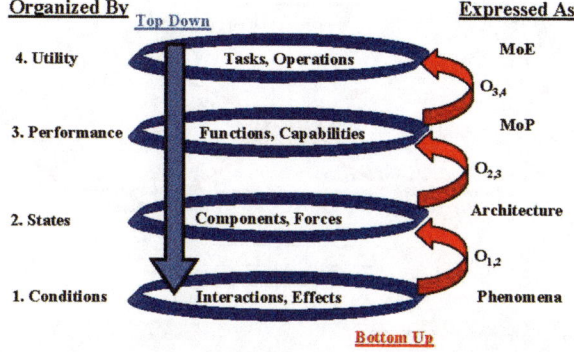

Fig. 5   The synthesis and employment processes.

situation that ultimately drives the need to plan and conduct an attack at the tactical level. The scenario is as follows.

A legitimate, pro-Western government of a Country of Interest is overthrown by radical elements and forced into exile. Radical elements, supported by a neighboring Hostile Country, form a new government and seek to force United Nations recognition. A majority of Country of Interest conventional military forces are loyal to a new government.

The United States believes its national interests and regional stability are threatened. Thus, it takes military action to deter the Hostile Country involvement in the Country of Interest, remove radical elements from power, restore the legitimate, pro-Western government, and stabilize the region and protect U.S. vital interests.

Deployed forces, organized as a Joint Task Force, have started offensive ground operations in the Country of Interest to establish conditions to return the legitimate government by defeating conventional forces loyal to the radical government and isolating the rebel government leadership inside the capital. The Joint Force Land Component Commander (JFLCC) intends to help isolate the rebel government by defeating conventional forces defending the approaches to the capital and preventing them from retreating inside the capital before the capital is surrounded.

Figure 6 shows the synthesis part of the MMF to perform top-down analysis via the planning and decision-making process. Starting at the top left, we identify a set of mission objectives for each level of war in terms of the overall mission Task

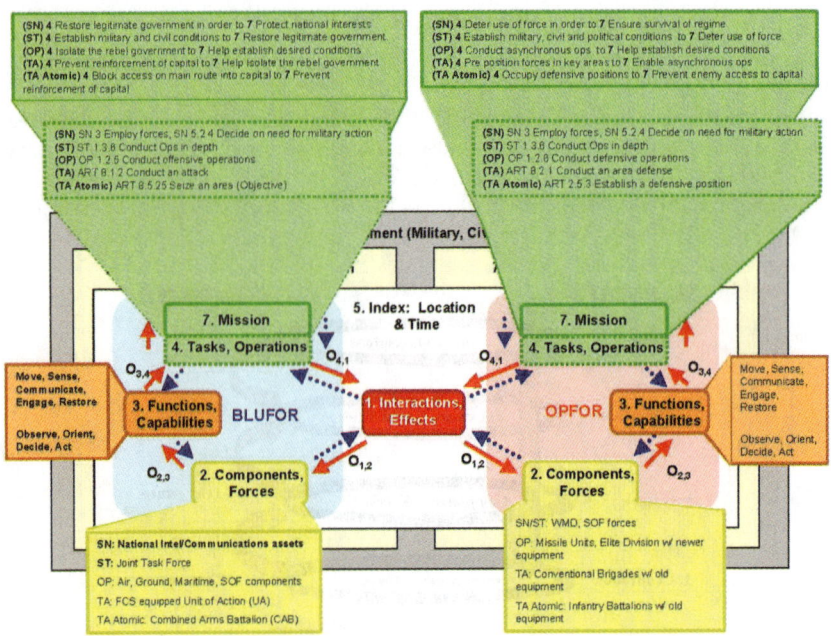

Fig. 6 Illustration of the synthesis component.

[or what is to be done (**Level-4**)] along with the mission purpose [or why it is to be done (**Level-7**)]. The why is expressed as the desired effect at each level from strategic national to tactical for both OWNFOR and OPFOR. In this case, the overall strategic effect desired by the OWNFOR is to restore the legitimate government to power. Conversely, the opposing force's desired strategic effect is to keep the rebel regime in power. Note that the mission objectives for each succeeding level of war are derived from and support the mission objective(s) at the preceding level of war. For example, the tactical-Joint objective Task of "prevent reinforcement of capital" for the UA becomes the objective purpose at the tactical-Service level for the Combined Arms Battalion (CAB). This "nesting" of Task and Purpose is precisely what military planners try to accomplish to ensure that operations are properly synchronized and focused on meeting the overall commander's intent for accomplishing the mission.

Supporting Tasks that must be performed to achieve the mission objectives are derived from authoritative task lists by applying the commonly used and accepted mission-to-operations-to-tasks decomposition process described in CJCSM 3500.03A, to determine the appropriate Tasks for each operation. Sample Tasks are illustrated in the box immediately below the mission objectives box in Fig. 6. Because these Tasks are derived in a specific mission context, taking the Environment (**Level-6**) and Location/Time (**Level-5**) into account, each objective Task and supporting Task can be associated with a relevant set of conditions, measures, and standards. Conditions are determined using Intelligence Preparation of the Battlefield (IPB) products and the results of wargaming during Course of Action (COA) analysis. Standards are likewise determined through wargaming and commander's guidance and may reflect both the qualitative purpose or the desired effect (**Level-7**) of the Task (MOE) and the quantitative performance (MOP) required to achieve the purpose under the anticipated interaction conditions (**Level-1**). The Task with associated conditions and MOPs drives the determination of the capability (**Level-3**) required to achieve the Task MOE (**Level-7**).

The combatant command headquarters employs the components (**Level-2**) that provide the strategic-level capabilities (**Level-3**) required to achieve the desired strategic effect. Included here are some of the broad categories of capabilities that would be needed at each level, such as move, engage, and sense to achieve the desired effects for that level.

How the mission is executed can be described by depicting the complex combination of Operations and individual Tasks, which must be performed in a logical and doctrinally correct sequence. One of the challenges inherent in this process is the management of the proliferation of Tasks that are generated during Task decomposition. When these Tasks are strung together, the number and complex relationship of the Tasks to each other and time can be overwhelming. For example, Fig. 7 represents an early attempt to graphically represent the string of Tasks required to conduct one operation from this vignette.

To help resolve this issue in situations where individual Tasks are habitually performed in sequence as part of a process, we have formalized the sequence in the form of an Operations Package (Fig. 8) (see Haddix, 2003). Some Tasks (e.g., passage of lines) may stand alone and be represented as a process because they are not habitually connected to other individual Tasks. Operations Packages help

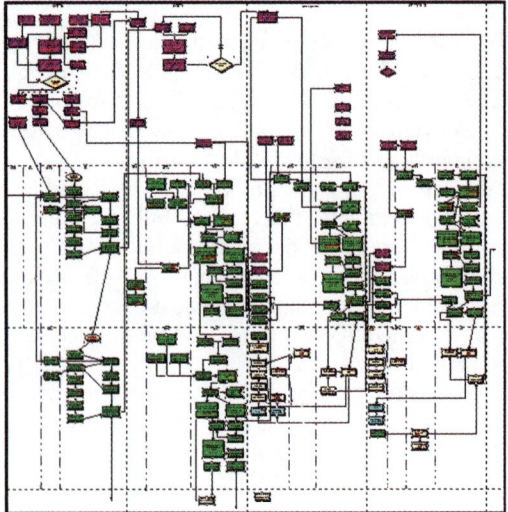

Fig. 7 Task explosion.

model the derived behavior without resorting to complex wiring charts of individual Tasks such as the one in Fig. 6. In fact, the entire derived specification can be modeled as one larger Operations Package that fits into the larger scenario. The modularity of Operations Packages facilitates editing the model and depicting continuous or iterative processes.

Another advantage of the Operations Package construct is the ability to save Operations Packages in the form of Formalized Data Products containing the

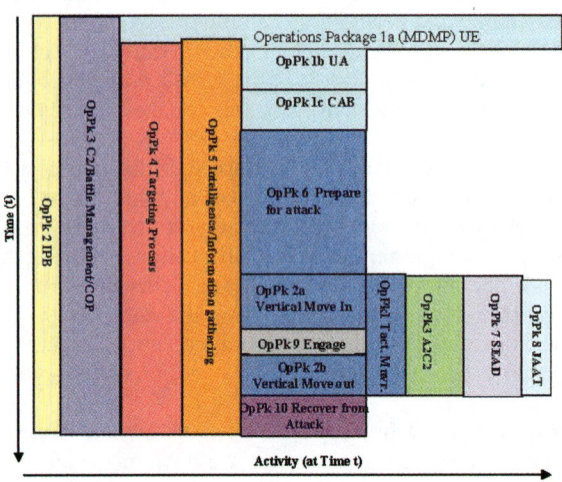

Fig. 8 Attack operations package.

Operations Package description, Tasks (with associated conditions and standards) included in the Operations Package, and a detailed description of their event or time-driven relationship to each other. These saved packages now become modular data packets that can be used as building blocks in the rapid development of "machine-parsable" scenarios/vignettes.

As previously noted, the MMF employs the DOD Architecture Framework views to define the concepts, rules, constraints, and interfaces to assemble Stocking Perspective parts (in this case, Tasks) into Assembly Perspective packages (in this case, Operations). The specific semantics and syntax for the assembly is drawn from the FDMS (Haddix, 2003).

As shown in Fig. 8, the Operations Package construct allows us to illustrate the major components (depicted as subordinate processes or Operations Packages) of the specification of behavior at a glance as well as the relationship of the major components to each other over time and to the desired effect—or MOE (the why)—for the top-level mission objective.

As seen in Fig. 9, the assembly of these Operation Packages from their component Tasks defines the relations needed to construct the Task-based fault trees that are essential to achieving the aim of relating performance to effectiveness by mapping MOPs to MOEs. By measuring the execution of each Task and subordinate Operations Package against associated MOPs, we ultimately reach the point where we can trace the performance of each subordinate Operations Package to the overall Operations Package's success or failure in achieving its associated MOE.

As illustrated in Figs. 10–12, the framework's iterative synthesis procedure explicitly links MOPs to MOEs via METL-based tasks, conditions, and standards. Within the Attack Operations Package, Fig. 10 illustrates the Task-based linkage between MOPs and MOEs for one constituent package (Operations Package 3, C2/Battle Management).

The Attack Operations Package represents one piece of the overall military scenario puzzle. By assessing the execution results of this one attack against the given MOE, we can determine the contribution this particular instance of the attack makes to overall mission utility. To further illustrate, we begin this example with Fig. 11 at the lowest level by seeing how the successful performance (as measured by a MOP) of a lower-level Task of the Conduct Surveillance Operations Package contributed to the ability to achieve the attack MOE.

In this portion of the scenario illustrated by Fig. 11, a UAV is in position and observes the Targeted Area of Interest (TAI) along the enemy's line of communication back to the capital. The CAB is able to observe the enemy moving out of defensive positions and returns to the capital without exposing friendly forces to enemy observation and fires. Thus, the CAB could engage with direct and indirect fires without exposure and prevent the enemy from reaching a key bridge on the way back to the capital.

Consequently, a significant enemy force is unable to return to the capital to reinforce defenders there. This in turn allowed other forces to surround the capital with less enemy resistance, contributing to the ability of the JTF to isolate the capital, which is one of the key conditions required by the combatant commander to restore the legitimate government to power.

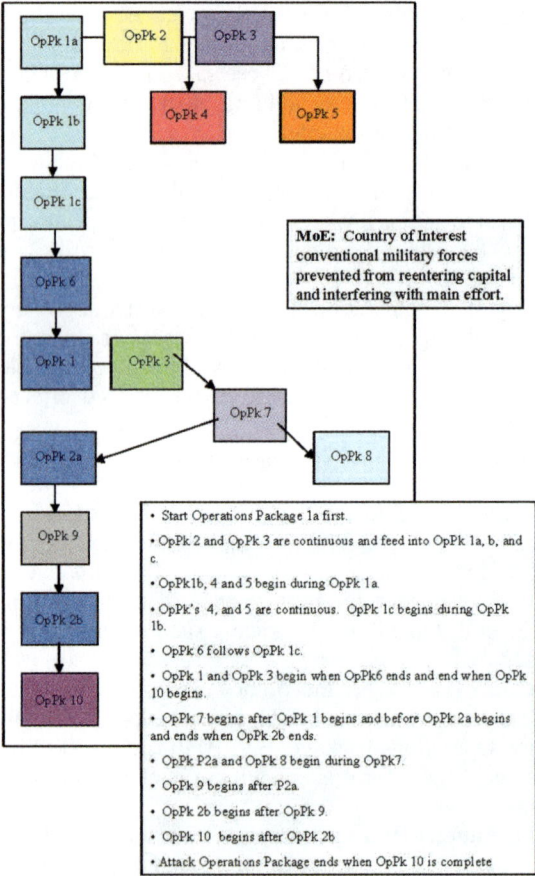

Fig. 9   Attack operations package sequencing.

As Fig. 12 illustrates, the impact of failure can also be traced. Failure to achieve the desired effect results in an undesirable set of conditions that can potentially start a chain of events leading to mission failure unless the situation is recognized and action is taken in the form of a new course of action to establish more favorable conditions.

Transformation Support

Rumsfeld (2003), Wolfowitz (2003a,b,c), Myers (2003), and Chu (2003) provide transformation guidance and promulgate acquisition reform for the DOD. The MMF provides a disciplined procedure to execute this guidance and reform as follows.

Mission Definition: At the Joint level, JCIDS requires a formal statement of national security strategy, strategy and overall concept for accomplishing, and

Sequence #	Task #	Task Title	MoP	Unit
MoE: Attack planning, coordination and execution is not adversely affected by inaccurate or outdated information concerning environmental conditions, friendly unit location and status or reported enemy activity, location, strength and intentions. (Y/N)				
3.8	ART 7.3.2.3	Conduct Risk Management	1) No offensive tasks executed that exceed maximum residual risk established by commander. 2) No casualties as a result of failure to manage risk.	All
3.9	ART 7.6.3	Make adjustments to resources, concept of ops or mission	Adjustments are made to exploit opportunities or resolve problems occurring during execution effectively. (Y/N)	Commander
3.10a	ART 7.5.4	Revise and refine the plan	Revision and refinements to the plan completed in less than one third of time available before execution	Commander and Staff
3.10b	ART 7.6.1.2	Adjust Graphic Control Measures	1) Adjustment of graphic control measures accurately reflected changes in METT-TC (w/in 100 meters) 2) Lag time between operations and adjustment of graphic control measures. (< 5 minutes)	Operations and Intelligence Cell

Fig. 10 Operations package 3 (C2/battle management).

Joint operational concepts. The traditional process does this in an ad hoc, implicit manner but does not explicitly structure the reference mission sets or the operational scenario. The MMF records this in the multi-sided (OWNFOR and OPFOR) specifications of **Level-7** Purpose/Mission referencing **Level-6** Context/Environment and **Level-5** Index/Location/Time. For the Army's FCS-equipped UAs, this would be the top-level definition of the reference mission sets and scenarios (e.g., Caspian Sea, Balkans, Northeast Asian, etc.) in the context of national security objectives.

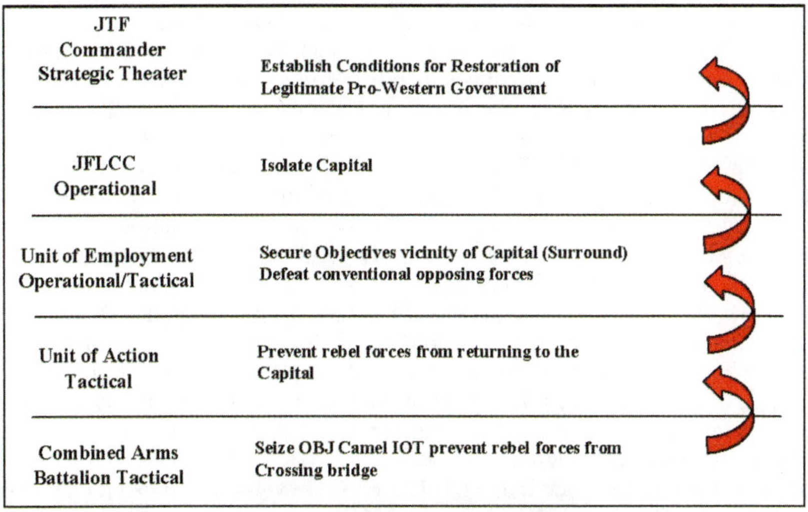

JTF Commander Strategic Theater	Establish Conditions for Restoration of Legitimate Pro-Western Government
JFLCC Operational	Isolate Capital
Unit of Employment Operational/Tactical	Secure Objectives vicinity of Capital (Surround) Defeat conventional opposing forces
Unit of Action Tactical	Prevent rebel forces from returning to the Capital
Combined Arms Battalion Tactical	Seize OBJ Camel IOT prevent rebel forces from Crossing bridge

Fig. 11 Mapping effects to utility.

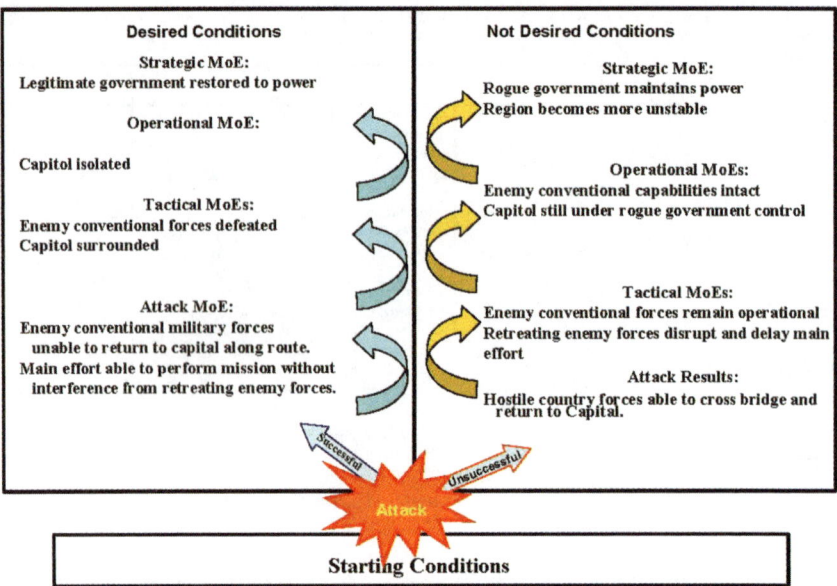

Fig. 12 Relating MOPs to MOEs.

Mission Analysis: JCIDS terms this FAA. The JTS calls this METL-based readiness requirements. The MMF employs the Military Decision Making Process (MDMP) to derive the specified and implied tasks, identify conditions, select measure, and assign standards. MMF records this as **Level-4** Tasks/Operations decomposition [e.g., see the mission decompositions referenced in the FCS Operational Requirements Document (ORD)] (U.S. Army, 2003).

Capability Assessment: JCIDS requires a functional concepts decomposition. Training Transformation calls this capability (to do what) based readiness. The MMF employs the $O_{3,4}S$ synthesis operator to derive a "catholically agnostic" (specifies what, not who/how) decomposition of **Level-3** Functions/Capabilities based on essential Level-4 Tasks/Operations. JCIDS uses an FNA to determine the ability of current/programmed/proposed capabilities to accomplish the METL under given conditions and standards. This is forward execution of the $O_{3,4}E$ employment operator. JCIDS records the **Level-3** capability requirements in an Initial Capabilities Document (ICD).

Integrated Architectures: JCIDS and the recently promulgated DOD 5000.2 require integrated Operational, Systems, and Technical standards architecture views. The traditional approach derives the solution (the System of Systems specification) directly from the ORD (with implicit rather than explicit mission analysis, capability assessment, and architecture) and then documents the de facto solution in architecture views.

This is backwards twice removed. The FAA (**Level-4** Task/Operation decomposition) and FNA (**Level-3** Function/Capability gap analysis) defined the required operational capabilities. The purpose for architecture is to design-in required composability/interoperability from the start. Hence, the Architecture Perspectives should be derived from the **Level-4** and **Level-3** content and associated $O_{3,4}E$ and $O_{3,4}E$ operator relationships. Then the solution (systems of systems specification) should be derived to comply with the All, Operational, Systems, and Technical standards view defined description of architecture concept of operations and systems engineering decomposition.

The current practice of documenting the full solution in architecture views overloads the intent of architecture. In the spirit of Occam's razor, for the same definition of function/form, a concise architecture is best.

Recall that the MMF has two names each for the seven Levels. The Stocking Perspective part-types are Purpose, Context, Index, Task, Function, Component, and Interaction. The Assembly Perspective package-types are Mission, Environment, Location/Time, Operations, Capabilities, Forces, Effects. And the MMF has two versions of each operator: Synthesis (planning and decision-making perspective) and Employment (execution and adjudication perspective). Within the JCIDS construct, MMF employs integrated architectures (as expressed in All, Operational, Systems, and Technical standards views) to provide the concepts, rules, and technologies to assemble Stocking Perspective parts into Assembly Perspective packages.

When designing/developing a plan, alternative, or solution, the MMF Synthesis operators use architectures to enable/constrain the assembly of Tasks into Operations, Functions into Capabilities, Components into Forces, and Interactions into effects. When employing a plan, alternative, or solution, the MMF Employment operators determine state changes imposed on Stocking Perspective parts by the execution of Assembly Perspective packages and use architectures to determine the state change effect on packages constituted from the affected parts.

Mission Evaluation: JCIDS requires an FSA to determine the degree to which alternative DOTMLPF solutions do/do not remove the FNA-identified capability gaps. In the MMF, the FSA is conducted as follows. The $O_{1,2}E$ Employment operator provides the degraded (or enhanced) states generated by **Level-1** Effects packages on **Level-2** Component parts. The $O_{2,3}E$ Employment operator uses architectures to determine **Level-3** Capability package performance based on **Level-2** Component part states and the appropriate architecture rules and constraints.

Summary

The MMF provides the necessary rigor to successfully define and execute a process to evaluate the capabilities and associated mission utility of alternative DOTMLPF solutions. This rigor is required for the DOD to successfully transform from a forces-based, materiel-centric Cold War posture to a capabilities-based, mission-centric asymmetric warfare posture.

References

Chu, D.S.C. (2003). Improved DOD Readiness Reporting, Office of the Under Secretary of Defense for Personnel and Readiness, 11 April 2003.

Deitz, P.H., & Ozolins, A. (1989). Computer Simulations of the Abrams Live-Fire Field Testing. BRL-MR-3755, U.S. Army Ballistic Research Laboratory, May 1989.

Deitz, P.H., Sheehan, J.H., Harris, B.A., Wong, A.B.H. (2002). Testing, Training, and Analysis: Relating Force Capabilities to Mission Utility. NDIA 5th Annual Testing and Training Conf., August 21, 2002.

Deitz, P.H., Sheehan, J.H., Harris, B.A., Wong, A.B.H., Bray, B.E., & Purdy, E.M. (2003). The Nexus of Military Missions and Means. 71st Military Operations Research Society Symposium, June 11, 2003.

Deitz, P.H., & Starks, M.W. (1999). The Generation, Use, and Misuse of "PKs" in Vulnerability/Lethality Analyses. *The Journal of Military Operations Research*, Vol. 4, No. 1, 1999, pp. 19–33.

Haddix, F. (2003). The Functional Descriptions of the Mission Space (FDMS) Data Interchange Format (DIF) version 2.0. Defense Modeling and Simulation, May 2003.

Hughes, W.J. (2003). FCS Acquisition and Evaluation Teaming. National Experimentation, Testing, Training and Technology (NET3) Conference & Exhibition, July 23, 2003.

Klopcic, J.T., Starks, M.W., & Walbert, J.N. (1992). A Taxonomy for the Vulnerability/Lethality Analysis Process. BRL-MR-3972, U.S. Army Ballistic Research Laboratory, May 1992.

Myers, R. (2003). Joint Capabilities Integration and Development System. CJCSI 3170.01C, June 2003.

Rumsfeld, D. (2003). Transformation Planning Guidance. Office of the SECDEF, April 2003.

U.S. Army (2003). Operational Requirements Document for the Future Combat System. Change 3 (JROC approved), 14 April 2003.

Wolfowitz, P. (2003a). Operation of the Defense Acquisition System. DODI 5000.2, 12 May 2003.

Wolfowitz, P. (2003b). Department of Defense Readiness Reporting System. DOD Directive 7730.65, 3 June 2003.

Wolfowitz, P. (2003c). DOD Architecture Framework. Version 1.0, draft DOD Instruction 8800.x.

Paper 2: The MMF Formal Process

M. L. Minchew and B. E. Bray
Dynamics Research Corporation

The MMF Formal Process

The interactions and relationships between missions and means are omnipresent in military operations. Through *experience and professional military education*, warfighters learn to apply a logical, cognitive, detailed thought process to the planning and execution of any mission. The Missions and Means Framework (MMF) Formal Process complements the Military Decision-Making Process (MDMP) and the planning and operations processes used within the warfighter community, and it provides the link between military art and science when determining capabilities that address mission requirements.

The MMF Formal Process (see Fig. 1) provides the hierarchy of steps used to properly apply the MMF and implement MMF analyses. Each step is an important and necessary requirement in the overall application of the process and must be applied in a disciplined manner to elicit the key data requirements needed for

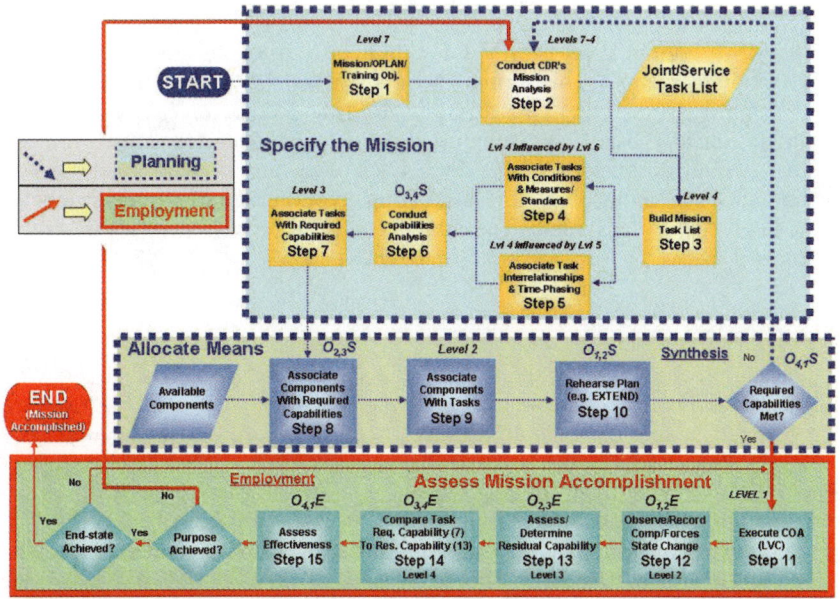

Fig. 1   MMF analysis process.

traceability and documentation of MMF results. When executed properly, MMF analytical results document and demonstrate the traceability of mission requirements to required capabilities and obtain comparisons between required and available capabilities and their potential effect on mission accomplishment. Application of the MMF process documents the top-down planning process and shows *linkage of the means to the mission* through the execution and adjudication process. The MMF Formal Process can be applied in support of analysis requirements for military planning, training, testing, readiness, or validation of current or future force requirements.

Operational Concept

The concept of the MMF Formal Process is to drive a methodical and disciplined process of explicitly capturing the inputs, cognitive processes, and outputs associated with planning, execution, and assessment of military operations iteratively and recursively. The MMF provides the organizing principle to capture critical information and interactions needed to understand all factors affecting the mission. Experienced warfighters apply the doctrinal planning and operations processes to generate planning products and ensure the operation remains on track. Many, if not most, of the inputs, cognitive processes, and even outputs of these planning and operations processes are not explicitly documented or retained; they remain implicit. The purpose of the MMF Formal Process then is to drive the explicit capture of the inputs, processes, and outputs involved in planning, executing, and adjudicating operational packages under various scenario contexts.

Methodology

The MMF Formal Process provides additional instructions and procedures on MMF implementation. The formal process breaks the framework down into three subprocesses; *Specify the Mission, Allocate Means, and Assess Mission Accomplishment.* These three subprocesses encompass the top-down synthesis and bottom-up employment of the MMF. Each of the process steps relates to one or more of the 11 elements (7 levels and 4 operators) of the MMF and is labeled accordingly above the symbol. The subprocesses and process steps are described as follows:

Specify the Mission

This subprocess includes the steps necessary to describe the What and Why (MMF Level 7) of the Mission as well as the When and Where (MMF Levels 5 and 6) that describes the Operational Environment. The steps involved are:

Step 1. Mission/OPLAN/Training Objective – The process begins with a mission statement of some kind. Mission statements can be conveyed in almost any form and typically specify the Who, What, When, Where, and Why, but not the How. This step relates to directly to MMF Level 7, the externally imposed Mission.

DETAILS AND DEVELOPMENTS IN THE MMF 321

Step 2. Conduct Commander's Mission Analysis – Chapter 3 of Army Field Manual (FM) 5-0, dated January 2005, contains a detailed explanation of the MDMP process. This step includes most of the tasks included in Step 1 (receipt of mission) and Step 2 (mission analysis) of the MDMP. The information derived from conducting this step is used to populate Levels 4 through 7 of the MMF.

Step 3. Build Mission Task List – One of the critical mission analysis tasks performed in Step 2 is the determination of specified, implied, and essential tasks required to accomplish the mission. This task typically results in a document or chart containing a written list of tasks in plain English form. The purpose of Step 3 is to convert the tasks from their plain English form to task numbers and titles derived from an authoritative source, such as the approved Joint and Service Task Lists. This step is one of the most important in the Formal Process because tasks represent the formal building blocks around which the MMF is based. Converting to authoritative task numbers and titles provides both a common language and a format that is much better suited to database applications. The information derived from conducting this step is used to further refine and populate MMF Level 4 data.

Step 4. Associate Tasks with Conditions and Measures/Standards – The purpose of this step is, for each task, to:
- Describe the purpose of the task being performed in terms of the desired effect/set of conditions to be created.
- Describe the environmental conditions that could impact on the ability to perform the task.
- Describe the measures of effectiveness that would indicate whether the task's purpose is being achieved.
- Describe the minimum measures of performance in terms of speed, accuracy, lethality, protection, etc., needed to perform the task to the required level.

The actions and information required to perform this step are described in Step 2 (Mission Analysis), Step 3 (COA Development), and Step 4 (COA Analysis) of the MDMP process. The information derived from conducting this step is used to further refine and populate MMF Level 4 data.

Step 5. Associate Task Interrelationships and Time Phasing – The purpose of this step is to capture task linkages and dependencies as well as the planned sequence of tasks for the selected Course of Action. The actions and information required to perform this step are described in Step 4 (COA Analysis) of the MDMP process. The information derived from conducting this step is used to further refine and populate MMF Level 4 data.

Step 6. Conduct Capabilities Analysis – The purpose of this step is to analyze the results of Steps 4 and 5 to determine the types and level of capabilities and functions needed to perform each task. The actions and information required to perform this step are described, and in some cases implied in Step 4 (COA Analysis) of the MDMP process. The process used to conduct this step represents the MMF $O_{3,4}S$ operator.

Step 7. Associate Tasks with Required Capabilities – The purpose of this step is to describe and document the required types and levels of capabilities and functions for each task. For automated modeling and simulation purposes, the format selected should match the format used to describe and document types and levels of capabilities and functions available given the current state of MMF Level 2 resources. The information derived from conducting this step is used to populate MMF Level 3.

Allocate Means

This subprocess includes the steps necessary to determine and describe the How of the Mission. Here we are literally associating the assigned mission with the means to be used to accomplish it. A pre-requisite for beginning this subprocess is gathering the set of components (units and associated people and equipment) available for planning purposes. The steps involved are:

Step 8. Associate Components with Required Capabilities – The purpose of this step is to describe and document the process used to match the required capabilities determined in Step 6 with the set of components available for planning purposes for this mission. The process used to conduct this step represents the MMF $O_{2,3}S$ operator.

Step 9. Associate Components with Tasks – The purpose of this step is to describe and document the selection of components to perform tasks. Components are selected from the set of components determined in Step 8 that deliver the required capabilities and functions for each task. The information derived from conducting this step is used to populate MMF Level 2.

Step 10. Rehearse Plan – The purpose of this step is to conduct a controlled execution of the selected COA In order to confirm or deny whether the required capabilities and functions for each task are satisfied by the available capabilities and functions delivered by the assigned component(s) at that moment in time and space. The process used to conduct this step represents the MMF $O_{1,2}S$ Operator. The intent is to identify capability gaps that could prevent successful accomplishment of the task's purpose. The process of identifying these capability gaps represents the MMF $O_{4,1}S$ Operator.

Assess Mission Accomplishment

This subprocess includes the steps necessary to execute the selected COA and assess whether the ongoing operation is satisfying the commander's intent for mission accomplishment. The steps involved are:

Step 11. Execute COA – The purpose of this step is to initiate the uncontrolled execution of the COA and associated tasks in a Live, Virtual, Constructive (LVC) Environment. Examples of this include closed loop simulations for analysis, constructive simulations for experimentation or training, map exercizes, Scenario War Games, live force-on-force training events, and actual operations. The information derived from conducting this step is used to populate MMF Level 1.

Step 12. Observe/Record Component/Units State Change – The purpose of this step is to capture the impact of the Level 1 interactions previously described on the Level 2 components (forces, equipment, people). The impact is described in terms of changes in the state of the Level 2 Components. The process of identifying these state changes represents the MMF $O_{1,2}E$ Operator, and the resulting information is used to refine the MMF Level 2.

Step 13. Assess/Determine Residual Capability – The purpose of this step is to describe and document the process used to determine the level of functionality/capability remaining for a particular component as compared to the full, or designed level of functionality/capability. The process of determining residual capability as a result of state changes represents the MMF $O_{2,3}E$ Operator, and the resulting information is used to refine the MMF Level 3.

Step 14. Compare Task Required Capability to Residual Capability – The purpose of this step is to describe and document the process used to determine the delta between required and residual capability at a given point in time and space. The process of determining this delta represents the MMF $O_{3,4}E$ Operator.

Step 15. Assess effectiveness – The purpose of this step is to describe and document the process used to determine whether the task that was just executed accomplished the desired purpose. The process of conducting this assessment represents the MMF $O_{4,1}E$ Operator.

The Decision Box immediately following Step 15 leads to a determination of whether to continue to execute the COA as planned or to adjust the plan to reflect the new set of conditions.

The final Decision Box determines when the mission is considered complete.

Paper 3: The Connection Between Functions and Capabilities

M. L. Minchew
74th Barchi Prize Entry
November 30, 2006

Introduction

War is a complicated business. The multitude and magnitude of tasks and operations executed simultaneously on a battlefield is staggering. From the performance of a single piece of equipment, to the actions of an infantry platoon, to the decisions made by military commanders engaged in an Area of Operations, all the disparate pieces and parts of equipment and personnel on a battlefield must be combined and synchronized to achieve maximum effect and devastation on the enemy. As the beginning of the 21st century unfolds, it is clear that the enemy and the way we wage war is changing. Staying ahead of this change will be a great challenge and will require a complete understanding of the threats that seek to harm us as well as the regions and conditions where they exist. It has become increasingly necessary to research, design and test new systems and concepts on shorter timelines while simultaneously fighting new threats. This pace of change in recent years also does not allow for the luxury of multiple year research, development and testing of new systems and force structures to meet these threats. The use of models and simulations to assist with this task is increasing but in order for models and simulations to be valid they must be able to describe, depict and replicate the intricate workings of military operations so that the resulting data can be used to study new Doctrine, Organizations, Training, Material, Leadership, Personnel and Facilities (DOTMLPF) solutions to counter future threats. How do we study the planning and execution of military operations, tracking hundreds of different types of equipment and personnel, performing myriad tasks that synchronously are able to achieve success in combat? Is it possible to replicate the chaos of a battlefield and capture the rigor of military operations in order to accurately predict the performance of new doctrinal concepts and capabilities so they can be quickly integrated into current doctrine and future force structures?

The 2001 Quadrennial Defense Review (QDR), the Joint Defense Capabilities Study (JDCS), and the Joint Capabilities Integration and Development System (JCIDS), all cited the need for a common framework to articulate a capabilities-based approach across strategy development, operational planning, resourcing, and execution. The Department of Defense (DOD) needed to transform itself from a forces-based, material centric Cold War posture to a capabilities-based, mission centric asymmetric-warfare posture. It is currently working hard to accomplish this task; however, in some areas DOD is falling short of the mark. Acquisition timelines remain too long from concept through incorporation on the battlefield. Modeling and simulation improvements have not kept pace with technology growth, and capabilities-based planning is still not fully understood across the DOD. There is no standard format on how to apply capabilities-based planning and analysis within the DOD and amongst the Services. JCIDS describes the three analysis phases: Functional Area Analysis (FAA), Functional Needs Analysis (FNA) and Functional Solutions Analysis (FSA), but these analyses do not provide a roadmap for how to apply them. Furthermore, it is unclear how to gather, capture, interpret and analyze data from these various analyses so as to extricate meaningful results that can be used for the basis of resourcing, acquisition, and future force decisions. To assist the JCIDS process and enable DOD transformation, from concept formulation through actual combat, a framework is required to enable the warfighter, engineer, and comptroller to specify a *common understanding* of military operations, systems, and information and to provide quantitative mission assessment of alternative DOTMLPF solutions.

The Missions and Means Framework

The Missions and Means Framework (MMF) was developed by a collective group of authors who possessed a desire to find a better way to think about and analyze military problems. They sought to describe the fundamental elements of military operations by designing a framework that comprehensively organized and rigorously specified all aspects of military operations with the intent to explicitly relate and map operational purposes and goals to the technical means needed to achieve military success.

The development of the MMF was borne from the military need to reduce the inefficiencies in capabilities development. The development of new technologies and systems is oftentimes done in a vacuum. *Warfighters* identify requirements based on new and emerging threats. *Engineers* assume the understanding of these requirements and develop new capabilities. *Comptrollers* and decision makers promote the need for new systems and obligate funds for programs before justifiable/quantifiable capabilities analysis is presented. Because warfighters, engineers and comptrollers speak in different vernaculars, clear insight into the cause and effect relationships between the entities is never reached. Without a full understanding or translation of requirements and capabilities, the nemesis between parties causes the understanding of requirements to literally become "lost in translation." This failure in understanding can be traced to the inability to *translate or frame* what warfighters need into *terms* engineers can understand. Alternately, materiel developers or warfighters fail to *translate or frame* their requirements and *quantifiably show proof* in terms comptrollers can understand in order to *justify* resourcing decisions.

In its military application, the purpose of the MMF is to provide a disciplined, comprehensive process that unifies warfighters, engineers and comptrollers in understanding cause and effect relationships of military *missions* and the *means* to accomplish those missions. Through its ability to integrate disparate communities, the MMF provides the basis for a common language across the warfighting, engineer and comptroller domains. One of the MMF's key contributions is in closing this lexicon gap so that the entire joint community can articulate, as a unified force, the needs of the warfighter. The MMF provides a disciplined and quantitative basis for justifying the *means* in terms of capabilities based development and funding supported by the linkage of those *means* to perform the *mission*. The establishment of the MMF as a universal framework to enable DOD transformation has proven valuable in building on standard practices for military operations analysis and has facilitated more effective communications between mission owners and various external support domains.

Fundamental Elements

The MMF prescribes an organizing principle that provides a disciplined way to examine military operations. Its 11 fundamental elements allow for the repeatable and scalable dissection of military problems providing a way to organize, document and capture top-down planning and bottom-up execution. The MMF provides the framework to reason about forces, materiel, and other resources in terms of the operational mission and its purpose and endstate.

The MMF uses 11 fundamental elements to organize and specify military operations. As shown in Fig. 1, mission content is organized into seven Levels:

Level-7. Purpose, Mission
Level-6. Context, Environment
Level-5. Index, Location/Time
Level-4. Tasks, Operations
Level-3. Functions, Capabilities
Level-2. Components, Forces
Level-1. Interactions, Effects

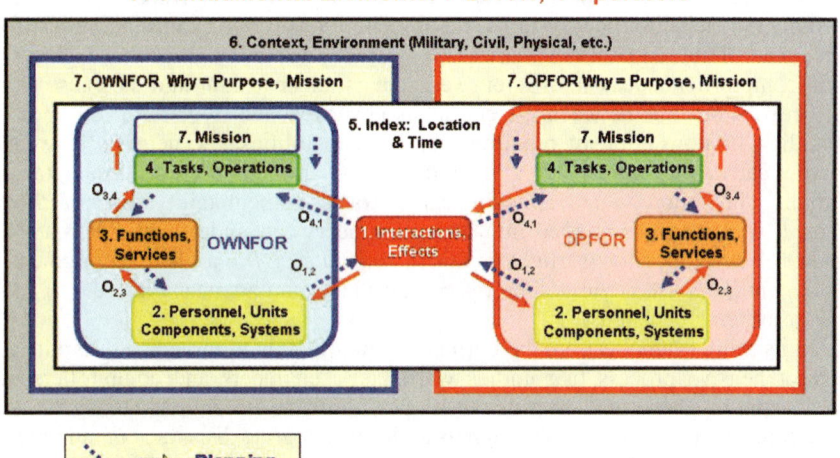

Fig. 1 Missions and means framework.

DETAILS AND DEVELOPMENTS IN THE MMF

In addition, the following four transformation Operators are included:

$O_{1,2x}$: transforms Level-1 interaction specifications into Level-2 component states.
$O_{2,3x}$: transforms Level-2 component states into Level-3 functional performance.
$O_{3,4x}$: transforms Level-3 functional performance into Level-4 task effectiveness.
$O_{4,1x}$: transforms Level-4 task sequences into Level-1 interaction conditions.

Also note in Fig. 1 the MMF's multi-sided nature. The OPFOR coalition influences the outcome of OWNFOR's mission prosecution. The MMF is a symmetric representation of an asymmetric (perhaps decidedly asymmetric) conflict.

Describing Scalability, Parts and Packages

The MMF is a brilliantly designed analysis framework. It was designed to break down and view military problems within the context of seven levels so that each level could be described separately from one another. This ability to view the content of each level in its pure state allows for rigorous analysis within each level. As shown in Figs. 2 and 3, the MMF is also scalable in that level content can be broken down into individual parts and parts can be combined to form larger packages. Parts combine and packages decompose as necessary to describe and relate level content. Rigor and scalability of the MMF are two of its unique and important characteristics.

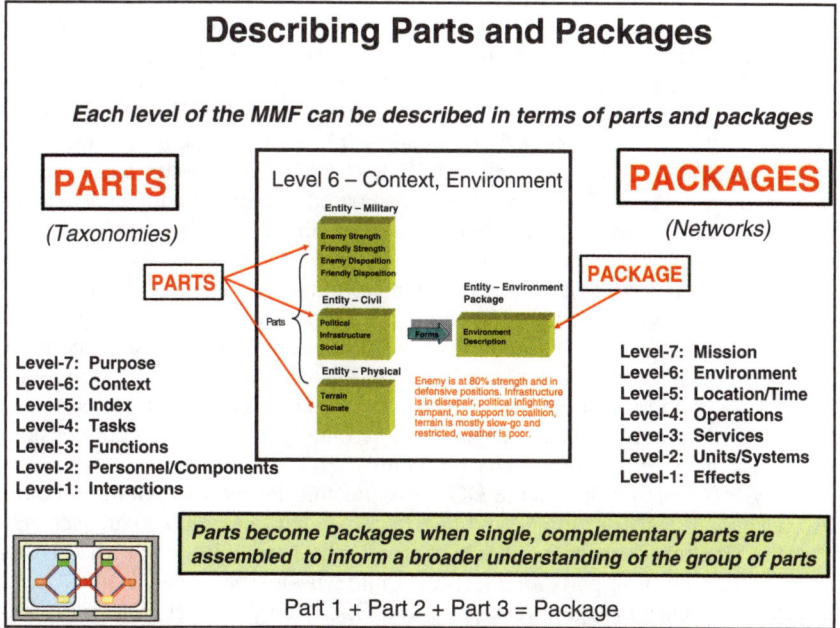

Fig. 2 Describing parts and packages.

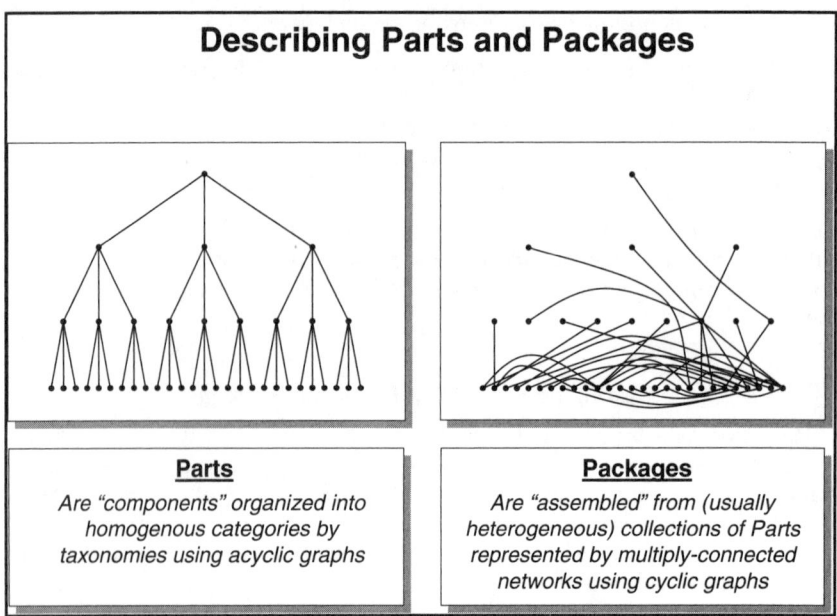

Fig. 3 Graphic depic on of the differences between parts and packages, Arthur K. Cebrowski, Director, Force Transformation.

The MMF structure breaks down military problems into **parts** so that they can be examined and studied in their homogenous state before being grouped or reassembled into **packages** that detail their heterogeneous nature. Scalability, the ability to expand something in various directions, allows analysis of the level content in relationship to other things that affect it. Parts can be broken out separately, atomically even, and compared to similar parts or to parts that make up other packages. Parts of one type can be compared to parts of another. They can be compared in scale vertically, horizontally or dimensionally. Parts combined into packages are compared to other packages in a similar way only at an aggregated level. This scalability allows for the continual aggregation (composition) and de-aggregation (de-composition) of parts and packages for analysis purposes and provides the rigorous structure upon which to analyze relationships between parts and packages across vertical, horizontal or dimensional paths. This scalability also provides traceability of effects and relationships between parts and packages and among parts and packages horizontally, vertically and dimensionally across the spectrum of military problems.

The MMF design employs the DOD Architecture Framework views to define the concepts, rules, constraints, and interfaces needed to assemble parts selected from the Stocking Perspective into packages in the Assembly Perspective. Using architectures, Purpose parts are assembled into Mission packages, Context parts into Environment packages, Task parts into Operation packages, Function parts into Capability packages, Component parts into Force packages, and Interaction parts into Effect packages.

Parts viewed as taxonomies are organized into similar groupings of content with open-ended outcomes. Taxonomies are structures that provide a way of classifying things into series of hierarchical groups that make them easier to identify. Think of the classification of musical instruments or musicians as an example. Strings, horns, percussion instruments and musicians are grouped and classified according to their like characteristics. They provide the individual parts that make up a band or orchestra.

Packages viewed as networks represent a multi connected entity whose *collective* actions work in concert with one another and provide a higher level capability that achieves a desired effect. Think of the individual musicians and instruments that make up a symphony orchestra. Musicians spend years of study learning to play and master a particular instrument. They practice with their instrument and hone their skills. They learn to read music and give solo performances. When combined or aggregated into a larger group of musicians and given a piece of music to play we begin to see how their disparate individual parts combine into a package that produces sound that is totally different. At the beginning of a concert, a guest in the audience can experience the orchestra warming up, musicians tuning their instruments or playing bits and pieces of the score. The musicians' acts are random. They aren't in synch with one other. It sounds like a bunch of noise. Now introduce the guidance of a skillful conductor and a studied piece of music and the combinations of the disparate musicians playing instruments execute Mozart's Requiem and achieve the overwhelming affect of delivering a beautifully synchronized piece of concert music.

Stocking and Assembly Perspectives

The MMF uses the semantics and syntax of the Functional Descriptions of the Mission Space (FDMS) to organize Stocking Perspective part specifications into Assembly Perspective package specifications (Haddix, 2003). The framework has two names for the mission content specified by each **Level**; the "stocking" (part) perspective and the "assembly" (package) perspective. Military problems are broken down into "parts" so that they can be examined and studied in their homogenous state. When assembled into "packages" the grouping of parts shows the synchronous, heterogeneous nature of the package.

	Stocking (Parts)	**Assembly (Packages)**
Level-7	Purpose	Mission
Level-6	Context	Environment
Level-5	Index	Time/location
Level-4	Tasks	Operations
Level-3	Functions	Capabilities
Level-2	Components	Forces
Level-1	Interactions	Effects

The Stocking (Part) Perspective

In the stocking (Parts) perspective, when the MMF names, records, and references content, the organization within the level is an orthogonal decomposition into **homogeneous** collections of similar content. Stocking *parts* are organized into similar groupings or taxonomies with open ended outcomes. The taxonomies serve as a method for logically categorizing information so it can be easily referenced. For example, consider a building construction metaphor. A buyer employs a builder to construct a home suitable for a resident. The builder constructs the home by tailoring and assembling standard building materials from a supplier. Using this metaphor, *parts* would be considered the standard building materials found in a supply warehouse. Similarly, musicians and instruments would be considered the building materials of a band or orchestra.

When mission content in a **Level** is viewed like standard *parts* in a reference library the MMF uses the first name for the **Level** content: *Purpose, Context, Index, Task, Function, Component and Interaction.* For example, Level 2 – Components may list a variety of types of military personnel or equipment in the library such as types of wheeled ground vehicles, armored ground vehicles or types of personnel. A decomposition of Level 2 Parts would reveal various types of equipment and personnel each described in a separate **Level-2** branch, and both would be considered Level 2 Component parts. Each reflects a "stock" of individual components (pieces of equipment or personnel) with virtually identical characteristics in each branch, not directly related to task or mission accomplishment. A few examples follow:

- Wheeled Ground Vehicles
  - M977 Heavy Expanded Mobility Tactical Truck
  - M99A2 HEMTT Cargo Truck
  - M998 Hummer
- Tracked Ground Vehicles
  - M1 Series Abrams Tank
  - M2A3 Bradley Infantry Fighting Vehicle (IFV)
  - M3A3 Cavalry Fighting Vehicle (CFV)
  - M4 Command and Control Vehicle (C2V)
- Personnel
  - Officers
  - Non-Commissioned Officers
  - Warrant Officers
- Officer Branches of Specialty
  - Infantry
  - Field Artillery
  - Quartermaster
- Radios
  - Single-Channel Ground Air Radio System (SINGARS)
  - Enhanced Position Location Reporting System (EPLRS)
  - AN/GRC-240 Have Quick (HQ) II UHF-AM Radio Set
- Small Arms Weapons
  - M16A4 5.56 mm Rifle
  - M4 5.56 Carbine
  - M107 .50 Semi Automatic Long Range Sniper Rifle (LRSR)
  - M249 5.56 mm Squad Automatic Weapon (SAW)

The Assembly (Package) Perspective

When mission content in a **Level** is viewed as an assembled package that satisfies a mission requirement (by analogy, a completed room or feature in the home under construction or the assembled orchestra), the framework uses the second name for the **Level** content: Mission, Environment, Location/Time, Operation, Capability, Force, and Effect. Just as in building construction (where diverse materials are assembled to provide a useful kitchen), when the MMF names, records, and references content from this perspective (hereafter called the Assembly Perspective), the organization within the **Level** is a decomposition into **heterogeneous** packages of diverse content. In the Assembly Perspective, a combined arms ground combat team would be specified in one **Level-2** branch, an aviation strike warfare package would be described in a distinct **Level-2** branch, and both would be described as Forces. Each reflects the characteristics of an "assembled" package created by appropriately combining components selected with an eye toward accomplishing a specific mission or task. For example, an Infantry Battalion, the 1st Brigade Combat Team or the 1st Armored Division all could be described as a package of disparate vehicles, equipment and personnel whose parts, when combined and employed in combat, perform synchronously to execute a particular military mission.

An NBC Part to Package Use Case

In Fig. 4, parts are combined at Levels 4, 3, 2 and 1 of the MMF to describe a Nuclear, Biological, and Chemical (NBC) use case. At Level 4, a number of separate NBC tasks describe separate activities such as Perform Decontamination, Conduct NBC Surveillance, or Occupy an Observation Point (OP). At an aggregated level these tasks are described as NBC Operations, particularly Decontamination, Detection and Smoke Operations.

The Level 3 Capabilities required to execute this Operations Package require a *chemical defense capability package*. These capabilities, broken down into Function parts require the individual functions of *protection* from NBC agents, preventing observation of friendly force movements specifically by *obscuring* and *sensing* (requires two different types of sensing capabilities: sensing enemy movements and sensing NBC agents within the area). Thus, executing the NBC Operations Package requires the individual functions of sensing, obscuring and protecting. Level 2 Components and Forces are required to provide these Level 3 Functions and Capabilities. These Level 2 Personnel/Component parts might consist of equipment components such as an M93 NBCRS, smoke generators, and M12A1 decontamination apparatus. These pieces of equipment would be operated by personnel most likely with an NBC military occupational specialty (MOS) such as 74D, Chemical Operations Specialist. When these individual soldiers and equipment are aggregated into a Level 2 Force Package, they are described as an NBC Decontamination Squad or a NBC Reconnaissance Platoon. When these forces execute their Level 4 NBC Tasks as part of the Level 4 Chemical Defense Operation, they engage in Level 1 Interactions at a time and place with enemy forces within a particular environment that produces Level 1 Effects. During the planning process the required/desired Effects are described as NBC contamination is detected and

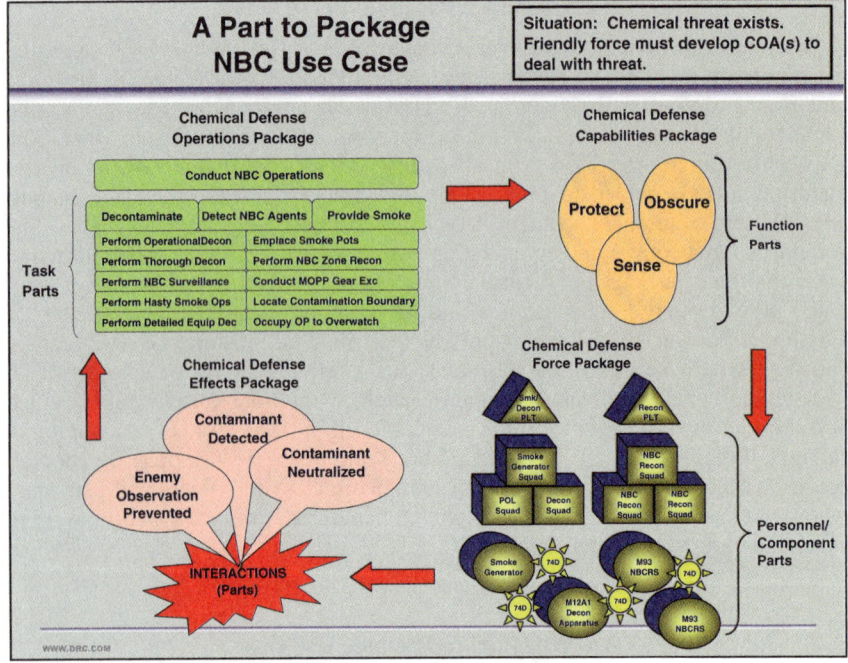

Fig. 4   Part to package NBC use case.

neutralized, and observation from the enemy was prevented. Once the interaction occurs the actual effects can be compared to the desired effects to determine success or failure of the operation.

The Varying Definitions of Capabilities

The term *capability* has many definitions and descriptions and is often used to describe military requirements in broad, overarching terms. Use of the term *capability* in military literature is extensive and often used in conjunction with Transformation requirements. Oftentimes, mention of a *capability* conjures up thoughts of a panacea to a military ailment or describes the next greatest military invention. The following excerpts from various military related articles demonstrate superficial or even overuse of the term *capability* as it intends to describe something the military needs to increase or improve.

> Current UAVs have a wide range of capabilities from the large Global Hawk high-altitude system to the hand-launched Raven ...... Establishment of new training opportunities in the Philippines will enhance U.S. force capabilities .... As this participation increases so does the effects on their logistics capabilities to maintain operational tempo become evident that there is a shortfall .... The US should improve its standing SRO capability .... The Standing Joint Force Headquarters (SJFHQ) provides a planning multiplier by bringing both joint and regional expertise

to the Joint Task Force Headquarters (JTFHQ), thereby increasing its capabilities. These capabilities assist in the rapid establishment of a fully functional JTFHQ, allowing for quicker reaction and crisis action planning.

It is difficult to determine exactly what the capability refers to in these excerpts. They seem to relate the *capability* to a piece of equipment or force structure, inferring that the equipment or force structure *is* the capability. In the first excerpt the term *capability* is linked to unmanned aerial vehicles (UAVs) and their wide range of *capabilities* from the Global Hawk to the Raven. Are the UAVs, as individual pieces of equipment, the *capabilities*? Another excerpt refers to increasing the capabilities of the Joint Force Headquarters. How do you increase the capability of a headquarters or a piece of equipment? Do you simply provide more of it by increasing the number of pieces of equipment in the inventory or personnel to the headquarters staff? Does changing the equipment or force structure in some way change its *capability*? Maybe the problem isn't really how the term *capability* is defined but in how it is used in the context of describing military operations.

To answer these questions requires a more in depth look at the meaning of the word capability and what it really is. This task would be helpful if there were a common definition of the term but unfortunately, there is not. For as often as the term *capability* is used in conjunction to anything military, one would think that a common definition could be found. Not the case and confusion abounds among and within the Services on what the term *capability* really means and more importantly, what it describes. Here are a few authoritative sources and definitions of the word.

The *CJCSI 3500 Joint Training Manual* describes a capability as: "The ability to conduct a Task under specified Conditions to (unit commander defined) Measures and Standards." This usage of capability embodies the mission essential task list (METL) approach to training and readiness that is organic to Training Transformation. This METL-centric approach is directed in JCIDS to document the requirements generated under the Functional Area Analysis (FAA).

The *CJCSM 3170, JCIDS* describes a capability as: "The ability to achieve a desired effect under specified standards and conditions through combinations of means and ways to perform a set of tasks." Examples of this description include: The capability to generate a specified effect with X probability given that the decision to employ the effect; the capability to develop the situation out-of-contact and close with and destroy the enemy at a time and place of our choosing; or, the capability to remain combat effective for follow-on missions after an engagement.

The *Joint Publication JP 1-02*, DODAF FCS Glossary describes it as: "The ability to execute a course of action (COA)."

Webster's Dictionary describes capability as: "Capability: *noun*, The quality of being capable" and, "Capable: *noun*, 1. Endowed with the physical power sufficient for an act. 2. Possessing mental powers: intelligent; able to understand." Inserting the definition of capable in the definition of capability, the definition becomes "Capability: the quality of being endowed with the physical power sufficient for an act." This definition corresponds to common usage in test and evaluation for traditional one-materiel-end-item at a time acquisition programs. It corresponds to the physical, tangible, objectively measurable capability of the system to deliver performance that is visible, relevant to the warfighter that actually operates the materiel.

These varying definitions of the term *capability* all touch on a common theme. They all include, and in some way focus on, the *action words such as execute, conduct, power or achieve*. Though there is indeed variance and confusion with reference to the meaning of the word *capability*, all these definitions indicate that to *possess a capability* means to have the ability to *execute an act, perform a function or provide some sort of physical power to achieve a desired effect*. What that act or function is, is unclear but is probably linked to the task needing to be performed. In the first definition it is *the action of executing a task*. In another definition, it is *executing a course of action*. In still another, it is *the ability to achieve a desired effect* at a time and place of our choosing. Collectively, the word *capability* infers that to possess it (a *capability*), the object must be able to *perform particular actions or perform particular functions to a specified standard in order to achieve a desired effect*. Function is action and action is accomplished by function.

Also unclear is whether a capability refers to one action or multiple actions. Is it practical to say that one action equals one capability or do multiple actions add up to a capability? This interpretation doesn't seem logical when you examine the word in context to the excerpts just provided. For example, take the JP 1-02 definition of a capability, *the ability to execute a course of action (COA)*. A course of action consists of myriad tasks that when executed synchronously, achieve a desired effect on the battlefield in order to accomplish a particular mission. This interpretation of the word *capability* means that the capability requires the execution of a *number of tasks or actions*. The JCIDS definition links capability to achieving a desired effect. Achieving a desired effect might require the execution of one or many actions which would require the performance of one or more *functions*.

These interpretations of the word *capability* infer a collective nature. To execute a course of action requires the execution of a collective number of tasks or actions performing various functions. A Joint Force Headquarters (JFH) is a collective group of individuals performing a variety of tasks to execute its mission. But is the JFH a *capability* in and of itself? We know a Global Hawk is a UAV but what *capability* does it possess? The answer lies in the description of what this piece of equipment or group of personnel collectively *do*. What functions do they perform? Within the JFH, each person executes particular tasks and performs certain functions to accomplish those specific tasks. The Global Hawk in the accomplishment of the task *Collect Information* performs the functions of flying, sensing, and communicating to satisfy the particular needs of a task. Therefore, it is safe to say that a capability is made up of a number of combinations of acts or functions and that *the capability is the method of describing the collective packaging of the number of actions or functions the piece of equipment or force can perform and execute*.

So, in order to increase the capability of a piece of equipment or force would require that it be able to accomplish either more actions or functions than it presently does or accomplish those actions or functions to a different standard. To increase the capability of the JFH would require that either more tasks (actions requiring function) be added to the staff or the ways or means by which those tasks are executed (improve the way the functions are executed) change in some way. This is where the term *function* must be succinctly described. The individual *functions* of the piece of equipment or force are the link to describing its overall capability. So when senior leaders ask for an increased capability to deploy forces or sense enemy activity, they are asking us to look deeper into the particular "do what" functions required to perform tasks.

Why Focus on Level 3 Functions and Capabilities

Functions and Capabilities are the performance-centric "how well" specifications for tasks that enable forces to conduct operations. The fundamental purpose of Level 3 entities within the MMF is to organize parameters for the purpose of evaluating task outcome and to facilitate the task organization, task force design and strike force package development process. The input for Level 3 comes from the MOEs and MOPs determined after wargaming and sequencing Level 4 tasks. Those MOPs and MOEs map to and help determine the Level 3 Functions and Capabilities needed to conduct the mission. The outputs from Level 3 are defined, quantifiable and conclusive Functions and Capabilities.

Unlike Level 4 Tasks and Operations and Level 2 Components and Forces, Level 3 Functions and Capabilities are not well understood or defined within the military community. In terms of Level 4 Tasks and Operations, there are a variety of already well established and defined authoritative task lists used by the Services to describe their particular military activities. Task lists such as the Universal Joint Task List (UJTL), the Army Universal Task List (AUTL), the Air Force Task List (AFTL) and the Naval Tactical Task List (NTTL) have become the cornerstone and building blocks of training and readiness programs. Additional task lists such as Mission Training Plans (MTPs) and Individual soldier task lists help to further define tasks at lower echelons. These task lists have been well vetted and describe the myriad tasks that the various Services are required to perform in order to accomplish their missions. They are the authoritative sources from which to construct training and operations plans (OPLANs) and build critical mission essential task lists (METLs) used as the basis of mission and readiness assessment. With the use of these task lists, military planners plan military operations from the Strategic National level down to the Tactical/soldier level. Level 4 Tasks form the stocking part of the MMF. Tasks, when combined and grouped collectively form the basis of Level 4 Operations, annotating the assembly or package perspective of the MMF.

Over the years, the Services have become very good at describing what they do using authoritative task language. Authoritative task lists provide the semantic descriptions of military operations. They supply detailed task descriptions, definitions and references. They include MOPs and MOEs with descriptive and quantifiable metrics that serve as the *benchmark for task performance*. Task MOPs and MOEs can be modified based on the mission and situation and change dynamically based on the time and place where and when the task is performed on the battlefield. Using authoritative task lists, the Services can discreetly and explicitly describe military operations down to the lowest level of individual performance.

Like Tasks and Operations, Level 2 Components and Forces are equally well understood. Level 2 Components, equipment and personnel, are well documented. Numerous references are available that list and specify numbers and types of military equipment. Detailed compilations of equipment specifications can be found in publications such as the Army Green Book, Jane's, RAND, Defense Weekly or other government database sources such as the Global Status of Resources and Training System (GSORTS), which contains personnel, equipment and training data on every DOD unit both Active and Reserve.

The Services employ thousands of pieces of equipment providing the necessary elements of air, sea and land power to fight and win war, and each piece of equipment performs to pre-determined specifications that are measurable and quantifiable. The Services also track and document the life cycle details of each piece of equipment in their inventory, noting every change or improvement made to the end item so that they know at any given time the exact specifications of a piece of equipment in the military inventory. Personnel are responsible for the command, control and operation of military equipment. They are responsible to execute the tasks within their respective Services. Each Service, to accomplish its military mission, requires the unique skill sets of Soldiers, Sailors, Airmen and Marines, enlisted or commissioned. Service training and educational institutions are built on the premise of training and educating men and women in the particular skill sets of their Service. For instance, within the Army, branches such as Armor, Infantry, Field Artillery, Quartermaster, Ordnance, Signal, etc., are responsible for training and educating soldiers on the skills necessary to accomplish the specific tasks within their Branch. The same holds true of the other Services. Naval, Marine Corps and Air Force personnel each have their own distinctively designed education and training system.

Equipment and Personnel, Level 2 Components within the MMF, form the individual *parts* of our military Forces. Combinations of equipment and personnel can be aggregated and described as Forces. When appropriate numbers and types of personnel and equipment are assembled into groups that meet specific military requirements, they become known as a Force package. Within the Army, for instance, documented force structures known as Tables of Organization and Equipment (TOEs) or Modified Tables of Equipment (MTOEs) detail the personnel and equipment elements of the assembled force and make up the basis of Corps, Divisions, and Modular Brigades. OPLANS contain detailed descriptions of the types and numbers of forces necessary to execute OPLANS.

Articulating and describing Level 4 Tasks/Operations and Level 2 Components/ Forces is not difficult. As has been discussed, Tasks/Operations and Components/ Forces are implicitly and explicitly understood and documented. Articulating Level 3 Functions and Capabilities is a different matter. Level 3 Functions and Capabilities play a key role in systemically linking required to available capabilities. Within the MMF construct, Level 3 Functions and Capabilities serve as the link between Level 4 Tasks/Operations and Level 2 Components/Forces.

> Level 3 Functions describe the specific *actions* required to execute tasks and acts as the conduit or interface between the two levels.

The relationships between Level 4 Tasks/Operations and Level 3 Functions/ Capabilities are made possible through the O3,4 Operator within the MMF. The relationships between Level 3 Functions/Capabilities and Level 2 Components/ Forces are made possible through the O2,3 Operator within the MMF. However, linking Level 4 Tasks/Operations to Level 2 Components/Forces is difficult due to a lack of understanding about the role Level 3 Functions and Capabilities play within the MMF construct.

Level 3 Functions and capabilities are probably the least understood of all the elements within the MMF. Yet, they provide the key link between two critical elements within the framework: Tasks/Operations and the Components (Equipment and Personnel)/Forces that conduct those operations. Level 3 Functions describe the specific *actions* required to execute tasks and act as the *conduit or interface* between Level 4 and Level 2. Level 3, in essence, is the Golden Spike linking Tasks/Operations to the appropriate Components/Forces best suited to accomplish them.

The Level 3 Function entity (part) within the MMF specifies the internal necessities required to deliver capabilities. It represents the actions or functions required to execute tasks. Examples of functions might include speed, navigate, transmit, shoot, hear, see, etc. The Capability entity (package) within the MMF reflects the collective grouping of actions or functions required to execute tasks or operations. Complementary functions grouped accordingly form the basis of a capability which must be present to conduct successful operations. Examples of capabilities might include move, sense, communicate, destroy, etc.

Connecting Tasks to Functions and Capabilities

Determining the "how well" specifications of Level 3 Functions and Capabilities is not an intuitive process. Because Level 4 Tasks and Level 2 Component data elements are better understood and defined, solutions to military problems often focus on satisfying task requirements by determining a force package solution without consideration of the impartial "do what/how well" functions required for task satisfaction and mission accomplishment. This isn't surprising. Planning military operations involves a great deal of implicit knowledge and understanding of military operations. Experienced military planners understand the identification of critical tasks, development of distinguishable and feasible Courses of Action, wargaming and task sequencing, task organization and force package development. This ability to conduct efficient and effective military planning evolves through years of experience and practice, but while experience can lead to better force package solutions, it is still based on an intuitive and implicit understanding of military tasks and operations. For instance, we *know* how an M1A1 tank will perform in open terrain, but what if it was employed in an urban environment. Inherently, a solution is determined based on what a component/force package historically has been able to do in the past without real consideration given to *how much of what is required now or in the future*. We instinctively move straight from the task to an equipping or manning solution because of an implicit understanding of the tasks *needs* and the components *abilities*. There is little thought given to *what functions are actually* required and the *amount of function* necessary to accomplish the task. This habitual, implicit and unconditional linkage between Level 4 Task data elements (MOPs/MOEs) and Level 2 Component specifications may not be good enough to distinguish between the best solution and just a good one. There is no *explicit* demonstration of linkage between task MOPs and component specification data to a level of detail needed to illustrate the discriminating differences between solution sets. There is no consideration given to what is actually required, *the actions or functions* to satisfy task MOPs.

Consider the variables in executing military operations and how every conflict is different from the last. How do we know that the force package chosen to fight the last conflict is the best one to satisfy the next operation in a different environment? What if no equipping or manning solution existed to satisfy execution of a task or military operation? Maybe the military operation is one never attempted before or being attempting in a foreign environment that could change the affects or outcome of task execution? To choose the best solution, the right tasks have to be selected and then broken down into quantifiable and measurable elements, MOPs/MOEs. Furthermore, Level 3 *functions and actions* necessary to execute each task MOP must be considered in order to determine and closely examine what actions have to take place to satisfy the task. *Level 3 Functions help to associate Level 4 Task (MOPs) with an appropriate Level 2 Component specification and describe what can be measured and depicted explicitly between the 3 Levels.*

Level 3 Functions inhabit the space between Level 4 Task MOPs/MOEs and Level 2 Component specifications. This is the space where task requirements and equipping solutions meet. Experienced military planners understand this space and consider, even if inadvertently, the functions needed for task satisfaction. They understand "how much" of a function is needed for task accomplishment and are able to analyze military problems by quantifying and *proving* whether an equipping/manning solution will satisfy a task requirement. Level 3 Functions/Capabilities within the MMF construct identify the common denominator between tasks and components. By reducing the level elements to common, quantifiable properties, expressed using a common unit of measure the three level elements can be compared to one another to determine whether the Level 2 Component/Force will provide the amount of Level 3 Function/Capability necessary to satisfy the Level 4 Task/Operations requirement. The common denominator identifies the *shared functions* and their accompanying, quantifiable *units of measure* between the three MMF levels making it possible to compare task MOPs with component specifications. *The importance of Level 3 is to identify the Functions required of Level 4 Tasks, determine the unit of measure and how much function the task requires and then compare that to the amount of function the Level 2 Component has available.* Units of measure, as the Level 3 mathematical common denominators, systematically link Level 4 Tasks to Level 2 Components.

Figure 5 provides an example of this concept. The Level 4 task is to Conduct Target Acquisition and the MOPs identify a need to conduct this acquisition task outside of the range of enemy weapon systems. The MOPs indicate that the target must be acquired without the acquisition component being detected and at least 50 kilometers away from friendly forces; the grid location of the target must be determined to an accuracy of 2.5 meters; the target must be 100 percent positively identified; and the information acquired must be transmitted in as near real time as possible so that a decision can be made on how best to service the target. The Level 3 Capabilities required to execute this task would include a capability to acquire and

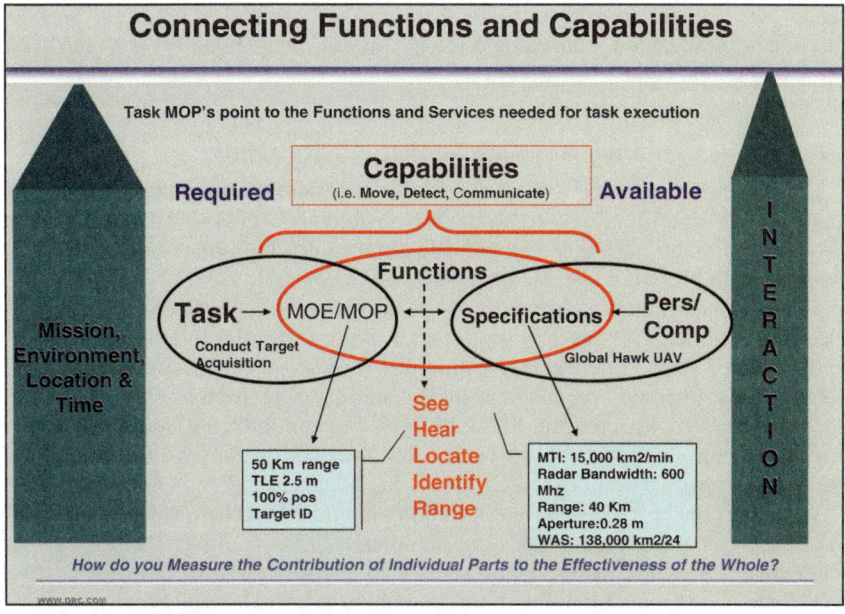

Fig. 5 Connecting functions and capabilities.

detect the target and communicate information. The Functions needed to execute this task and achieve its performance measures include the ability to *see* the target from a distance (magnification factor, kilometers, MPH), *locate* the target (grid location, GPS), *record* data on the target (radar bandwidth in MHz, aperture, shutter speed, RAM, loiter time), *transmit* data to another location (BoD rate, bps, kilometers/miles to receiver), *identify* the target as friendly or enemy (MTI) and *range*, distance to the target (km, miles). These particular functions might identify a number of components that satisfy task execution for any of the required functions. Components such as UAVs, Special Reconnaissance Forces, aerial or ground Forward Observers (FOs), etc., might provide the needed functionality, to satisfy the task. What is required is a component that provides *all* the functionality, however. The example in Fig. 5 indicates that a Global Hawk UAV was the component selected to execute this task, but the UAV does not have the range to be able to get to the target location so would be ineffective in executing the task.

This is just one example of how the identification of Level 3 Functions helps focus analysis efforts and make the comparison of required to available functions and capabilities more efficient and exact. Identifying functions helps to narrow and define the common quantifiable evaluation criteria with their associated UoMs required for task execution. Determining these critical functions provides a way to quantitatively and objectively compare required to available amounts of functionality so components that are unable to satisfy task execution can be eliminated from consideration and better choices can be made on those that best suit the tasks needs. This type of analysis provides realistic and rigorous proof for evaluating net-centric warfare concepts for potential capability gap solutions.

It provides decision makers a way to unambiguously articulate required capabilities that are traceable to current and future missions which can serve as the basis for resource and acquisition decisions.

Understanding Nesting and Scalability

Understanding nesting and scalability are important in conducting MMF analysis because they help to frame the problem, highlight the relationships between parts and packages and provide a traceable chain of events that can be studied. Scalability is a critical attribute of the MMF because it reveals the depth and breadth at which effects can be traced. Understanding nested constructs and the scale at which they reach is important in linking Level 1 Interactions/Effects to the Level 7 Purpose/Mission. Nesting, the repeated combination of individual parts into complementary packages and scalability, how high or low the nesting construct will go with respect to the analysis being conducted, must be considered when conducting MMF analysis so that the relationships within and among the nested constructs and the MMF levels can be described and analyzed.

When the concept of nesting and scalability is applied to other levels of the MMF, the context of the problem grows in complexity. Like 3-dimensional chess the relationships and interactions within and among the various MMF levels are extensive. Parts as individual building blocks combine in complementary ways to form packages. At a higher echelon in a nested construct, packages can be viewed as individual parts and again combined with other complementary parts to form another higher level package. Take one example of a nesting construct using Level 2 Components/Forces. An individual soldier is a single Component part of an Army Force at the lowest level in a Level 2 nested construct. Nine soldiers as individual Component parts with like characteristics, Military Occupational Specialty (MOS) or rank, combine to form a Force package, a Squad. At a higher echelon within this nesting hierarchy, the Squad can be considered as an individual Component part and when combined with other like Squads forms to become another Force package of complementary parts, a Platoon package. Continuing on the Platoon can also be viewed as a part and when combined with other like or complementary Platoon parts forms a Company package. Companies, as individual parts combine to form an assembled Battalion package and so on. At every echelon there are parts being combined and formed into packages. This nesting example could continue up through the highest levels within the Army or a Joint Task Force and illustrates the reach and scale at which the nesting could occur.

Applying this nesting principle to other levels within the MMF reveals the scope an analysis effort could take and the multitude of relationships to be expected within and among the levels. Content in one MMF level affects content in another level and the scope of analysis becomes more complex in comparison with the degree of scale within each level. In order to conduct analysis to this degree of detail it is imperative to be able to explicitly describe the relationships of the parts and packages at each level. Level 1 Interactions involving Level 2 Components within a nested construct may affect the actions of other Level 2 Components at higher or lower echelons. These interactions may influence the outcome of Level 3 Functions at the various nested echelons which in turn influence the performance of Level 4 Tasks. From an aggregated point of view the Effects of Level 1

Interactions influence Level 2 Forces who provide Level 3 Capabilities to execute Level 4 Operations. The conduct of Level 4 Operations ultimately affects whether the Level 7 Purpose/Mission is achieved. Analyzing mission content thus requires breaking the problem apart bit by bit during the planning process and then tracing the effects of an interaction up through the nested constructs of each level to determine the result on the overall Mission. The scale of the analysis effort could be far reaching and yield countless permutations of interactions and results.

Scalability of Functions and Capabilities

Nesting and scalability are fairly easy to describe when discussing Level 2 Components and Forces or Level 4 Tasks and Operations. The example just provided how Level 2 Components and Forces can be scaled and nested using personnel and force structure concepts. Individual soldiers combine to form variations of force structures that grow in complexity and capability. Level 4 Tasks and Operations can be described similarly. Level 4 individual Tasks combine to form packages of collective tasks and collective tasks can also be grouped to describe an Operations package. Tasks are derived from the examination of the higher headquarters mission and are analyzed by examining the higher headquarters mission two levels up. All the individual task parts group and combine to form collective tasks and operations packages which contribute to the higher level operation or mission.

Describing the scalability and nesting of Level 3 Functions and Capabilities is similar. Since Level 3 Functions and Capabilities link to Level 4 Tasks and Operations and Level 2 Components and Forces, it is important to describe Level 3 Functions and Capabilities in sufficient detail in order to determine the relationships between the level parts and packages and find the common denominator between the three levels.

In Fig. 6, Capabilities are broken down into individual *Function parts* so that the Capabilities' *functions* can be described more acutely. From the example below Fire Support as a higher level Capability is broken down and described by analyzing its parts, what makes up the *capability* of Fire Support. Field Manual 3-90, Doctrine for Fire Support, describes what individual parts contribute to providing a Fire Support capability. Based on doctrine, the parts or functions that combine to form a Fire Support capability include Target Acquisition, Command and Control and Attack Resources. Fire Support is the Capability package where Target Acquisition, Command and Control and Attack Resources are the individual Function parts. Now apply the concept of nesting, and any one of these individual parts can be described as a package. Target Acquisition, by the example, becomes a Target Acquisition Capability package, and it too can be broken down further into individual *function* parts that describe what makes up the Target Acquisition capability. These functions include Detect, Locate, Identify, Track, and Classify.

Taking it a step further, each of these individual functions can be considered a *capability* and broken down further to describe each in more detail. The capability to Detect could consist of the individual functions of *seeing, hearing, tasting, smelling or feeling*. The capability to locate might require the functions of *positioning, triangulation, distance, and timing*. But why go to this level of detail and scrutiny in describing Level 3 Functions and Capabilities? Because it is here, at the lowest point of dissection, that the common denominators are revealed and to

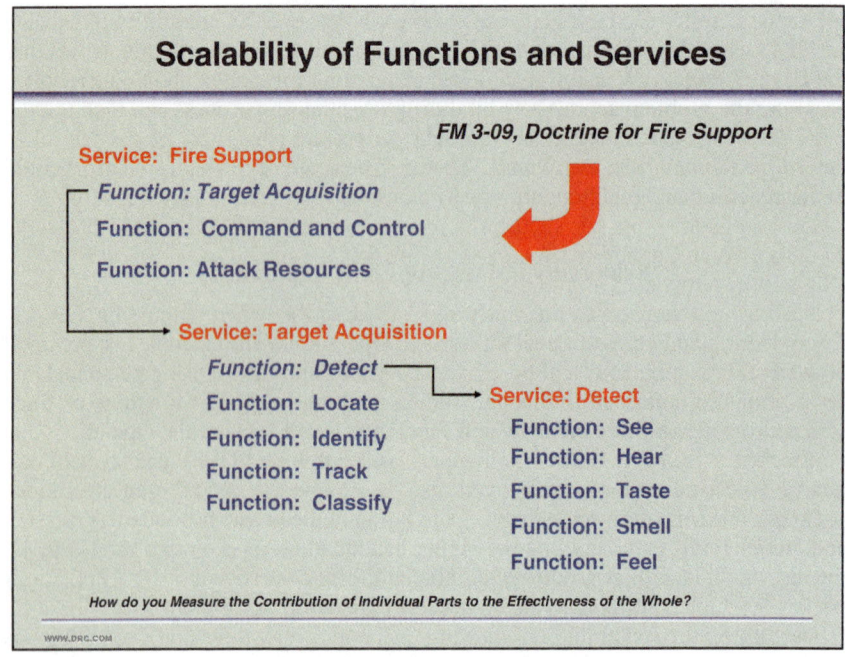

Fig. 6 Scalability of functions and capabilities.

which links Level 4 Task (MOPs) to Level 3 Functions (UOMs) and Level 2 Components (Specifications).

The Need for a Functions and Capabilities Database

As discussed throughout this paper, Functions and Capabilities are difficult to pinpoint or describe because they are uncommon to the normal lexicon of military language and discussion. They occupy a space beyond what is customarily discussed in military planning. Yet, if Functions and Capabilities in their pure state were better understood and considered more frequently in military planning, the quality of analysis efforts would improve greatly. What is missing is a reference library or database of Capabilities and Functions that can be readily accessed for use in analysis efforts.

MMF analysis requires access to a variety of sources that provide the necessary data content for the various levels within the MMF. For instance, the CJCSM 3500.04D, UJTL, Enclosure C provides a substantial reference of military, physical and civil Conditions data that satisfy Level 6 Context and Environment content. References such as the UJTL, the Service Task Lists or Mission Training Plans (MTPs) provide excellent sources of task information for Level 4 Task content. There are also databases and information sources that provide Level 2 Component and Forces data content such as GSORTS. What does not exist though is a comprehensive database of definitive and quantifiable Functions and Capabilities that can be readily accessed to provide MMF Level 3 data content. Such a database would provide access to reference materiel for Level 3 data content necessary to compare Level 4 Task data elements to those of Level 2 Components specifications. A comprehensive Functions and Capabilities resource library would fill the void that exists when trying to bridge the gap between these two critical MMF levels.

A Functions and Capabilities reference would require a structured taxonomy, classifying Capabilities and Functions accordingly. It should include descriptions of capabilities and functions in a nested construct with their associated units of measure. A start point for such a reference has already begun within the DOD. The Joint Force Capabilities Assessment Sub Study under Operational Availability (OA)-05 developed 21 Joint Capability Areas (JCAs) representing the beginning of a common language to discuss and describe capabilities across every related [DOD] Departments activities and processes. The JCAs represent an evolving capabilities lexicon. Tier 1 of the JCAs represents the top level of the capabilities lexicon and includes definitions found in Joint Publication 1-02. Tier 2 refines the lexicon to provide more detail to enhance the JCA's usefulness. These JCAs will be integrated into future joint doctrine, the UJTL and JCIDS where appropriate. See Fig. 7 for an example taxonomy.

While the Joint community is well on its way to developing a common language and understanding of Joint capabilities via establishment of the JCAs it is possible that these JCAs will require further drill down to provide the level of detail useful

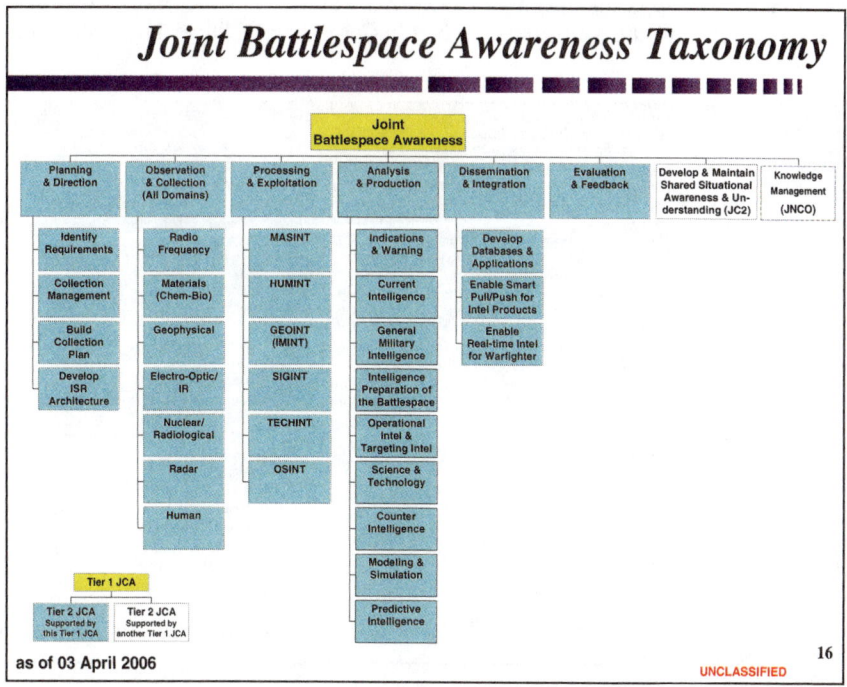

Fig. 7 Excerpt from CJCS JCA progress report.

for MMF analysis efforts. The JCAs, in their present form, refer to a very high level of capability and do not provide the degree of detail necessary for more careful and systematic analysis using the MMF.

What would be applicable to future MMF analysis efforts is access to a more comprehensive Functions and Capabilities reference library that captures the multitude of functions that describe military actions as well as the functions that describe what components are able to deliver as defined by component specification data. Such a database should include functional descriptions with associated quantitative units of measure that can be easily related to both task MOPs and component specification data.

Figure 8 illustrates a sample version of a notional Functions and Capabilities database. Beginning with the Tier 1 JCAs as the highest level of capability, each JCA was broken down further to identify underlying individual functions. In some cases these functions were decomposed further until the lowest level of functionality was identified.

At its lowest point of functionality, each function was then associated with a function scale and a variety of accompanying units of measure.

Figure 8 serves as a starting point for development of a reference library of Functions and Capabilities using the JCAs as an entry point. While this type of reference library can be developed to describe semantically, functions and capabilities the second piece of the database must contain associated units of measure. Units of measure are the universal keys that link task MOPs to equipment specification data. Dictionaries of units of measurement, such as the one available by Russ Rowlett and the University of North Carolina at Chapel Hill, document

Capability	Lower Level Capability	Function	Function_Scale
Battlespace Awareness	Plan and direct resources	Develip Intel requirements	# PIRs linked to CCIRs
Battlespace Awareness	Plan and direct resources	Develip Intel requirements	Time to answer PIR (Mins)
Battlespace Awareness	Plan and direct resources	Coordinate Intel assets	Time to engage assets
Battlespace Awareness	Plan and direct resources	Coordinate Intel assets	Error rate on orders to subordinates
Battlespace Awareness	Plan and direct resources	Coordinate Intel assets	Fraction of assets collection
Battlespace Awareness	Plan and direct resources	Position Intel assets	# NAI's covered
Battlespace Awareness	Plan and direct resources	Position Intel assets	% of targets identified
Battlespace Awareness	Plan and direct resources	Manage collection assets by type	SIGINT
Battlespace Awareness	Plan and direct resources	Manage collection assets by type	HUMINT
Battlespace Awareness	Plan and direct resources	Manage collection assets by type	IMINT
Battlespace Awareness	Plan and direct resources	Manage collection assets by type	ELINT
Battlespace Awareness	Plan and direct resources	Manage collection assets by type	Time to redirect assets
Battlespace Awareness	Plan and direct resources	Build Collection Plan	Time to approve
Battlespace Awareness	Plan and direct resources	Develop Intel architecture	Level of Classification
Battlespace Awareness	Plan and direct resources	Develop Intel architecture	# alternative mean of communication
Battlespace Awareness	Detect	See	Passive Electro Optic Sensor
Battlespace Awareness	Detect	See	Visual acuity 20/20 (Magnification)
Battlespace Awareness	Detect	See	Infrared
Battlespace Awareness	Detect	See	SAR/GMTI
Battlespace Awareness	Detect	See	STW
Battlespace Awareness	Detect	See	Distance meters
Battlespace Awareness	Detect	See	Distance kilometers
Battlespace Awareness	Detect	Hear	Decibels
Battlespace Awareness	Detect	Taste	Sweet
Battlespace Awareness	Detect	Taste	Sour
Battlespace Awareness	Detect	Taste	Salty
Battlespace Awareness	Detect	Taste	Bitter
Battlespace Awareness	Detect	Smell	Olfactory - decipols
Battlespace Awareness	Detect	Feel	Pressure - Psi

Fig. 8 Sample functions and capabilities data set.

measurements for things such as length, area, speed, torque, volume, mass, temperature, density, energy or work, force, power, fuel consumption, etc. These units of measure are relevant when measuring task requirements for military operations and equipment performance specifications. Coupling a Functions and Capabilities database with Unit of Measurement database could be a very powerful tool and data source for MMF Level 3 content.

Conclusion

The MMF clearly provides the rigor to evaluate transformation requirements and the utility of future DOTMLPF solutions. Disciplined, repeatable and scalable analysis of future DOLTMLPF solutions is within reach and rests on the application of MMF principles coupled with a clear and true delineation of the role Functions and Capabilities play in the MMF process. Future force transformation efforts would be significant if required and available capabilities could be examined in detail within the context of task execution and component availability and specification. Planning, resourcing and fielding decisions could be made with more confidence based on analysis that scientifically traces the linkage between equipment and forces and the military tasks they are required to execute. Establishment of a Capabilities and Functions reference library and database is imperative in providing this linkage. Understanding the role Functions and Capabilities play in future analysis efforts is recommended and undoubtedly required for successful quantitative assessment of alternative solutions. More work is needed toward the establishment of a useable Functions and Capabilities database, but there is no doubt that the development of such a data source could have an extraordinary effect on future force transformation analysis efforts.

References

The Army Green Book, 2006.
Cebrowski, A. (2004). Transforming Defense, Fear of Flying... Into The Future.
CJCSI 3500.01, (2006). Joint Training Policy of the Armed Forces.
CJCSM 3170, (2005). Operation of the Joint Capabilities Integration and Development System (JCIDS).
CJCS JCA Progress Report, (2006).
FM 101-5, (1997). Staff Organization and Operations.
FM 3-0, (2001). Operations.
FM 3-90, (2002). Doctrine for Fire Support.
Joint Publication 1-02, (2001). Department of Defense Dictionary of Military and Associated Terms.
The Joint Staff Officers Guide, (2000).
Haddox, F. (2003). The Functional Description of the Mission Space (FDMS) Data Interchange Format (DIF) version 2.0 Defense Modeling and Simulation.
Kearley, (2006). Joint Mission Decomposition Framework, A Missions and Means to Train Framework (MMTF).
Montague Institute, Managing Taxonomies Strategically, (2001).

Minchew, M. L. (Nov. 2006), "The Connection Between Functions and Capabilities," submitted for the Barchi Prize at the *74th Military Operations Research Society Symposium*.

Minchew, M. L., and Bray. B. E. (unpublished), "The MMF Formal Process," Dynamics Research Corporation.

Rowlett, Russ, (2000). A Dictionary of Units of Measure.

Sheehan, J. H., Dietz, P. H., Bray, B. E., Harris, B. A., and Wong, A. (1–4 Dec. 2003), "The Military Missions and Means Framework," *Proceedings of the Interservice/Industry Training, Simulation, and Education Conference,* IITSEC, Orlando, FL.

Appendix F

Acronyms and Abbreviations

ABC	Activity-Based Costing
ACAT	Acquisition Category
AFATL	Air Force Armament Test Laboratory
AFSC	Air Force specialty code
AFV	armored fighting vehicle
AIS	Abbreviated Injury Scale
	Abbreviated Injury Severity
AJEM	Advanced Joint Effectiveness Model
A-Kill	Catastrophic kill within 5 min (ground vehicles); attrition kill within 5 min (aircraft)
AMSAA	Army Materiel Systems Analysis Activity
AP	Anatomic Profile
AP	armor-piercing
APC	armor-piercing capped
APDS	armor-piercing discarding-sabot
APFSDS	armor-piercing, fin-stabilized, discarding sabot
APFSDS-T	armor-piercing, fin-stabilized, discarding sabot – tracer
APG	Aberdeen Proving Ground
API	armor-piercing incendiary
ARA	Applied Research Associates, Inc.
ARL	Army Research Laboratory
ATB	articulated total body
ATEC	Army Test and Evaluation Command
ATGM	anti-tank guided missile
AURA	Army Unit Resiliency Analysis
AVVAM	Armored Vehicle Vulnerability Analysis Model
BAD	behind-armor debris
BAE	behind-armor effects
BDAR	battle damage assessment and repair
BDR	battle damage repair
BFV	Bradley Fighting Vehicle
B-Kill	Catastrophic kill within 20 min (ground vehicles); attrition kill within 30 min (aircraft)
BLAST	Battlefield Laser Acquisition Sensor Test

BLOS	beyond line-of-sight
BRASS	Ballistic Research Laboratory Anti-armor Sensor Simulation
BRL	Ballistic Research Laboratory
BRL-CAD	Ballistic Research Laboratory – computer-aided design
BRLESC	BRL Electronic Scientific Computer
C4I	command, control, communications, computers, and intelligence
CAD	computer-aided design
CAIV	Cost As an Independent Variable
CAM	computer-aided manufacturing
CAMAD	computer-aided modeling, analysis, and design
CARDE	Canadian Armament Research and Development Establishment
CARS	Chemical Response Simulation
CASE	computer-aided software engineering
CASTFOREM	Combined Arms and Support Task Force Evaluation Model
CCWG	Crew Casualty Working Group
CD	charge diameter
CE	chemical energy
C-Kill	Attrition kill within 40 min (ground vehicles)
CJ	Chapman–Jouguet
COMPKIL	Component Kill
COVART	Computation of Vulnerable Areas and Repair Times
CP	cumulative probability
CSG	constructive solid geometry
CVC	combat vehicle crewman
DAL	damage assessment list
DCU	degradation of combat utility
DDT	deflagration-to-detonation transition
DE	directed energy
DIS	distributed interactive simulation
DMSO	Defense Modeling and Simulation Office
DOD	Department of Defense
DODI	Department of Defense Instruction
DPICM	dual-purpose improved conventional munition
DRI	Denver Research Institute
DS	degraded state
DSVM	Degraded-States Vulnerability Methodology
DU	depleted uranium
ECV	elemental capability vector
EDVAC	Electronic Discrete Variable Computer
EFP	explosively formed penetrators
EKE	Expected Kinetic Energy
E-Kill	Attrition kill due to additional damage upon landing (aircraft)
EM	evaluation model
ENIAC	Electronic Numerical Integrator and Computer
EOD	explosive ordnance disposal
EVA-3D	Effectiveness/Vulnerability Assessment in 3-D
FATE	Fast Air Target Encounter

FATEPEN	Fast Air Target Encounter Penetration
FDMS	Functional Descriptions of the Mission Space
F-Kill	Firepower kill (ground vehicles)
F-LoF	firepower loss of function
FMEA	Failure Modes and Effects Analysis
FMECA	Failure Mode, Effects, and Criticality Analysis
FPIIM	Fragment Penetration and Impact Initiation Model
FUSL	full-up, system-level
GPS	gunner's primary sight
GUI	graphical user interface
HE	high-explosive
HEAT	high-explosive anti-tank
HEI-T	high-explosive incendiary-tracer
HEP	high-explosive plastic
HVAC	heating ventilation and air conditioning
HVAP	high-velocity armor-piercing
HVAPDS	high-velocity, armor-piercing, discarding sabot
ICE	Interactive Criticality Estimator
ICM	Improved conventional munition
IDA	Institute for Defense Analysis
I-Kill	Interdiction kill (ground vehicles)
ILS	integrated logistics system
IM	interaction module
ISS	Injury Severity Score
JASP	Joint Aircraft Survivability Program
JASPO	Joint Aircraft Survivability Program Office
JLF	Joint Live Fire
JMEM	Joint Munitions Effectiveness Manual
J–R	Jacobs–Roslund
JTCG	Joint Technical Coordinating Group
JTCG/AS	Joint Technical Coordinating Group on Aircraft Survivability
JTCG/ME	Joint Technical Coordinating Group for Munitions Effectiveness
KE	kinetic energy
K-Kill	Catastrophic kill (ground vehicles); attrition kill within 30 s (aircraft)
KK-Kill	Immediate attrition kill (aircraft)
LF	live fire (or Live Fire)
LFT	Live Fire (or live fire) test (or testing)
LFT&E	live fire test and evaluation
LoF	loss of function
LOS	line-of-sight
LVD	low-velocity detonation
MAGI	Mathematical Applications Group, Inc.
MANPRINT	manpower and personnel integration
MAVEN	Modular Aircraft Vulnerability Estimation Network
MDMP	military decision-making process

MEVA	Modular Effectiveness Vulnerability Assessment
M/F-LoF	mobility/firepower loss of function
MILES	Multiple Integrated Laser Equipment System
M-Kill	Mobility kill (ground vehicles)
M-LoF	mobility loss of function
MMF	Missions and Means Framework
MOE	measures of effectiveness
MOP	measures of performance
MOPP	mission-oriented protective posture
MOS	military occupational specialties
MRMC	Medical Research and Materiel Command
M&S	modeling and simulation
MTM	model-test-model
MTOS	Major Trauma Outcome Study
MUVES	Modular UNIX-based Vulnerability Estimation Suite
NATO	North Atlantic Treaty Organization
NBC	nuclear, biological, and chemical
NEC	Navy Enlisted Code
NISS	New Injury Severity Score
NLOS	non-line-of-sight
NOL	Naval Ordnance Laboratory
NRL	Naval Research Laboratory
NSWCDD	Naval Surface Warfare Center Dahlgren Division
OCO	Office of the Chief of Ordnance
OPFOR	opposing forces
ORCA	Operational Requirements-based Casualty Assessment
ORDVAC	Ordnance Variable Automatic Computer
OSD	Office of the Secretary of Defense
OWNFOR	friendly forces
$P_{CD|H}$	probability of component dysfunction given a hit
P_{DET}	probability of detection
$P_{D|H}$	probability of damage given a hit
P_H	probability of hit
$P_{H|DET}$	probability of hit given detection
$P_{I|H}$	probability of incapacitation given a hit
P_K	probability of kill
$P_{K|H}$	probability of kill given a hit
P_{SURV}	probability of survival
PATRIOT	Phased-Array Tracking, Ranging, and Intercept of Target
PHD	profile hole diameter
RAM	reliability, availability, and maintainability
RCC	right circular cylinder
RCC	Residual Combat Capability
R&D	research and development
RDEC	Research, Development, and Engineering Center
RHA	rolled homogeneous armor
RISS	Revised Injury Severity Score
RPG	Recommended Practices Guide

FATEPEN	Fast Air Target Encounter Penetration
FDMS	Functional Descriptions of the Mission Space
F-Kill	Firepower kill (ground vehicles)
F-LoF	firepower loss of function
FMEA	Failure Modes and Effects Analysis
FMECA	Failure Mode, Effects, and Criticality Analysis
FPIIM	Fragment Penetration and Impact Initiation Model
FUSL	full-up, system-level
GPS	gunner's primary sight
GUI	graphical user interface
HE	high-explosive
HEAT	high-explosive anti-tank
HEI-T	high-explosive incendiary-tracer
HEP	high-explosive plastic
HVAC	heating ventilation and air conditioning
HVAP	high-velocity armor-piercing
HVAPDS	high-velocity, armor-piercing, discarding sabot
ICE	Interactive Criticality Estimator
ICM	Improved conventional munition
IDA	Institute for Defense Analysis
I-Kill	Interdiction kill (ground vehicles)
ILS	integrated logistics system
IM	interaction module
ISS	Injury Severity Score
JASP	Joint Aircraft Survivability Program
JASPO	Joint Aircraft Survivability Program Office
JLF	Joint Live Fire
JMEM	Joint Munitions Effectiveness Manual
J–R	Jacobs–Roslund
JTCG	Joint Technical Coordinating Group
JTCG/AS	Joint Technical Coordinating Group on Aircraft Survivability
JTCG/ME	Joint Technical Coordinating Group for Munitions Effectiveness
KE	kinetic energy
K-Kill	Catastrophic kill (ground vehicles); attrition kill within 30 s (aircraft)
KK-Kill	Immediate attrition kill (aircraft)
LF	live fire (or Live Fire)
LFT	Live Fire (or live fire) test (or testing)
LFT&E	live fire test and evaluation
LoF	loss of function
LOS	line-of-sight
LVD	low-velocity detonation
MAGI	Mathematical Applications Group, Inc.
MANPRINT	manpower and personnel integration
MAVEN	Modular Aircraft Vulnerability Estimation Network
MDMP	military decision-making process

MEVA	Modular Effectiveness Vulnerability Assessment
M/F-LoF	mobility/firepower loss of function
MILES	Multiple Integrated Laser Equipment System
M-Kill	Mobility kill (ground vehicles)
M-LoF	mobility loss of function
MMF	Missions and Means Framework
MOE	measures of effectiveness
MOP	measures of performance
MOPP	mission-oriented protective posture
MOS	military occupational specialties
MRMC	Medical Research and Materiel Command
M&S	modeling and simulation
MTM	model-test-model
MTOS	Major Trauma Outcome Study
MUVES	Modular UNIX-based Vulnerability Estimation Suite
NATO	North Atlantic Treaty Organization
NBC	nuclear, biological, and chemical
NEC	Navy Enlisted Code
NISS	New Injury Severity Score
NLOS	non-line-of-sight
NOL	Naval Ordnance Laboratory
NRL	Naval Research Laboratory
NSWCDD	Naval Surface Warfare Center Dahlgren Division
OCO	Office of the Chief of Ordnance
OPFOR	opposing forces
ORCA	Operational Requirements-based Casualty Assessment
ORDVAC	Ordnance Variable Automatic Computer
OSD	Office of the Secretary of Defense
OWNFOR	friendly forces
$P_{CD|H}$	probability of component dysfunction given a hit
P_{DET}	probability of detection
$P_{D|H}$	probability of damage given a hit
P_H	probability of hit
$P_{H|DET}$	probability of hit given detection
$P_{I|H}$	probability of incapacitation given a hit
P_K	probability of kill
$P_{K|H}$	probability of kill given a hit
P_{SURV}	probability of survival
PATRIOT	Phased-Array Tracking, Ranging, and Intercept of Target
PHD	profile hole diameter
RAM	reliability, availability, and maintainability
RCC	right circular cylinder
RCC	Residual Combat Capability
R&D	research and development
RDEC	Research, Development, and Engineering Center
RHA	rolled homogeneous armor
RISS	Revised Injury Severity Score
RPG	Recommended Practices Guide

SAFE	Stochastic Analysis of Fragment Effects
SAMSMAE	Surface-to-Air Missile Site Mean Area of Effectiveness
SC	shaped charge
SCJ	shaped-charge jet
SDAL	Standard Damage Assessment List
SDT	shock-to-detonation transition
SFF	self-forging fragment
SME	subject-matter expert
SoS	systems of systems
SPARC	Sustainability Prediction for Army Requirements for Combat
SPH	smooth particle hydrodynamics
SPRAE	Stochastic Processor for Artillery Effectiveness
SQuASH	Stochastic Quantitative Analysis of System Hierarchies
SRS	Shock Response Spectrum
SURVIAC	Survivability/Vulnerability Information Analysis Center
TEMP	Test and Evaluation Master Plan
THAAD	Theater High-Altitude Air Defense
TIS	thermal imaging sight
TOW	tube-launched, optically tracked, wire-guided
TRADOC	Training and Doctrine Command
TTP	tactics, techniques, and procedures
UAV	unmanned aerial vehicle
V/L	vulnerability/lethality
VAMP	Vulnerability Analysis Methodology Program
VAST	Vulnerability Analysis for Surface Targets
VR	vulnerability reduction
V&V	verification and validation
VV&A	verification, validation, and accreditation
WFS	warhead-function surface
WRAIR	Walter Reed Army Institute of Research
XDT	unknown detonation transition

Index

Abbreviated Injury Scale (AIS), 88
   score, 203
ABC. *See* Activity-Based Costing.
Aberdeen Proving Ground (APG), 5
Accreditation process, 208, 214–215
   application-specific, 215
   class, 215
Acronyms, 347–351
Activity-Based Costing (ABC), 225
   cost determinations, 225
Advanced joint effectiveness model
      (AJEM), 199
AFV. *See* armored fighting vehicle.
Aircraft damage categories
   A-kill, 75, 76
   B-kill, 75, 76
   C-kill, 76
   E-kill, 75
   K-kill, 75
   KK-kill, 75
Aircraft
   armored vehicles vs, re measures of
      vulnerability, 74–76
   criticality analysis and, 44–45
   fuel system failure, 47–48
   vulnerability testing, Joint Technical
      Coordinating Group for Munitions
      Effectiveness, 8
AIS. *See* Abbreviated Injury Scale.
AJEM. *See* advanced joint effectiveness model.
A-kill, 75, 76
Ammunition components, 50–51
   ballistic impact prediction on, 134–138
   booster explosives, 51
   depth charges, 137
   explosively loaded warheads, 50
   fuzes, 51
   gun propellants, 137
   liquid gun propellants, 51
   liquid-propellant rocket motors, 50–51
   primers, 51
   pyrotechnics, 51
   solid gun propellants, 51
   solid-propellant rocket motors, 50
Ammunition damage, 48–53
   initiation mechanisms, 51–53
   terminology, 49–50
Ammunition terminology, 49–50
   deflagration, 49
   detonation, 49
   energetic material, 49

explosion, 48–49
explosive, 49
explosiveness, 50
failure diameter, 49
high explosive, 49
initiation, 49
low-velocity detonation, 49
propellant, 49
pyrotechnic, 49
sensitivity, 49
shock, 49
AMSAA. *See* U.S. Army Material Systems
   Analysis Activity.
Anti-tank guided missile (ATGM), 11, 36
   damage mechanism, 37
      blast-induced shock, 37
      SCJ, 37
AP shot. *See* full-bore shot.
APC. *See* armor-piercing capped projectiles.
APDS. *See* armor-piercing discarding-sabot
   round.
APG. *See* Aberdeen Proving Ground.
API. *See* armor-piercing incendiary.
Applications, 217–240
   concept refinement, 218–219
   demonstration, 219–221
   doctrine, 240
   future trends, 237–239
   live fire testing, 221–226
   requirements documentation, 217–218
   system development, 219–221
   system life cycle, 226–232
      BDR example, 231–232
      RAM analyses, 228–231
      SPARC/BDR methodology, 226–228
   tactics, 239–240
      performance requirements,
         239–240
   technology development, 218–219
   vulnerability reduction, 232–239
Application-specific accreditation, 215
ARL. *See* U.S. Army Research Laboratory.
Armored fighting vehicle (AFV), 39
Armored vehicles, 67
   aircraft vs, re measures of
      vulnerability, 74–76
   early vulnerability studies, 168–170
   testing, 69, 72–73
   vulnerability, 6
      armor-piercing capped projectiles, 6
      armor-piercing projectiles, 6

353

INDEX

Armored vehicles (*Continued*)
  catastrophic kill, 6
  firepower kill, 6
  high-explosive anti-tank, 6
  high-explosive plastic projectiles, 6
  high-velocity armor-piercing projectiles, 6
  Joint Technical Coordinating Group for Munitions Effectiveness, 8
  mobility kill, 6
Armor-piercing (AP) projectiles, 6
Armor-piercing capped (APC)
  projectiles, 6
  design, 22–25
Armor-piercing discarding-sabot round (APDS), 22–25
Armor-piercing incendiary (API), 25
Army Unit Resiliency Analysis (AURA), 8
Army V/L requirements, 9
  acquisition, 10
  development, 10
  Live Fire Law, 10
  logistical burden, 10–11
  military occupational specialties, 10
  mission rehearsal, 11
  research, 10
  Research, Development and Engineering Centers, 9
  Training and Doctrine Command, 9
  training, 11
  U.S. Army Material Systems Analysis Activity, 9
  U.S. Army Research Laboratory, 9
  U.S. Army Test and Evaluation Command, 9
Artillery damage mechanism, 36
  blast shock, 37
  fragments, 37
ATEC. *See* U.S. Army Test and Evaluation Command.
ATGM. *See* anti-tank guided missile.
AURA. *See* Army Unit Resiliency Analysis.
AURA model, 206

BAD. *See* behind-armor debris.
Ballistic dose, 88
Ballistic impact prediction, ammunition components, 134–138
  depth charges, 137
  detonation, 134
  dual-purpose improved conventional munition, 137
  gun propellants, 137
Ballistic penetration, 20
  behind-armor debris, 21
  chemical energy munition, 20
  chemical energy, 20
  explosively formed penetrators, 20
  perforation, 20
  self-forging fragments, 20

Ballistic Research Laboratory (BRL), 4, 40, 139, 221
Ballistic shock, 20, 35–38
  analysis of, 36
  damage, 36–38
    artillery, 36
    ATGM, 36
    gun-fired anti-armor projectiles, 36
    mines, 36
    thin-skinned vehicles, 38
  effects of, 138–139
    loading function, 138
    sensitive components, 138
  low-frequency, 36
Ballistic threats, 1
  high explosive warheads, 1
  kinetic energy penetrators, 1
Battle damage assessment and repair (BDAR), 80
Battle damage repair (BDR)
  analyses, 177
  definition of, 2
  example of, 231–232
BDAR. *See* battle damage assessment and repair.
BDR. *See* battle damage repair.
Behind-armor debris (BAD), 21, 31–34
  characterization of, 267–274
    witness pack, 267–269
    witness plates analysis, 269–270
  debris cloud formation, 33
  fragments, 31
  materials, 33
    copper, 33
    rolled homogeneous armor, 33
    tungsten, 33
    penetrators, 33
  mathematical modeling, 270–274
    calibration, 274
    fragment angular distribution, 273
    fragment presented area, 272–273
    fragment shape, 272
    fragment type, 271–272
    fragments, numbers of, 271
    stochastic, 271
    velocity field, 273–274
  target vulnerability, 33
Beyond line of sight (BLOS), 124, 125
BFV. *See* Bradley Fighting Vehicle.
Binary damage, 78–79
B-kill, 75, 76
Blast, 20
Blast analysis, 43
Blast damage mechanism, 34–35
Blast-fragment impingement, 43
Blast-induced shock, 37
Blast methodology, 47
  mobile ground target components, 47

INDEX

Blast overpressure injuries, 55–58
  Bowen's free-field injury curves, 55–56
  ears, 55
  effective peak pressure technique, 56
  gastrointestinal tract, 55
  INJURY code models, 57
  injury curves, 56
  Kevlar body armor, 57
  lungs, 55
  Medical Research and Materiel Command, 57
  study advancements in, 57
  sub-lethal blast injuries, 56
Blast shock, 37
Blast wave, 34–35, 42, 55
Blast-waves in air, 131
Blocked shotline, 157
BLOS. *See* beyond line of sight.
Booster explosives, 51
Bourrelet, 23
Bowen, curves, 56
Bowen's free-field injury curves, 55–56
  Lovelace Foundation, 56
Bradley Fighting Vehicle (BFV), 7, 8, 25, 40, 60, 221
  MUVES-SQuASH VV&A, 283–296
BRASS. *See* BRL Anti-armor Sensor Simulation model, 158
BRL. *See* Ballistic Research Laboratory.
BRL ΔKE casualty criteria, 87
BRL Anti-armor Sensor Simulation model (BRASS), 158
BRL-CAD software, 148–149, 154
Bullet impact
  solid gun propellants, 53
  solid rocket motors, 52
  solid-explosive charges, 53
  warheads, 53
Bullet incapacitation analysis
  ballistic-pendulum method, 87
  BRL ΔKE casualty criteria, 87
  Expected Kinetic Energy, 87
Burn injury assessment, 41
  free-air temperature, 51
  heat-flux calorimetry, 51
Burn rates, 83
Bursting indirect-fire munitions modeling, 189–195
  fragment characterization, 193–194
  historical analysis process, 194
  stochastic analysis of fragment effects model, 195
  stochastic processor for artillery effectiveness, 194–195

CAIV. *See* Cost as an Independent Variable Strategy.
Calibration, fragment modeling and, 274
CAMAD. *See* computer-aided modeling, analysis and design.
Canadian Armament Research and Development Establishment (CARDE), 7, 132–133
Canted warheads, 177
Capability state, 17
CARDE. *See* Canadian Armament Research and Development Establishment.
CASTFOREM. *See* Combined Arms and Support Task Force Evaluation model.
Casualty, definitions of, 55
Casualty assessment, 83–89
  correlations, 86
  human vulnerability data, 83
  process of, 83–85
    Crew Casualty Working Group, 83
    defining, 85
    elemental capability vector, 85
    Joint Aircraft Survivability Program Office, 83
    Operational Requirement-based Casualty Assessment, 84
Casualty correlations
  ballistic dose, 88
  bullet incapacitation analysis, 87
  ComputerMan model, 88
  formulas, 86
  injury type, 88
  Major Trauma Outcome Study, 88
Catastrophic (K-Kill), 6, 69
CCWG. *See* Crew Casualty Working Group.
CD. *See* charge diameter.
CE. *See* chemical energy.
Charge diameter (CD), 30
Chemical effects, 20
Chemical energy (CE)
  munitions, 20
  shaped-charge jets, 20
C-kill, 76
Class accreditation, 215
Closed-form algorithms, 59
Closed-form phenomenological modeling, 129–131
  SCJ penetration, 130
Code verification, 213
Combat analysis, 11–12
  Joint Munitions Effectiveness Manual, 11
  measures of effectiveness, 12
Combat utility, 67–68
  level analysis, 13
Combat utility of tank, 93–110
  control of power, 106–107
  engine power, 94–105
  suspension system, 107
  transmission of power, 106

Combined Arms and Support Task
    Force Evaluation model
    (CASTFOREM), 205
Communication problems in armored tank, 111
Compartment-level analysis, target modeling
    and, 150–151
Compartment vulnerability modeling, 170–176
  experimental database, 170–171
  future use of, 174–175
  implementation of, 174–175
  model development, 171–174
    profile hole diameter, 173
Component
  critical, 44
  definition of, 44
Component capability, 288–290
  crew incapacitations, 290
  critical component
    hits, 289
    kills, 289
  vehicle perforation, 288–289
Component dysfunction, 44–58
  ammunition damage, 48–53
  analysis, 46
  blast methodology, 47
    mobile ground target components, 47
  criticality analysis, 44
  electrical component failures, 45–47
  formulas, 45
  fuel system, 47–48
  mechanical failures, 45–47
  penetrators, 46
  personnel vulnerability, 53–58
  probability, 45
  soldier survivability, 53–58
  structures, 58
  vulnerability analysis, 46
  point burst, 153
  target modeling and, 152–153
    parallel ray, 152–153
Computation of Vulnerable Areas and
    Repair Times, 152
Computer-aided modeling, analysis and
    design (CAMAD), 145–149
  constructive solid geometry, 145–149
ComputerMan injury modeling,
    88, 141–143
  Abbreviated Injury Scale, 88
Constructive solid geometry (CSG), 145
  interactive modeling, 148–149
  modeling system, 147
  raytracing technique, 147
  tools, 155–167
Contents, enemy access to, 91
Control of power, 106–107
Control problems in armored tank, 111
Cooling system, 100–103
Copper, 33

Correlation curves, 83
Cost as an Independent Variable Strategy
    (CAIV), 225
Cost determinations, 225
COVART
  model, 178–179, 199
  program, 152
Crew Casualty Working Group (CCWG), 83
Crew incapacitations, 290
Critical component, 44
  hits, 289
Critical energy, 82
Criticality analysis, 44, 89, 182
  aircraft, 44–45
  failure modes and effects analysis, 44
  fault trees, 45
  functional, 44
  mission, 44
  reliability, availability and maintainability
    studies, 45
CSG. See constructive solid geometry.

DAL. See damage assessment list.
Damage assessment list (DAL), 176
Damage mechanisms, 19–44
  ballistic penetration, 20
  ballistic shock, 20, 35–38
  behind-armor debris, 31–34
  blast, 20, 34–35
  chemical effects, 20
  directed energy, 20
  explosively formed penetrators, 29–30
  kinetic energy, 21–27
  secondary effects, 20, 40–41
    electronic failures, 40
    fire, 40, 41
    software failures, 40
  shaped charges, 27–28
  synergistic effects, 20, 41–44
  vaporific effect, 20, 38
    armored fighting vehicle, 39
Damage probability, 275–281
  graphical method, 278–279
  multiple fragments, 279–281
Damage states, 17, 188, 189
Damage to ammunition, 48–53
DDT. See deflagration-to-detonation
    transition.
Deactivation diagrams. See fault trees.
Debris cloud, 26–33, 254, 259–264
Deflagration, 49
Deflagration-to-detonation transition (DDT), 52
Degraded states, 92–93
  examples, 110–117
    loss of communication, 111
    loss of control, 111
    mobility loss, 113–117
    transmission problems, 112

INDEX

Stochastic Quantitative Analysis for System Hierarchies model, 92
Vulnerability Analysis Methodology Program, 92
Vulnerability Methodology (DSVM), 92, 227
Dehn model, 133
Depth charges, TORPEX-loaded, 137
Detonation,
   definition of, 49
   hot embedded fragments, 136
   hydrocodes, 134
   Jacobs-Roslund (J-R) equation, 134–135
   low-velocity, 49
   residual projectile modeling, 135
   SDT, 134
   warhead, 30, 136
   wave, 30, 33
   XDT, 52, 135
Differential thermal analysis, 82
Differential-scanning calorimetry, 82
Directed energy, 20
DIS. *See* distributed interactive stimulation simulators.
Distributed interactive stimulation (DIS) simulators, 11
Doctrine, 240
   Organizations, Training, Material, Leadership, Personnel and Facilities (DOTMLPF), 324
Documentation, 217–218
DOTMLPF. *See* Doctrine, Organizations, Training, Material, Leadership, Personnel and Facilities.
DPICM. *See* dual-purpose improved conventional munition, 137
Drop-weight impact test, 82
DSVM. *See* Degraded-States Vulnerability Methodology.
Dual-purpose improved conventional munition (DPICM), 137
Dysfunction of components, 44–58

Ears, injuries to, 55
ECV. *See* elemental capability vector.
Effective peak pressure technique, 56
Effectiveness/Vulnerability assessment in 3-D (EVA-3D) program, 200
   damage accumulation, 200
EFPs. *See* explosively formed penetrators.
EKE. *See* Expected Kinetic Energy.
E-kill, 75
Electrical component failures, 45–47
   definition of, 45
Electrical system, 103–104
Electronics, failure of, 40
Elemental capability vector (ECV), 85, 204
   elements, 204

Embedded fragments, 136
Empirical modeling, 128–129
   Thor equations, 128
Energetic materials, 49
Energetic materials sensitivity, 80
   burn rates, 83
   critical energy, 82
   differential thermal analysis, 82
   differential-scanning calorimetry, 82
   drop-weight impact test, 82
   friability test, 83
   gap test, 81
   pop plot, 81
Energy-based analysis, 68
Engineering model, 59, 143–144
   judgment, 144
   physical damage, 143
Engineering simulations, 129–131
   blast-wave in air, 131
   finite-element codes, 131
   hydrocodes, 131
Environmental factors, mission effectiveness and, 17
Eulerian technique, 42
EVA-3D program. *See* Effectiveness/Vulnerability assessment in 3-D.
Execution environment, 201
Expected Kinetic Energy (EKE), 87
Expected-value point-burst modeling, 176–179
   COVART, 178
   VAST, 176–177
Explosion, 48–49
Explosive, definition of, 49
Explosively formed penetrators (EFP), 20, 29–30
   charge diameter, 30
   components, 29
   lethality analysis, 177
   liner designs, 30
Explosively loaded warheads, 50
   improved conventional munitions, 50
Explosiveness, definition of, 50

Failure diameter, 49, 137
Failure modes and effects analysis (FMEA), 44, 46
Failures
   electrical components, 45–47
   mechanical, 45–47
Fast Air Target Encounter (FATE), 254
Fast Air Target Encounter Penetration (FATEPEN), 128–129, 253
   model overview, 260–264
      initial debris characteristics, 262
      residual debris characteristics, 262
FATE. *See* Fast Air Target Encounter.

358 INDEX

FATEPEN. *See* Fast Air Target Encounter Penetration.
Fault trees (deactivation diagrams), 45, 62, 89–92
  cargo, 91
  common functions, 91
  contents, 91
  criticality analysis, 89
  damage state to capability state, 89
  development
    degraded states methodology example, 110–117
    traditional metrics example, combat utility of tank, 93–110
  nonbinary forms, 92
  parallel path diagram, 90
  series diagram, 90
  situation recognition, 91
  time considerations, 91
Finite-element analysis, 58
Finite-element codes, 42, 131
  eulerian technique, 42
  Lagrangian technique, 42
Fire, 40, 41
  prediction of, 132–134
    CARDE tests, 132–133
    fuel fire start algorithms, 133–134
  probability of, 48
  vehicle fuel, 41
F-kill, 6, 68
FMEA. *See* failure modes and effects analysis.
Force-level modeling, 205–206
  Combined Arms and Support Task Force Evaluation, 205
  unit reconstruction, 206
Formulas, component dysfunction, 45
FPIIM. *See* Fragment Penetration and Impact Initiation Model.
Fragment, 31, 37
  angular distribution, 273
  characterization, 193–194
  high-penetration velocities, 31
  impact on
    solid gun propellants, 53
    solid rocket motors, 52
  number of, 271
  presented area, 272–273
  projectiles, 25–27
  shape, 272
  size, 32
  spall ring, 31
  type, 271–272
  velocity field, 273–274
Fragment Penetration and Impact Initiation Model (FPIIM), 254
Fragmentation munitions, penetration from 253–264
Fragmentation penetration
  characteristics of, 257–260

Fast Air Target Encounter Penetration, 253
  model overview, 260–264
  Fragment Penetration and Impact Initiation Model, 254
  low-impact velocities, 259
  simple methods, 255–257
  Thor model, 253
Free-air temperature burn injury assessment, 51
Friability test, 83
Friedlander blast waves, 56
Fuel, damage to, 78–79
  hypergolic, 78
Fuel fire damage assessment, 80
  battle damage assessment and repair teams, 80
  personnel, 80
    free air temperature, 80
    heat-flux calorimetry, 80
  vehicle, 80
Fuel fire start algorithms, 133–134
  Dehn model, 133
Fuel system, 95–97
  dysfunction, 47–48
    aircraft, 47–48
    diesel, 48
    JP-4, 48
    tanks, 47–48
  model, 155
  vulnerability, 237
Full-bore shot (AP), 21
  rotating bands, 21
  tracer, 21
Full-up, system-level (FUSL) testing, 222
Functional analysis, 44
Functions and capabilities, 341–345
Fuze, 51, 158

Gap test, 81
Gastrointestinal tract injuries, 55
Geometric modeling, 182
Geometric representation of targets, 144–145
  computer-aided modeling, analysis and design, 145–149
  target modeling practices, 150–155
Geometry interrogation, 156–160
  shotlines, 157–158
Granularity, 16
  importance of data, 17
  infinite, 16
  level of, 17
  state vectors, 16
Graphic software tools, 155–167
  BRL Anti-armor Sensor Simulation model, 158
  geometry interrogation, 156–160
  pix weight, 163–164
  polar-fb, 164–167
  presentation graphics, 161–167
  probability of kill given center of impact, 161–163
  scientific visualization, 161–167

INDEX

Graphical user interface (GUI), 201
GUI. *See* graphical user interface.
Gun-fired anti-armor projectiles, 36
Gun propellants, 137
   failure diameters, 137
   nitramine-based, 137

HE. *See* high explosive.
HEAT. *See* high-explosive anti-tank projectiles.
Heat-flux calorimetry, 80
   burn injury assessment, 51
HEP. *See* high-explosive plastic projectiles.
High-explosive (HE), 34, 49
   anti-tank (HEAT), 6
   plastic (HEP) projectiles, 6
   warheads, 1
High-resolution injury modeling, 141
   ComputerMan, 141–143
High-resolution modeling, 83
High-velocity armor-piercing (HVAP)
   discarding sabot (HVAPDS) design, 23
   projectiles, 6
High-yaw SC warheads, 177
Hit point, 181
Holes
   spatial distribution of, 269–270
   witness plate analysis and, 269–270
Hollow-charge projectile, 27
Hot embedded fragments, 52, 136
Human vulnerability data, 83
   correlation curves, 83
   high resolution modeling, 83
HVAP. *See* high-velocity armor piercing projectiles.
HVAPDS. *See* high-velocity, armor-piercing discarding design.
Hydrocodes, 131, 134
Hydrodynamics, 43
Hypergolic fuels, 78

ICMs. *See* improved conventional munitions.
ILS. *See* integrated logistics system.
Improved conventional munitions (ICMs), 50
Incapacitation, 54
   degrees of, 56
   potential, 86
   wound ballistics, 139
Initial debris, characteristics of, 262
Initiation, definition of, 49
Initiation mechanisms, 51–53
   bullet impact on
      solid gun propellants, 53
      solid rocket motors, 52
      solid-explosive charges, 53
      warheads, 53
   deflagration-to-detonation transition, 52
   fragment impact on solid
      gun propellants, 53
      rocket motors, 52
   hot embedded fragments, 52
   projectile impact on warheads, 53
   SCJ impact on solid gun propellants, 53
   shock-to-detonation transition, 52
   XDT, 52
Injuries, modeling of, 140–143
   high-resolution, 141
INJURY code models, 57
Injury Severity Score (ISS), 54
Injury
   curves, 56
   lethal, 88
   model, 140, 203
      Abbreviated Injury Severity score, 203
   serious, 88
   standard, definition of, 54
   See also casualty, personnel vulnerability.
Institute for Defense Analysis (IDA), 295
Insult, definition of, 54
Integrated logistics system (ILS), 10
Interactive CSG geometric modeling, 148–149
   BRL CAD package, 148–149
Interactive synergistic effects, 42
Intermediate air defense systems, 78
   Phased-Array Tracking, Ranging and Intercept of Target system, 78
ISS. *See* Injury Severity Score.

Jacobs-Roslund (J-R) equation, 134–135
   Ledsham, 134–135
JLF. *See* Joint Live Fire.
JMEM. *See* Joint Munitions Effectiveness Manual.
Joint Aircraft Survivability Program Office (JASPO), 83
Joint Live Fire (JLF) program, 7
Joint Munitions Effectiveness Manual (JMEM), 11, 193
Joint Technical Coordinating Group for Munitions Effectiveness (JTCG/ME), 8, 193
JP-4 fuel, 48
J-R equation, 134–135
JTCG/ME. *See* Joint Technical Coordinating Group for Munitions Effectiveness.
Judgment, engineering modeling and, 144

KE. *See* kinetic energy.
Kevlar body armor, 57
Kill categories, aircraft, 75
Kill criteria, 68–69
   catastrophic, 69
   firepower, 68
   mobility, 68
Kill probability, 74

Killer-victim scoreboard, 10
Kills, 289
Kinetic energy, 21–27
  munition, 20
  penetrators, 1, 253
Kinetic energy projectiles, 21–25
  APC design, 22
  armor-piercing
    discarding-sabot round, 22
    incendiary, 25
  fragments, 25–27
    low-impact velocities, 25–27
  full-bore shot, 21
  high-velocity, armor-piercing discarding sabot, 23
  tank, 21
K-kill, 6, 69, 75, 295
KK-kill, 75

Lagrangian technique, 42
  smooth particle hydrodynamics, 43
Ledsham equation, 134
Lethality, definition of, 2
LFT. *See* live fire testing.
Line of sight (LOS), 123
Lined-cavity charges, 27–28
  hollow-charge, 27
Liner designs, 30
Liquid gun propellants, 51
Liquid-propellant rocket motors, 50–51
Live Fire Law, 10
Live fire testing (LFT), 207, 221–226, 286–287
  Activity-Based costing, 225
  Cost as an Independent Variable strategy, 225
  full-up, system-level (FUSL) testing, 222
  Law, 10, 222
  legislation, 7
  Test and Evaluation Master Plan, 222
Loading function, 138
LoF. *See* loss of function.
Logical verification, 213
Logistics, 10–11
  integrated logistics system, 10
  killer-victim scoreboard, 10
  vehicles, 76–77
    payloads, 77
Long-rod KE, 33
LOS. *See* line of sight.
Loss of function (LoF), 37
  value calculation, 72
Low-frequency ballistic shock spectrum, 36
  von Mises technique, 36
Low-impact velocities, 25, 259
  debris cloud, 26
Low-velocity detonation (LVD), 49
Lubrication system, 97–100

Lungs, injuries to, 55
LVD. *See* low-velocity detonation.

Major Trauma Outcome Study (MTOS), 88
Manpower and personnel integration program (MANPRINT), 53
MANPRINT. *See* manpower and personnel integration program.
Mathematical basis, vulnerability/lethality (V/L) taxonomy and, 18
Measures of effectiveness (MOE), 12, 62, 306
  mission-centric, 63
Measures of mission effectiveness, history, 65–66
Measures of performance (MOP), 306
Measures of vulnerability
  aircraft damage categories, 75–76
  armored vehicles, aircraft vs, 74–76
  casualty assessment, 83–89
  energetic materials sensitivity, 80
  fuel fire damage, 80
  logistics vehicles, 76–77
  mobile missile-launching systems, 78–79
  self-propelled artillery, 77
  vulnerable areas, 79–80
Mechanical components, 104–105
Mechanical failures, 45–47
  definition of, 45
Medical casualty, 55
Medical Research and Materiel Command (MRMC), 57
MEVA. *See* Modular Effectiveness Vulnerability Analysis.
MILES. *See* Multiple Integrated Laser Equipment System.
Military occupational specialties (MOS), 10
Mine damage mechanism, 36, 38
Mission analysis, 44
Mission centric measures of effectiveness, 63
Mission effectiveness analysis, 13
  capability state, 17
  damage state, 17
  environmental factors, 17
  granularity, 16
  levels, observability, 15
  measures of, 65–126
  n-tuple, 18
  opposing forces (OPFOR) coalition, 15
  outcome of the friendly forces, 15
  quantifying movement, 17
Mission-oriented protective posture (MOPP), 206
Mission rehearsal, 11
Missions and Means Framework, 13–19, 63, 297–345
  application of, 309–317
  capabilities, 332–340
  construction of, 302–309

INDEX

formal process, 319–323
  allocations, 322
  methodology, 320
  mission accomplishment assessment, 322
  mission specifics, 320–322
  operational concept, 320
functions and capabilities, 324–345
  Doctrine, Organizations, Training, Material, Leadership, Personnel and Facilities, 324
fundamental elements, 326–327
mapping to V/L Taxonomy, 15
military, 299–317
need, 300–301
nesting, 340–341
objective, 301
parts and packages, 328
radiation, 19
related efforts, 301–302
scalability, 327–329
stocking perspective, 329–330
M-kill, 6, 68
MMF. *See* Missions and Means Framework.
Mobile ground target components, 47
Mobile missile-launching systems, 78–79
  damage to, 78–79
    binary, 78–79
    fuels, 78–79
    propellant events, 79
    multi-vehicle systems, 79
    surface-to-air, 78
    surface-to-surface, 78
Mobility/firepower LoF, 294–295
Mobility kill (M-kill), 6, 68
Mobility loss of armored tank, 113–117
Modeling
  advanced joint effectiveness, 199
  empirical vs semi-empirical, 128–129
  engineering, 143–144
  force-level, 205–206
  modular effectiveness vulnerability analysis, 199–202
  MUVES, 195–199
  ORCA program, 202–205
  phenomenological, 129–143
  verification, validation and accreditation, 206–207
Model-test-model (MTM) approach, 214
Modular Effectiveness Vulnerability Analysis (MEVA), 8, 199–202
  Effectiveness/Vulnerability assessment in 3-D program, 200
  execution environment, 201
  graphical user interface, 201
Modular Unix-based Vulnerability (MUVES), 8
MOE. *See* measures of effectiveness.

MOP. *See* measures of performance.
MOPP. *See* mission-oriented protective posture.
MOS. *See* military occupational specialties.
MRMC. *See* Medical Research and Materiel Command.
MTM. *See* model-test-model.
MTOS. *See* Major Trauma Outcome Study.
Multiple fragment effects, 279–281
Multiple Integrated Laser Equipment System (MILES), 11
Multi-vehicle missile launching systems, damage to, 79
MUVES, 195–199
  functionality, 283–285
MUVES-SQuASH VV&A
  Bradley fighting vehicle, 283–296
  component capability, 288–290
  Institute for Defense Analysis, 295
  live fire test, 286–287
  model validation, 286
  VAST modeling, 295–296
  vehicle capability, 290–294
  vehicle utility, 294–295

Needs assessment, 9–10
Nesting, 340–341
Network configuration, 121
Networked combat systems, 117–126
  smart munitions, 118
  unmanned aerial vehicles, 119
  vulnerability, 119–121
    configuration, 121
    survivability of, 121–126
Networks, 117–126
New Injury Severity Score (NISS), 54
NISS. *See* New Injury Severity Score.
Nitramine-based gun propellants, 137
NLOS. *See* non-line of sight.
Nonbinary forms, 92
  Degraded-States Vulnerability Methodology, 92
Non-line of sight (NLOS), 124
n-tuple, 18, 19

Observability, 15
OCO. *See* Office of the Chief of Ordnance.
Office of the Chief of Ordnance (OCO), 5
Office of the Secretary of Defense (OSD), 7
Office of the Secretary of Defense (OSD), 83
Operational casualty, 55
Operational Requirement-based Casualty Assessment (ORCA), 84, 202–205
  elemental capability vector, 204
  injury models, 203
OPFOR. *See* opposing forces coalition, 15
Opposing forces (OPFOR) coalition, 15
ORCA. *See* Operational Requirement-based Casualty Assessment.

OSD. *See* Office of the Secretary of the Defense.
Outcome of the friendly forces (OWNFOR), 15
Output validation, 214
OWNFOR. *See* outcome of the friendly forces.

Parallel path diagram, 90
Parallel ray, 152–153
  COVART program, 152
PATRIOT system, 78
Payloads, vehicles and, 77
Penetration from fragmentation munitions, 253–264
Penetrators
  analysis, 21, 43
  areal density, 80
  component dysfunction and, 46
  KE munition, 20
  long-rod KE, 33
  materials, 33
  M-kill, 175
Perforation, definition of, 20
Personnel injury, fuel fire and, 80
Personnel vulnerability, 53–58
  blast overpressure injuries, 55–58
  casualty, 54
  incapacitation, 54
  insult, 54
  manpower and personnel integration program, 53
  medical casualty, 55
  operational casualty, 55
  standard injury, 54
  trauma severity scoring, 54
Phased-Array Tracking, Ranging and Intercept of Target (PATRIOT) system, 78
PHD. *See* profile hole diameter.
Phenomenological models, 129–143
  ballistic impact prediction, 134–138
  ballistic shock, 138–139
  closed-form, 129–131
  engineering simulations, 129–131
  predicting fire, 132–134
  wound ballistics, 139–143
Physical damage, engineering modeling and, 143
Pix weight, 163–164
Platform capability level, 13
Platform survivability, 61
Point burst, 153
  model, 176–179
    stochastic, 179–189
    VAST, 177
Polar-fb, 164–167
Pop plot, 81
Preconditioning synergism, 42
  modeling of, 42

Presentation graphic software, 161–167
Probability of damage, 275–281
Probability of kill given center of impact, 161–163
Profile hole diameter (PHD), 173
Projectiles, 21–25
  impact on warheads, 53
Propellant, 49
  events, missile damage and, 79
  liquid gun, 51
  solid gun, 51
Pyrotechnics, 49, 51

Radiation, 19
RAM studies. *See* reliability, availability and maintainability studies.
Raytracing technique, 147
RCC. *See* Residual Combat Capability.
RDECs. *See* Research, Development and Engineering Centers.
Reliability, availability and maintainability (RAM) studies, 45, 228–231
Research, Development and Engineering Centers (RDECs), 9
Residual Combat Capability (RCC), 206
Residual debris, characteristics of, 262
Residual penetrator deflection, 181
Revised Injury Severity Score (RISS), 54
RHA. *See* rolled homogeneous armor.
RISS. *See* Revised Injury Severity Score.
Rocket motors
  liquid-propellant, 50–51
  solid-propellant, 50–51
Rolled homogeneous armor (RHA), 33, 268
Rotating bands, 21

Sabot, 22–23
SAFE. *See* stochastic analysis of fragment effects.
SC. *See* shaped charges.
Scientific visualization software, 161–167
SCJs. *See* shaped-charge jets.
SDAL. *See* Standard Damage Assessment List.
SDT. *See* shock-to-detonation transition.
Secondary effects, 20, 40–41
Secondary kill phenomena, 181–182
Self-forging fragments (SFFs), 20
Self-propelled artillery, 77
Semi-empirical modeling, 128–129
  Fast Air Target Encounter Penetration, 128
Sensitive components, 138
Sensitivity, 49–50
Series diagram, 90
SFF. *See* self-forging fragments.
Shaped charges (SC), 27–28
  lined-cavity, 27–28
  modeling of, 28
  tests, 132–133

INDEX

Shaped-charge jets (SCJs), 20, 29–33, 37
   impact on solid gun propellants, 53
   penetration, 46, 130
Shock, definition of, 49
Shock Response Spectrum (SRS), 280
Shock-to-detonation transition (SDT), 52
Shotlines modeling, 157–158
   ammunition, 157
   blocked, 157
   negative hit, 157
   wet hit, 157
Smart munitions, 8, 118
Smooth particle hydrodynamics, 43
Smoothed-particle-hydrodynamic numerical calculations, 68
Software failures, 40
Soldier survivability, 53–58
   blast overpressure injuries, 55–58
   casualty, 54
   incapacitation, 54
   insult, 54
   manpower and personnel integration program, 53
   medical casualty, 55
   operational casualty, 55
   standard injury, 54
   trauma severity scoring, 54
Solid explosive charges, bullet and projectile impact on, 53
Solid gun propellants, 51
   bullet or fragment impact on, 53
   SCJ impact on, 53
Solid rocket motors, bullet or fragment impact on, 52
Solid-propellant rocket motors, 50
SoS. *See* systems of systems.
Spall
   component interaction, 186
   production, 181
   rays, 177
   ring, 31
SPARC. *See* Sustainability Prediction for Army Requirements for Combat.
SPARC/BDR methodology, 226–228
SQuASH (Stochastic Quantitative Analysis for System Hierarchies model), 92
   application, 92, 180–189
      hit point, 181
      residual penetrator deflection, 181
      secondary kill phenomena, 181–182
      spall production, 181
      warhead performance, 181
   case study, 263
   inputs
      criticality analysis, 182
      geometric modeling, 182
      SDAL modifications, 185
      spall/component interaction, 186
      threat characterizations, 182, 185
   model, 179–180
   outputs
      damage states, 188, 189
      difference of, 186
SRS. *See* Shock Response Spectrum.
Standard Damage Assessment List (SDAL), 68–69
   armored vehicle testing, 69, 72–73
   kill
      criteria, 68–69
      probability, 74
   LoF value calculation, 72
   modifications, 185
   simulation models, 74
      degradation qualities, 74
   turreted tanks, 70–71
   weaknesses in, 73
Standard injury, 54
State vectors, 16
Stochastic analysis of fragment effects (SAFE) model, 195
Stochastic modeling, 18–19, 271
Stochastic point-burst model, 179–189
   SQuASH application, 179–189
   vulnerability analysis, 180
Stochastic processor for artillery effectiveness, 194–195
Stochastic Quantitative Analysis for System Hierarchies model. *See* SQuASH.
Stocking perspective, 329–330
Structural-response analysis, 58
   energy-based analysis, 58
   finite-element, 58
   smoothed-particle-hydrodynamic numerical calculations, 68
   time-marching analysis, 58
Structural validation, 214
Sub-lethal blast injuries, operational impact of, 56
   Bowen curves, 56
   degrees of incapacitation, 56
   Friedlander blast waves, 56
Surface-to-air missiles, 78
   intermediate air defense systems, 78
   Theatre High-Altitude Air Defense system, 78
Surface-to-surface missiles, 78
Survivability
   definition of, 2
   enhancement of, 233
   See also soldier survivability.
Susceptibility, definition of, 2
Suspension system of tank, 107
Sustainability Prediction for Army Requirements for Combat (SPARC), 177

Sustained fire, probability of, 48
Synergistic effects, 20
  blast-fragment impingement, 43
  interactive, 42
  modeling of, 42–44
    blast, 43
    finite-element codes, 42
    penetrators, 43
    preconditioning, 42
System life cycle, 226–232
System performance modeling, 167–170
Systems of systems (SoS), 61–62
  fault trees, 62
  platform survivability, 61

Tactical utility, 62–63
  measures of effectiveness, 62
  mission-centric MOE, 63
Tanks, 21, 70–71
  combat utility of, 93–110
  engine power, 94–105
    cooling system, 100–103
    electrical system, 103–104
    fuel system, 95–97
    lubrication system, 97–100
    mechanical components, 104–105
  fuel system failure, 47–48
Target modeling practices, 150–155
  compartment-level analysis, 150–151
  component-level analysis, 152–153
  philosophy, 153–155
Target
  definition of, 2
  geometric representation, 144–145
  response, 58–62
    closed-form algorithms, 59
    engineering aspects, 59–60
    systems of systems, 61–62
  vulnerability, 33
TEMP. *See* Test and Evaluation Master Plan.
Test and Evaluation Master Plan (TEMP), 222
THAAD. *See* Theatre High-Altitude Air Defense system.
Theatre High-Altitude Air Defense (THAAD) system, 78
Thor
  equations, 128
  model, 253
Threat, definition, 2
  characterizations, 182, 185
Time-marching analysis, 58
TORPEX-loaded depth charges, 137
TOW warhead, 39
Tracer, 21
TRADOC. *See* Training and Doctrine Command.
Training and Doctrine Command (TRADOC), 9, 206
Training, 11
  distributed interactive stimulation simulators, 11
  Multiple Integrated Laser Equipment System, 11
Transmission of power, 106
Transmission problems in armored tank, 112
Trauma severity scoring, 54
  Injury Severity Score, 54
  New Injury Severity Score, 54
  Revised Injury Severity Score, 54
Tungsten, 33
Turreted tanks, 70–71

U.S. Army Material Systems Analysis Activity (AMSAA), 9
U.S. Army Research Laboratory (ARL), 9
U.S. Army Test and Evaluation Command (ATEC), 9
UAVs. *See* unmanned aerial vehicles.
Unit reconstitution modeling, 206
  mission-oriented protective posture, 206
  Residual Combat Capability, 206
Unmanned aerial vehicles (UAVs), 119, 333

V/L. *See* vulnerability/lethality analysis.
Validation, 208
  output, 214
  structural, 214
VAMP. *See* Vulnerability Analysis Methodology Program.
Vaporific effect, 20, 38–40
  armored fighting vehicle, 39
  wire-guided SC warhead, 39
  spalling, 39
VAST model, 176–177, 290–294
  battle damage repair, 177
  canted warheads, 177
  high-yaw SC warheads, 177
  Sustainability Prediction for Army Requirements for Combat, 177
Vehicle capability, 290–294
Vehicle fuel fires, 41
  burn injury assessment, 41
  equipment damage assessment, 41
Vehicle perforation, 288–289
Vehicle utility, 294–295
  K-kill, 295
  mobility/firepower Lof, 294–295
Vehicles, ballistic shock damage to, 38
Verification, validation and accreditation (VV&A), 206–207
  benefits, 207
  case study, 283
  challenges to, 215–216

definitions, 208–209
policy documents, 207–208
process of, 210–216
   accreditation, 214–215
   validation, 214
   verification, 210. 213
resources, 207–208
users, 209–210
Von Mises technique, 36
VR. *See* vulnerability reduction.
Vulnerability, definition of, 2
Vulnerability Analysis for Surface Targets (VAST), 177
Vulnerability Analysis Methodology Program (VAMP), 92
Vulnerability analysis, 46
   stochastic, 180
Vulnerability assessment
   degraded states, 92–93
   fault trees, 89–92
   measuring, 66–80
      armored vehicles, 67
      combat utility, 67–68
      mission effectiveness, 65–126
      Standard Damage Assessment List, 68–69
   networks, combat systems, 117–126
Vulnerability/lethality (V/L) analysis, 1
See also Army V/L requirements.
   Bradley Fighting Vehicle, 7
   Canadian Armament Research and Development Establishment, 7
   combat, 11–12
   component dysfunction, 44–58
   damage mechanisms, 19–44
   history of, 4–5
      Aberdeen Proving Ground, 5
      armored vehicle vulnerability, 6
      Office of the Chief of Ordnance, 5
      Joint Live Fire program, 7
   large-scale systems, Army Unit Resiliency Analysis, 8
   missions and means framework, 13–19
   modeling, 7
   Modular Effectiveness Vulnerability analysis, 8
   Modular Unix-based Vulnerability Suite, 8
   roles of, 8–12
      fielding new material, 8–11
      needs assessment, 9–10
      technology opportunities, 9–10
   stages of, 3–4
   tactical utility, 62–63
   target response, 58–62
Vulnerability/lethality (V/L) taxonomy, 13
   casualty analysis, 85
   combat utility, 13
   initial state, 13
   mathematic basis, 18
   mission effectiveness, 13
   MUVES, 284
   platform capability, 13
   stochastic modeling, 18–19
Vulnerability reduction (VR), 232–239
   definition of, 233
   myths, 234–235
   principles, 235–237
   relation to MMF, 233–234
   survivability enhancement, 233
Vulnerable areas, measuring of, 79–80
VV&A. *See* verification, validation and accreditation.

Walter Reed Army Institute of Research (WRAIR), 41
Warheads, 136
   bullet impact on, 53
   explosively loaded, 50
   high-yaw SC, 177
   performance of, 181
Wet hit, 157
Wire-guided (TOW) SC warhead, 39
Witness pack, 267–269
Witness plate, analysis of, 269–270
   spatial distribution of holes, 269–270
Wound ballistics, 139–143
   modeling conventional injuries, 140–143
WRAIR. *See* Walter Reed Army Institute of Research.

XDT, 52

Supporting Materials

Many of the topics introduced in this book are discussed in more detail in other AIAA publications. For a complete listing of titles in the Progress in Astronautics and Aeronautics Series, as well as other AIAA publications, please visit http://www.aiaa.org.